PAPER LANDSCAPES

Cover: Tengger, East Java. Lithograph after a painting by Abraham Salm, ca 1870 (KITLV print collection 2219).

VERHANDELINGEN
VAN HET KONINKLIJK INSTITUUT
VOOR TAAL-, LAND- EN VOLKENKUNDE

178

PAPER LANDSCAPES
Explorations in the environmental history of Indonesia

Edited by

PETER BOOMGAARD
FREEK COLOMBIJN
DAVID HENLEY

1997
KITLV Press
Leiden

Published by:

KITLV Press
Koninklijk Instituut voor Taal-, Land- en Volkenkunde
(Royal Institute of Linguistics and Anthropology)
P.O. Box 9515
2300 RA Leiden
The Netherlands

Cover: Youetta de Jager and Rita DeCoursey

ISBN 90 6718 124 2

Printed in the Netherlands

Contents

PETER BOOMGAARD

Introducing environmental histories of Indonesia

Prologue

Ecological or environmental history of Indonesia is a new specialization.[1] It is true that one can name a few scholars who have been active in this field, *avant la lettre* so to speak, such as Clifford Geertz, James Fox, and Roy Ellen,[2] but two or three scholars do not make a discipline. Admittedly, environmental history has similarities with historical geography and other disciplines, such as hunter-gatherer studies as they were presented during the famous conference on Man the Hunter, now more than thirty years ago.[3] It also leans heavily on findings from such fields as agricultural history, demographic history, and economic history. We can learn much of these disciplines, and in view of what has been done already modesty becomes us. As regards Indonesia, however, most of these disciplines are not highly developed.

What is new in the field of environmental history of Indonesia, is a concerted effort at 'trawling' the archives and the published primary and secondary sources for relevant data. Too much of what has been presented as historical background is thinly disguised backwards projection of recent developments. These projections may or may not be correct, but, as a rule, the evidence is negligible. What is also new is that the environmental historian is looking for – theoretical – explanations of past developments that have hitherto escaped both the attention of historians, because they had neither the data nor the theoretical framework of the environmental historian at their disposal, and that of present-day environmentalist scholars, because their ideas about causation are entirely shaped by present-day concerns.

There are, alas, no shortcuts that may enable us to circumvent the often tedious and always time-consuming weeks or rather months spent on

1 In this essay and in this volume no distinction will be made between 'ecological history' and 'environmental history'. A case could be made, however, for distinguishing them (Arnold and Guha 1995:1-4).

2 See, for instance, Geertz 1963; Fox 1977, 1991; Ellen 1979.

3 The results were published in Lee and DeVore 1968.

research in archives and libraries. In interpreting the data we hope to find there, we will most certainly be guided by theoretical notions developed by other disciplines. Nevertheless, environmental history of Indonesia, thus perceived, should be regarded as a new specialization.

This volume presents a number of essays that reflect some major concerns of the environmental historian of Indonesia. It is, however, not a representative sample of environmental historical research. Some issues are not addressed here because they were recently dealt with elsewhere, sometimes by authors in this volume. Other topics had to be left out because no one could be found to cover them. This introduction was written to redress the balance by filling in the blanks left by the contributions to this book. In order to do so, it will be necessary to dwell a while upon the question of what environmental history is all about.

Briefly put, environmental history deals with the mutual influence of humans and the environment. Human beings exert influence on the environment, and the thus altered environment influences humans differently than before these changes occurred. These changes, and the feedback mechanisms they trigger, are the objects of environmental historical research.

Now although many historians may be inclined to see human beings as the 'prime movers' of (environmental) change, it is self-evident that the natural environment itself is of utmost importance to the environmental historian. One does not need to be in favour of environmental determinism to safely state that the environment, even in historical (as opposed to prehistorical) times, has to a large extent shaped many features of any given society. Moreover, autonomous changes in the natural environment, such as major and minor climatic fluctuations, have been 'prime movers' of change as much as humans have been. So natural givens, and the autonomous changes to which they are prone, are to be regarded as an environmental-historical topic *par excellence*.

Given the definition of environmental history presented above, it will not come as a surprise that population growth, or the lack of it, is also one of the main concerns of this young discipline. Environmental history is interested in both causes and effects of population growth. Among the former, disease patterns rank high. As population increases, even at a low rate, so do forest exploitation, agriculture, animal husbandry, hunting, and mining. Generally speaking, in the long run each of these activities leads to deforestation, temporarily in some cases but permanently in most. This, in turn, often results in land degradation and eventually in soil erosion. Population growth will, as a rule, also cause the expansion of fishing, in rivers, in lakes, and in the sea. A most relevant but also vexing question, cutting right across the various ways in which the environment can be manipulated, is whether people ever succeeded in doing so on a sustainable basis, with or without population growth.

The topics mentioned so far could well be studied in a 'closed system'.

Closed systems, however, are rare, and in historical times Indonesia was not one of them. Outside forces, such as foreign trade, foreign overlords, migrants, new crops and new diseases have always had a profound influence on the ways in which Indonesia's natural environment was exploited. It is, therefore, imperative to study the effects of these outside forces as clearly distinguishable entities.

Finally, the new discipline is fascinated by what could be called people's attitude towards nature, or rather by all beliefs, norms, and values that have a bearing on the environment, and the way in which these beliefs change over time. This includes beliefs that are not, or not necessarily, of an explicitly 'environmental' nature, but do have important ecological consequences. Environmentalist movements represent the tail-end of this topic.

The history of pollution, particularly urban and industrial pollution, will not be discussed here. The contributions in this volume are focused on the period in which pollution, although certainly not absent (Nagtegaal 1995), was not a major problem.

Having listed the main features of the discipline, and the connections between them, I will now give an impression of what has been done so far regarding the various topics, and of the many gaps that are still to be found.

The environment

Within this topic, two broad themes can be distinguished, namely the natural background against which Indonesian society came into being, and climatic variability and change. In this volume the first theme is addressed by Boomgaard, Brookfield, and Kathirithamby-Wells; it is also mentioned in passing in other contributions. The issue is whether soil, climate, and original vegetative cover can explain differences in the – original – availability of (large) animals and uncultivated edible plants, and therefore in population densities. A corollary of this question is how we can distinguish man-made landscapes from 'original' ones. Usually, these questions take us back to the period predating the written sources, leaving the environmental historian at the mercy of archaeologists, prehistorians, and palaeobotanists.[4] Their findings suggest that it is unlikely that the entire Indonesian archipelago was covered with dense primary forests before humans made their presence felt. Some areas may have been too dry for such cover and in other areas large natural fires changed the original vegetation. There is also a rather neat coincidence between high population densities on the one hand, and on the other

[4] A good summary is to be found in Bellwood 1985. For more recent data, see Maloney 1995a and 1995b.

moderate rainfall, temperatures that were not too high, and young volcanic non-acidic soils. Examples are the Minangkabau highlands in Sumatra, the southern part of Central Java, the island of Bali, Southwest Sulawesi, and various islands of the Ambon group. The humid tropical rain forest does not seem to have been a preferred habitat to early man.[5]

The second theme, climatic variability and change, is addressed in this volume in the contributions of Brookfield, Henley, and Knapen, and it is also touched upon in Reid's paper. We know, of course, that Indonesia is, and was, from time to time hit by weather anomalies. A prolonged drought, a long period of heavy rains and floods, or, in the worst possible case, a combination of a very wet and a very dry season, would cause harvest failures and high food prices. Occasionally, such abnormal conditions would obtain during a number of years in a row, with disastrous consequences. What many of us may not be aware of is that these anomalies seem to occur in more or less regular patterns. In other words, it seems that there are very long, long, and short cycles.

This transpired in an analysis of tree-rings of a number of teak trees (*jati*), cut down in the Rembang area of Java around 1925, that were almost four hundred years old. The first analysis of these data, by the Dutch meteorologist H.P. Berlage, was published in 1931. The data are still being investigated, while researchers are employing more and more sophisticated statistical tools for their analysis. Lately, Brookfield has been using them for his ENSO research. The researchers found a 3.5-year cycle, called the Southern Oscillation, linked to what we now call the El Niño phenomenon. Accordingly, this is now often called an ENSO effect. Longer cycles were also found, namely one of 50 years and one of almost 90 years. The last one fits eight 11-year sunspot cycles. The 50-year one has no theory behind it, but every economic historian is tempted to cry: Kondratieff![6]

Such cycles, then, may be held responsible for food shortages, famines, and, perhaps, in the long run, changes in land use. For instance, a series of droughts may have led to repeated forest fires (probably often man-made) which, on unfavourable soils, could cause permanent deforestation, to the point of rendering these soils useless for agriculture. This may have happened in the Batak area in northern Sumatra, eastern Java, and Timor. This line of reasoning might also shed new light on the origins of large expanses of *alang-alang* (*Imperata cylindrica*). In both cases there could be a link with the importance of livestock breeding and hunting in the areas mentioned. Large floods could have similar effects. Finally it should be

5 The basic outline of this is already present in Mohr 1938.

6 Berlage 1931; De Boer 1951; Lamb 1982:45; Murphy and Whetton 1989; Brookfield and Allen 1991; Palmer and Murphy 1993. A 'long wave' of ca 50 years was first observed after World War I by the Russian economist Nikolai D. Kondratieff. Economists and historians refer to this as the 'Kondratieff wave' or 'Kondratieff cycle'.

mentioned that long spells of bad weather seldom came alone, but were often accompanied by epidemics and wars.[7]

It is self-evident that in an area where weather anomalies occurred as often as they did in Indonesia, a population that consisted largely of agriculturalists in the broad sense of the word, must have suffered repeatedly from food shortages and high mortality. It stands to reason that climatic variation influenced their behaviour as producers and 'reproducers'. It is one of the tasks of the environmental historians to find out just how it shaped their attitudes.

Population growth

Essays on this topic are lacking in this volume, partly because some of the participants (Boomgaard, Reid) not so long ago published their findings on this topic elsewhere, partly because so much research still has to be done. The contributions to this volume by Henley and Knapen deal with this topic in passing.

Owing to a lack of numerical data, nothing definite can be said about growth rates of the population in Early-Modern Indonesia. The few studies that hazard an opinion on this matter assume that population-growth rates were low to very low in Java between 1600 and 1750, and in all other areas between 1600 and 1850. Taking the Indonesian archipelago as a whole, it seems unlikely that the average annual growth rate of the population was much higher than 0.1% during the period 1600-1800. That compares unfavourably with Europe (without Russia), and Asia as a whole during the same period, with growth rates of 0.25 and 0.3% per annum respectively.[8]

That growth rates outside Java prior to 1850 were low was also the opinion of most contemporary observers. They often mentioned the following factors. In the first place, in many areas men had to pay such a high bride price that marriages were often concluded rather late, and people remaining single were not exceptional. Secondly, it was argued that female fertility was rather low because women had to carry out many physically taxing tasks.[9] In the third place, all these societies were familiar with means to prevent conception and to end pregnancies prematurely. That explained why even to women who married early few children were born and that women stopped giving birth at an early age. In the fourth place, infant, child, and maternal mortality were very high, owing to a lack of obstetrical knowledge and dangerous post-natal practices. Finally, general mortality was high because of the many

7 Reid 1988:11-8, 1993:291-8; Boomgaard 1996a.
8 Reid 1988:14-8; Boomgaard 1989a:175; Livi-Bacci 1992:31.
9 On this point see also Reid 1987:39-40; Cohen 1989:87-9.

endemic and epidemic diseases and constant warfare. Strangely enough, not much was said about the influence of famines.[10]

It is much too early to pass judgement on these opinions, as no one, to my knowledge, has attempted to really investigate this whole complex. However, some of the factors mentioned are fairly well documented. General mortality, indeed, seems to have been rather high, although precise figures are lacking. There is also sufficient proof for high infant and maternal mortality.[11] The widespread practice of abortion was also attested from the early seventeenth century onwards, although some caution with these observations might be in order. This information derives from sources written by outsiders, to wit the Dutch, to whom abortion was a grave offence against the divine order. It was not beyond them to exaggerate those features of a foreign civilization that were regarded as morally reprehensible.[12]

Much less well documented is the purported link between high bride prices, late marriages and people remaining single, and low percentages of children. There are, however, some indications, particularly as regards Sumatra, that there is some truth in all this. In the first place, there are population counts, dated 1819-20, of two districts near the town of Bengkulu, which seem to suggest that the district where the patrilineal and virilocal *jujur*-marriages (high bride price marriages) were usual, had higher proportions of females marrying late or remaining single and lower percentages of children than the districts where these marriages did not occur.[13] However, these counts seem to have been carried out right after a rather troubled period, which makes it questionable whether they are representative.

Furthermore, as the *jujur*-marriages slowly but surely disappeared during the nineteenth and early twentieth centuries, population-growth rates started to rise. These shifts in marriage patterns had been stimulated by Dutch colonial policy and by the ongoing spread of Islamic influence. This does not constitute proof of a causal link, but at least it supports the suggested connection between high bride prices and low population-growth rates. Finally, there seems to be a – not altogether unexpected – connection between the most densely settled areas (Minangkabau, Batak), where rice was grown on permanent fields (*sawah*), early marriages, and the absence of (high) bride prices, or at least the custom to waive the bride price in exchange for labour. Among the Batak, high proportions of children were observed around 1900, but as yet I have not found similar observations

[10] Marsden 1975:256-7; Heyne 1814:427; *Substance* 1820:1-16; Raffles 1830:470; Du Bois 1852:265; Bickmore 1868:496-7; Wallace 1869:100-1, 264-5; Jacobs 1883:5-7; Helfrich 1889:547-50, 561; Snouck Hurgronje 1893-94, I:73, 411-3; Jacobs 1894, I:111-7, 124-5.

[11] Reid 1988:11-18; Boomgaard 1993:88, 1996a:6-7.

[12] References to cases of abortion in *Generale Missiven* (*GM* for short) I, 88 (20.8.1618); *GM* I, 547 (8-1-1635).

[13] Halhed et al. 1820; MacKenzie et al. 1820; Moyer 1983:253.

regarding the Minangkabau.[14] The upshot of all this is that early and universal marriage was, outside Java, probably limited to specific areas, and that it may have been linked to *sawah* cultivation.[15]

The combination of permanently cultivated fields, high population density, and high population-growth rates also obtained in Java, probably since the 1750s, but certainly from the early nineteenth century onwards. Early and universal marriage was found here as well, although the age of marriage was perhaps not as low around 1800 as it would be later on. Population growth was stimulated by the so-called Pax Neerlandica – Dutch-imposed absence of internal wars – and by a successful, island-wide system of vaccination against smallpox, in place since the early nineteenth century. In addition, improved communications are supposed to have reduced the influence of harvest failures on the death rate (Boomgaard 1989a). I also have the impression – which, admittedly, at the moment I cannot yet back up with firm data – that droughts and other weather anomalies, although occurring as frequently as they did in the seventeenth century, were less often followed by famines or harvest failures than they were before. If this is true, I can think of a number of factors that may explain it, namely increased 'sawahization' and therefore more secure yields, increased reliance on other, less 'risky' foodcrops such as maize, a slightly lower proportion of the population being entirely dependent on agriculture, and fewer wars preceding or following weather anomalies.

Between 1850 and 1910, the Dutch colonial empire in Indonesia was rounded off, which meant that the factors responsible for higher population-growth rates in Java also started to operate in the Outer Provinces. Thus smallpox vaccination, Pax Neerlandica, and improved communications all started to make themselves felt in those areas.

One factor that locally may have been partly responsible for low population-growth rates prior to 1850 has not been dealt with so far, and is in fact hardly ever mentioned in a demographic context, namely slavery. Slavery was already present in the archipelago when the Europeans arrived. Their presence certainly boosted its existence, although it is doubtful whether the European share in the slave-trade was all that important. However that may be, it would seem that certain areas have been exporting their population 'surplus' for centuries, whereas other regions apparently always had a shortage of people. Notable exporters of slaves were Bali, Buton, and Nias, three islands of roughly the same size, though different in every other respect. It may be that the people of these

14 Van Hasselt 1882:290; Volz 1909, I:34, 79-80, 169, 355; Moyer 1983:249-50; Wuisman 1985:32-72. Dobbin (1983:16) mentions birth control among the Minangkabau prior to 1850, but that does not imply that percentages of children were not high. The institution of *merantau* (temporary migration) may have kept marital fertility rates low.

15 On a recent conference on population development in Asia in the past (Taiwan, January 1996), the notion of early and universal marriage was found to be wanting for other Asian areas as well.

islands succeeded in creating a stable population, without having to go through the trouble of 'family planning'.[16] This had the added advantage that they received money for their exported population surplus, whereas regions where a stable population was the result of birth-control had to forego such windfall profits.

Areas with a constant demand for slaves were places such as Banda and the Ambon islands, producers – for the world market – of nutmeg and cloves respectively. Mortality among slaves was high, both during the sea voyage to their destinations and in the areas where they were employed. Therefore, the loss of people in the exporting areas was not balanced by an increase of the same order of magnitude in the receiving areas. In fact, the receiving areas probably showed zero growth rates. Thus, both the exporting and the importing areas may have had more or less stable populations, with which they managed to establish a system of sustainable production during two or three centuries.[17]

Banda and Ambon were places where the Dutch had reshaped the original mode of production, but slavery was certainly not limited to areas producing for the world market. Although slavery in the archipelago is an under-researched topic,[18] I think it is fair to say that it was present in a great many areas. The question, then, remains to be answered why certain areas were exporting slaves, while other areas were net importers, and why in some regions, such as Java, slavery was probably restricted to the commercial centres (Banten, Batavia).[19] It is tempting to see slavery as a solution for areas where population growth was – consciously? – kept at low levels because agricultural yields were too uncertain for the upkeep of large families.[20] Slaves could be bought in good years in order to increase family income, whereas they could be left to their own devices in bad years.

In areas where the presence of *sawah* led to a higher predictability of yields but also to higher labour requirements per unit of land, an extra child would soon produce more than its upkeep. In a bad year production would be lower, but as a rule not so low that survival was threatened. This would explain why population growth was relatively high in *sawah* areas, and why slavery was unusual: under these circumstances an extra child was a better investment than a slave.

16 It seems that in some areas the export of slaves was practised in addition to other methods to keep the population from increasing. This was the case on the island of Sumba, where we find late marriages, abortions, constant warfare, and a tradition of exporting slaves (Bieger 1890:12-3).

17 For Bali, see Schulte Nordholt 1996; for Buton, see Schoorl 1994; for Ambon and Banda, see Knaap 1987 and Loth 1996.

18 Exceptions are, for instance, Reid 1983 and Knaap 1995.

19 Debt bondage was known in Java, but to what extent is unclear.

20 Adriani and Kruyt (1950-51, I:76) found that slaveholding Toraja groups in Central Sulawesi had fewer children than non-slaveholding groups, as a slave could take the place of a child.

Finally, it is difficult to see what slave-exporting areas such as Bali, Buton, and Nias had in common, except their size. The islands must have had a 'surplus' population big enough to be a dependable export commodity. On the other hand the islands may have been small enough to keep enslaved people from running away to 'empty' areas. At the same time, the islands apparently did not have a shortage of labour, because in that case the slaves would not have been exported but put to work. We may also expect that either internecine warfare, or a high incidence of debts (or both), were at the root of these continuous slave-exports. The question of debts is particularly interesting, as it may have been linked in some cases (Nias) to high bride prices (Raffles 1830:491). If this would be borne out by further investigations, high bride prices may have been responsible in more than one sense for low population growth.

This is not the whole story, and matters get complicated when we take into account that slaves were not held for economic purposes only, and that in densely settled areas the aristocracy could use, although probably to a limited extent, statute labour instead of slaves. Other densely settled areas (Madura, Minangkabau highlands) used temporary or permanent outmigration, rather than the export of slaves, as a solution to their problems.

It is clear that, given the state of research, we cannot do much more than formulate some possible connections between slavery, population growth, and levels of economic development, thereby giving an example of the kind of questions to be asked by environmental historians.

However, we can be certain of one thing. Slavery was a bottomless pit, into which not only large numbers of people disappeared, but also large amounts of capital. It may have been one of the mechanisms that enabled the Indonesian archipelago to participate in the world market without getting developed. In an economic sense this is bad, but judged from an environmental angle it was perhaps not bad at all.

Disease environments

No paper in this volume is entirely dedicated to diseases, but they are quite central in the contributions by Henley and Knapen. Interest in disease in Indonesian history has been growing recently (Owen 1987), but our knowledge of this field is still rather limited.

Disease patterns constitute a mix of non-human and human factors. For instance, parasites, as a rule, are more numerous in a tropical environment than in moderate climatic zones (Boomgaard, this volume). Some diseases, however, were introduced through the trade routes. Epidemics such as the plague, cholera, and smallpox came, often time and again, from the Asian mainland, while syphilis arrived, via Europe, all the way from the

Americas.[21] Some disease are so-called civilized diseases, in the sense that they do not bother low-density populations which are, moreover, often constantly on the move (Knapen, this volume).

Human activities other than trade can also influence the incidence and spread of diseases. The spread of malaria, for instance, seems to be related to the expansion of wet rice cultivation, irrigation, changes in water management in general, and the opening up of new lands. However, historical malaria research is fraught with many problems, not only because the sources prior to the second half of the nineteenth century do not use the term malaria, but also because so many different species of mosquitoes and several species of plasmodia are involved. Nevertheless, it should be attempted to get a better grip on the complex circumstances that have been instrumental in making malaria locally an important killer. On the positive side, smallpox eradication, a colonial product and already quite effective in Java around 1850, removed one of the important killers in the archipelago.[22]

On the other hand, large-scale availability of quinine and the eradication of malarial mosquitoes, no matter how laudable for the victims of malaria, removed an efficient protection from certain 'wild' areas. The absence or presence of certain diseases, and certainly changes in these patterns, may, therefore, have important ecological consequences.

An interesting hypothesis (Knapen, this volume) could be the following. It is known that certain rather isolated groups, such as the Kubu in Sumatra and certain Dayak tribes in Kalimantan, were mortally afraid to come near foreigners; hence the phenomenon of 'silent barter'. One is tempted to assume that disease experiences in the past had driven them into the interior, thereby possibly even substituting a hunting-and-gathering lifestyle for an agricultural one.

It should be mentioned in passing that diseases are not limited to humans. Epizootics, or epidemics among animals, both wild and tamed, a hardly researched topic among historians of Southeast Asia, could have an enormous impact locally. They may have followed some density-related cycle, and they were doubtlessly influenced by extreme weather conditions, as was the case with many human diseases.[23]

Land clearing and deforestation

Expansion of forest exploitation, agriculture, animal husbandry, hunting and mining all follow in the wake of population growth. All these activities lead to – local – deforestation and the depletion of some forest

21 Boomgaard 1987, 1993:78-9, 1996a:7-9; Reid 1993:291-8.

22 See, for instance, Dobbin 1983:7; Boomgaard 1987:55-9, 1989b; Van der Brug 1994; Knapen, this volume.

23 Dempster 1975:43-4; see also Knapen, this volume.

resources. This, in turn, often results in land degradation and eventually in soil erosion. Some of these themes are addressed in this volume, namely by Boomgaard, Colombijn, De Jong Boers, Kathirithamby-Wells, Nibbering, and Potter.

Of the topics mentioned here, only agriculture and forest exploitation have recently received some attention. In the case of agriculture these studies are focused, as a rule, on production rather than on expansion of arable lands.[24] Generally speaking, we are well informed about developments in Java after 1800, but on Java prior to 1800 and on the Outer Islands our information is rather meagre.

There are, to my mind, two themes that require our special attention, namely the connection between wet rice cultivation and population growth, already touched upon above, and the effects of slash-and-burn agriculture on forest cover. We have seen that wet rice cultivation may have been conducive to population growth, but it is also often assumed that population growth stimulates 'sawahization'. Nevertheless, the early origins of wet rice cultivation in Java may have had other roots as well, such as conscious attempts of the state to settle an otherwise still rather mobile peasantry (Dove 1985a). It will be worthwhile to test that assumption for other areas, applying it not only to indigenous rulers but also to the Dutch colonial state.

Slash-and-burn (*gaga, huma, ladang*) cultivation has been studied quite intensively over the last decades, but really historical studies have been few and far between.[25] Yet, if we leaf through monographs and articles dating from the nineteenth century, particularly concerning Sumatra, it is clear that we have sufficient information for a good historical analysis. Of particular interest to the environmental historian are questions related to the carrying capacity (Henley, this volume) of local shifting cultivation systems. We need a good description and analysis of (changes in) fallow regimes, of the choice between clearing a patch of *rimba* (old forest) or *belukar* (secondary forest), of early agroforestry practices, and of possible transitions from slash-and-burn to permanent field agriculture.[26] We assume that population densities influenced all these features, but we should not limit our search to demographic variables. The constant threat of war, for instance, often induced people to live in concentrated settlements which would be moved when new clearings had to be made too far away from the village, thus

24 On agriculture, see for instance Dobbin 1983; Booth 1988; Boomgaard 1989a; Van der Eng 1993. A strong historical component can be found in two recent studies that pay attention to land use and deforestation: Donner 1987 and Hefner 1990.

25 Regarding Borneo/Kalimantan, I can mention a number of studies that do try to place present-day developments in a historical perspective, namely Freeman 1970; Padoch 1982; Dove 1985b; Rousseau 1990.

26 On 'early' agroforestry, see for instance Pelzer 1945:24; Mary and Michon 1987; Wiersum 1996; on the transition from *ladang* to permanent fields, see Seavoy 1973.

New tobacco-fields in Deli, North Sumatra; photograph by M. Mazaraki, ca 1905 (KITLV photo collection 35740).

establishing rather clear boundaries between secondary growth areas and old forest. With the threat of war, or head-hunting raids, gone, smaller, dispersed settlements would become feasible, resulting in a quilt pattern of new and old forest patches (Rousseau 1990:120, 137). War, therefore, not only kept population-growth rates down, but it also may have made for concentrated land-use patterns, and therefore less disturbance of the 'original' forest cover. Large, unbroken tracts of old forest were probably less vulnerable to forest fires and to damage from grazing livestock.

The historiography on shifting cultivation has always had moral/political overtones. Colonial writers often accused the indigenous peasantry of irresponsible behaviour regarding the forests, arguing that their slash-and-burn agriculture, livestock-rearing practices, and hunting methods had created the vast expanses of *alang-alang*. Later writers were inclined to blame large-scale commercial, sometimes government-sponsored, agriculture (coffee, pepper) for this type of land degradation.[27] Nevertheless, it is also clear that there were many deforested regions, such as the Batak area, where no large-scale agriculture had ever been practised (see also Boomgaard, this volume). It must be assumed, therefore, that small-scale shifting cultivation, probably in combination with livestock keeping and hunting, may cause permanent soil degradation, even at rather low levels of population density. It is up to the environmental historian to find out when it does and when it does not. Regarding the Batak, my guess would be that a combination of infertile soils, high rainfall totals, fairly high livestock densities, and repeated droughts go a long way towards explaining land degradation under small-scale slash-and-burn conditions.

As slash-and-burn shades imperceptibly into deforestation, we leave the realm of agriculture and dedicate a few lines to forest exploitation.[28] This topic can be analytically subdivided into two themes, namely the collection of non-timber forest products, and the production of timber and firewood.

As we are trained to think in more or less clear-cut categories, we are inclined to regard wet-rice cultivators, slash-and-burn agriculturists, and hunter-gatherers as separate groups, or even as people (or peoples) on different rungs of the developmental ladder. There are, of course, groups like the 'wild' Kubu and the Punan whose existence supports such an approach, but it is questionable whether it also works for the majority of societies we want to study. To the environmental historian, this is a relevant question, because he or she runs the risk of misrepresenting the

[27] Pelzer 1945:16-24; Dove 1986; Reid 1995:101-2. Recently it has been argued that grasslands created by shifting cultivation may have a number of economic and environmental advantages (Umans 1993:29).

[28] A review of the recent literature on forest exploitation in the past is given in Boomgaard 1996b.

environmental impact of settled and semi-settled peasants, if hunting and gathering activities are relegated to the realm of specialized groups. Therefore, the collection of non-timber forest products should be studied in conjunction with other (economic) activities. I am inclined to think that, at least prior to let us say 1850, this type of gathering was almost always a sideline, and more often than not an arduous or even dangerous one (tigers, head-hunting, epidemics), which would be undertaken only if the crops had failed or if someone had to finance a life-cycle ceremony.[29] In most areas of the archipelago, this mechanism would have led to a system of sustainable collection of non-timber forest products.

Occasionally, however, such products had disappeared locally, perhaps after a number of very bad harvests, possibly in combination with fairly high population densities, or after a period of very high prices. Examples of overexploited forest products before 1800 – and probably much earlier – are sandalwood and sappan in Java. An example on a much smaller scale comes from the surroundings of Mount Dempo in Southern Sumatra, where the Englishman Presgrave visited the Bukit Kayu Manis. This hill was named after the cassia trees that had grown there but that at the time of the visit (1817) had vanished without a trace.[30]

During the nineteenth century, when collecting and transporting forest products had become less onerous a burden, when outside demand showed more peaks, and when population-growth rates became higher, sustainable 'production' became more problematic (Potter, this volume).

As many non-timber forest products are only to be found in old forests, substitution of secondary forest for *rimba* was a threat to sustainable collection of these products. However, this was not the only threat, as some products are easily overcollected, even if the old forest area is not decreasing. In such cases, the answer might be cultivation, and we should try and find out under what circumstances this response could be observed (Potter, this volume). Finally, it seems that after circa 1870, wage labour increasingly became an attractive alternative to foraging, whereas specialized foragers became increasingly 'commercialized' (Persoon 1994:41; Boomgaard 1996b:13-5).

The exploitation of non-timber forest products is seldom the cause of deforestation. The production of timber and firewood, however, often is. Neither deforestation, nor the exploitation of timber resources are recent phenomena. When the first survey of forested areas in the entire archipelago, dated 1927, became available, many regions had already lost

29 Even nowadays the vast majority of collectors are part-timers (De Beer and McDermott 1989:86). For increased gathering activities in South Sumatra after the great drought of 1877-78, see Forbes 1885:232-3.

30 Presgrave 1822:22; Boomgaard 1995a:5, 8. Sandelwood is *Santalum album*, sappan is *Caesalpinia sappan* (see also De Jong Boers, this volume), and (Padang) cassia is *Cinnamomum burmanni*.

large chunks of their forest cover (Boomgaard 1996b:27, 164-6). The process of deforestation accelerated somewhere around 1870, but even then much timber had vanished already. It seems likely that shipbuilding, both by the Indonesians and the Dutch, was one of the main culprits. Already in the sixteenth century shipbuilding was reported from Aceh (Sumatra), Java, Borneo (Brunei?), and Tidore (North Moluccas). In the seventeenth century, in addition to these areas, shipbuilding was not only mentioned in Riau (Sumatra) and Makassar (Sulawesi), but also, surprisingly enough, on the Kai Islands, where ships were even built for customers elsewhere in Eastern Indonesia.[31] I see it as one of the most challenging tasks for environmental historians to find out whether in any of these areas timber was being felled in a way that did not destroy the local forests.

Animal husbandry, hunting (Boomgaard and Kathirithamby-Wells, this volume), and mining as historical phenomena hardly have been studied at all, let alone in an environmental historical context. If these aspects of exploitation of the terrestrial natural environment have been badly studied, even less research has been done on the use of the marine environment in the past (Masyhuri, this volume). Therefore, the essays presented here on these topics cannot be regarded as summaries of mature investigations, but as attempts to spark off research in this direction.

Outside forces

Foreign trade, foreign people, new crops, and new diseases have, time and again, influenced the exploitation of Indonesia's natural environment. It is difficult to find a contribution to this volume in which this theme is entirely absent, but it is explicitly dealt with by Colombijn, Cribb, Dove, De Jong Boers, and Potter.

Most scholars would agree that the demand created by international trade began to exert a notable influence on the environment throughout Indonesia after 1870 or 1900 and even more so after the 1950s. However, in some areas this influence came much earlier, as was the case in the Moluccas (cloves), Banda (mace, nutmeg), Timor (sandal), Sumbawa (sappan), Java (rice, pepper, sugar, coffee, timber), and Irian (birds of paradise). The last-named commodity (Cribb, this volume) is perhaps the most intriguing. The trade in these feathers goes back at least a millennium, and yet Irian Jaya was until recently seemingly hardly touched by the usual forces that elsewhere seem to accompany the penetration of the world economy.

Therefore, when we try to establish the effects of demand stemming from international trade, we should investigate the buffer mechanisms operating between demand and supply. These mechanisms seem to have

31 Valentijn 1724-26, III:35; Dampier 1931:91; Pigafetta 1969:127; Andaya 1975:148; Teensma 1989:321; Boomgaard 1991:18-20; Heersink 1995:39.

been responsible for shielding (some) indigenous societies from trade-induced change. In the late nineteenth and twentieth centuries they started to function less effectively. One mechanism was already mentioned when non-timber forest products were dealt with, namely the probability that collecting these products was a sideline. Another factor is that every exported product had to pass several middlemen before it reached its destination. At each step some money was skimmed off as payment for services rendered and as interest for advances given, because without the lubricant of advances, trade in the archipelago was impossible (Boeke 1955:132). So net profits accruing to the producer/collector were but a shadow of the market value of any commodity. Finally, if some income remained after subsistence needs had been met, it was often spent on traditional 'conspicuous consumption' items, such as slaves, jewellery, bridewealth, and 'potlatch'-type ceremonies. After circa 1870 or 1900, the removal of a number of uncertainties (epidemics, pirates, head-hunting; Knapen, this volume), increased demand, a larger array of desirable imported commodities, improved communications, and the abolition of slavery all contributed to increased production or collection of export commodities for foreign trade, thereby inducing more social and economic change and an increased transformation of the natural environment.

The introduction of new crops is, of course, a trade-related phenomenon. Almost all new crops came from the New World, and the list is impressive. Chillies, peanuts, maize, sweet potato, and cassava fundamentally influenced both Indonesian cuisine and – subsistence – agricultural practice. Coffee and tea, which did not come from America, and tobacco and rubber, which did, revolutionized the demand from international trade and changed the agricultural landscape of some Indonesian regions almost beyond recognition. Both series of crops had a formidable impact on the environment. For some crops such as coffee, this is rather well known, but the influence of maize, sweet potatoes, and cassava, which became ubiquitous food crops, has hardly been studied. Maize and cassava influenced land reclamation and agriculture not only directly, but also indirectly, because the increasing area under these two crops with relatively high yields per unit of time and unit of land, facilitated higher population-growth rates in certain areas. It should be pointed out that the expansion of some old crops, present before 1500, could also have considerable environmental consequences, as witness sugar and pepper (Boomgaard 1995b; Reid 1995:98-103).

Foreign migrants (Colombijn, this volume) and internal migrants (Reid, this volume) have a good name as 'motors' of economic development, but a bad name in environmental matters. As Colombijn demonstrates, this applies particularly to foreign migrants, who arrive poor and burdened with debts, and who want to exploit their new environment to the hilt. Small wonder, therefore, that 'virgin' areas are quickly depleted of their natural resources after the arrival of outsiders.

The motives for migration of foreigners do not belong to the realm of environmental history of Indonesia, but to the history of their countries of origin. The causes of internal migration, however, should be a research topic for Indonesia scholars. This is a complex question, as Reid demonstrates in his contribution to this volume. The question becomes even more convoluted if we take into account that migrants did not only settle in lower-lying areas, but also in the less inviting uplands (Boomgaard 1995b).

Attitudes

I will start with a familiar straw man: 'Western' people have a destructive attitude towards nature, 'Eastern' people live in harmony with it. However primitive a notion this may be, it does serve to pose the 'attitude' problem that has to be faced by every environmental historian. In fact, there are two clusters of problems. The first one can be formulated as follows: Do people destroy their natural environment because they have a destructive mentality, or is a destructive mentality just something that is thrown in when a society goes through a phase of (rapid) economic development. The corollary of this question is, of course, whether people have succeeded in preserving their natural environment because they want to live in harmony with it, or because economic development was absent. The second problem complex is whether there is any sense in seeing Western or Eastern people as meaningful analytical categories. Even if the answer to this question is negative, it could still be argued that Indonesians, who share so many cultural characteristics, constitute a meaningful unit of analysis.

Starting with the last question, I would argue that only in a rather limited sense Indonesians are a helpful analytical category for the environmental historian who wants to study attitudes towards nature. I am not alone in thinking that hunters-gatherers among the Indonesians have, and had, other cultural notions about their relationship with nature than do, and did, settled and wet-rice-cultivating peasants. In this sense, Javanese peasants – although many Dutchmen called them 'children of nature' – were not much different from peasants and farmers throughout history, in whatever part of the world: they feared and often hated 'wild' nature, and they expected their rulers to do something about it. This is all rather straightforward and, to my knowledge, uncontroversial. It becomes more interesting if we take into account that the same Javanese peasant who wanted to get rid of 'wild' nature, was not willing to kill an animal as dangerous as the tiger (unless the tiger had killed his cattle or a relative). Killing a tiger was morally or spiritually dangerous, because a 'benign' tiger could very well be inhabited by the soul of an ancestor. Neither was the Javanese peasant willing to fell 'sacred' trees or 'haunted' forests.

Felling these trees and forests would evoke the wrath of the spirits dwelling there.[32]

These animistic beliefs the Javanese peasant had in common with most other people in the Indonesian archipelago, and in that respect it makes sense to study Indonesia as a whole. Intra-Indonesian comparisons are particularly interesting when remarkable differences between various groups can be observed in the exploitation of one and the same commodity. Sandelwood is a good example, given the fact that the Javanese cut all their sandal long before 1800, whereas the people of Sumba declared it taboo until circa 1880, and the Timorese had been exporting it on a sustainable level since way before 1500 until 1915. At least in the cases of Sumba and Timor we know that the inhabitants of both islands believed that the sandal tree was inhabited by the spirits of ancestors, and yet we observe such glaring differences in exploitation (Boomgaard 1995a). Furthermore, even within particular societies we should not expect uniform beliefs, as there are differences according to rank, income, occupation, and gender. We will have to be very specific as to whose attitude towards nature we are talking about.

Finally reverting to the problem complex mentioned first, I can only say that the 'primacy' of economic development or attitudes towards nature in reshaping the natural environment will have to be established empirically. In fact, we should consider a third factor that has to be taken into account, namely the vulnerability of the local environment. The Batak, mentioned above, can be taken as an example. Was their environment deforested at such an early point in time because, as I suggested above, it was a rather fragile one? Or was it just a case of economic growth forcing them to be less careful with their forests? Or was their attitude towards nature different from that of the surrounding groups in Sumatra? Of course, it may have been a combination of all three factors.

Returning to the more simple dichotomy I formulated at the beginning of this section, prima facie evidence suggests that outside forces often seem to bring changes both in ideology and in the pace of economic development, so that beliefs and exploitation itself may have been transformed at the same time. Remnants of the older attitudes do survive, but they no longer seem to have much importance for everyday life. Nevertheless, that still leaves us with the question of what caused the observed manipulation of nature prior to these influences. Owing to a lack of data, it will be a difficult point to clear up.

32 Lombard 1974; Schefold 1988, 1990:16-9; Boomgaard 1994:164-70, 1995c.

Epilogue

The report on the workshop that formed the basis of this collection of essays identified three major themes which between them run through the majority of the contributions and lend coherence to the collection. Those themes are: sustainable and unsustainable use of natural resources, natural hazards and uncertainties, and the importance of a long-term historical perspective when it comes to assessing current cases of environmental degradation (Colombijn et al. 1996:9). A long-term perspective, of which the importance is self-evident to most historians, and which has received due attention in this introduction, might be our most relevant contribution to present-day environmental debates. I expect that, within this perspective, the – interrelated – themes of sustainability (or the lack of it) and natural hazards and uncertainties will provide us with a promising framework for further analysis.

This introduction has shown that there are a large number of fascinating questions, and, as yet, very few answers. Given the tender age of environmental history of Indonesia as a discipline, this was to be expected. This volume hopes to make a small dent in our vast ignorance regarding this topic.

Research in this direction will be continued for the years to come by the KITLV, which started the EDEN project some years ago. The acronym stands for Ecology, Demography, and Economy in Nusantara, another word for Indonesia. The fact that ecology, demography, and economy are mentioned explicitly implies that these disciplines play an important role in the investigations carried out by the research team, but it does not imply that other fields of study are ignored. As this introduction demonstrates, other factors, such as politics and mentalities, are also taken into consideration.

Finally, these introductory remarks are also meant to convey a sense of plurality. Although there are good reasons to view the Indonesian archipelago as a meaningful unit of analysis, we are also confronted by an astounding array of variations in soil, micro-climate, flora and fauna, products, levels of economic and political development, and mentalities, such that, for the time being, we cannot do much more than concentrate on writing specific environmental histories. It is much too early for the one comprehensive environmental history of Indonesia envisaged by all of us.

References

Adriani, N. and A.C. Kruyt
1950-51 *De Bare'e sprekende Toradjas van Midden-Celebes (de Oost-Toradjas).* Amsterdam: Noord-Hollandsche Uitgevers Maatschappij. 3 vols.

Andaya, L.Y.
1975 *The kingdom of Johor 1641-1728.* Kuala Lumpur: Oxford University Press.

Arnold, D. and R. Guha
1995 'Introduction; Themes and issues in the environmental history of South Asia', in: D. Arnold and R. Guha (eds), *Nature, culture, imperialism; Essays on the environmental history of South Asia*, pp. 1-20. Delhi: Oxford University Press.

Beer, J.H. de, and M.J. McDermott
1989 *The economic value of non-timber forest products in Southeast Asia; With emphasis on Indonesia, Malaysia and Thailand.* Amsterdam: Netherlands Committee for IUCN/WWF.

Bellwood, P.
1985 *Prehistory of the Indo-Malaysian archipelago.* Sydney: Academic Press.

Berlage, H.P.
1931 'Over het verband tusschen de dikte der jaarringen van djatiboomen (*Tectona grandis* L.f.) en den regenval op Java', *Tectona* 24:939-53.

Bickmore, A.S.
1868 *Travels in the East Indian archipelago.* London: Murray.

Bieger, Ph.
1890 'Bezoeken op Soemba', *Mededeelingen van wege het Nederlandsche Zendelinggenootschap* 34:1-29.

Boeke, J.H.
1955 *Economie van Indonesië.* Haarlem: Tjeenk Willink.

Boer, H.J. de
1951 'Tree-ring measurements and weather fluctuations in Java from AD 1514', *Proceedings of the KNAW, Series B* 54:194-209.

[Bois, J.A. du]
1852 'De Lampongsche distrikten op het eiland Sumatra', *Tijdschrift voor Nederlandsch-Indië* 14, I:245-75, 309-33.

Boomgaard, P.
1987 'Morbidity and mortality in Java, 1820-1880; Changing patterns of disease and death', in: N.G. Owen (ed.), *Death and disease in Southeast Asia; Explorations in social, medical and demographic history*, pp. 48-69. Singapore: Oxford University Press. [ASAA Southeast Asia Publication Series 14.]

1989a *Children of the colonial state; Population growth and economic development in Java, 1795-1880.* Amsterdam: Free University Press. [CASA Monographs 1.]

1989b 'Smallpox and vaccination on Java, 1780-1860; Medical data as a source for demographic history', in: A.M. Luyendijk-Elshout et al. (eds), *Dutch medicine in the Malay archipelago, 1816-1942*, pp. 119-31. Amsterdam/Atlanta: Rodopi.

1991 'The non-agricultural side of an agricultural economy; Java, 1500-1900', in: P. Alexander, P. Boomgaard and B. White (eds), *In the shadow of agriculture; Non-farm activities in the Javanese economy, past and present*, pp. 14-40. Amsterdam: Royal Tropical Institute.

1993 'The development of health care in colonial Java; An exploratory introduction', *Bijdragen tot de Taal-, Land- en Volkenkunde* 149:77-93.

1994 'Death to the tiger! The development of tiger and leopard rituals in Java, 1605-1906', *South East Asia Research* 2:141-75.

1995a 'The VOC trade in Asian forest products in the 17th and 18th centuries'. Rewritten version of paper prepared for the conference on Environmental History of South and South-East Asia, New Delhi.

1995b 'Maize and tobacco in upland Indonesia, 1600-1940'. Paper prepared for the workshop on Agrarian Transformation in Upland Indonesia, Halifax.

1995c 'Sacred trees and haunted forests in Indonesia, particularly Java, nineteenth and twentieth centuries', in: O. Bruun and A. Kalland (eds), *Asian perceptions of nature; A critical approach*, pp. 47-62. Richmond, Surrey: Curzon Press.

1996a 'Fluctuations in mortality in 17th-century Indonesia'. Paper prepared for the conference on Asian Population History, Taipei.

1996b (with the assistance of R. de Bakker), *Forests and forestry 1823-1941*. Amsterdam: Royal Tropical Institute. [Changing Economy in Indonesia 16.]

Booth, A.
1988 *Agricultural development in Indonesia*. Sydney: Allen and Unwin. [ASAA Southeast Asia Publications Series 16.]

Brookfield, H. and B. Allen
1991 'Environmental and human responses to climatic variability in the West Pacific "maritime continent" region', in: J.E. Hay (ed.), *South Pacific environments; Interactions with weather and climate; Proceedings of a conference organized by Environmental Science, University of Auckland, 2-7 September 1991*, pp. 71-82. Auckland: University of Auckland.

Brug, P.H. van der
1994 *Malaria en malaise; De VOC in Batavia in de achttiende eeuw*. Amsterdam: De Bataafsche Leeuw. [PhD thesis, Rijksuniversiteit Leiden.]

Cohen, M.N.
1989 *Health and the rise of civilization*. New Haven/London: Yale University Press.

Colombijn, F., D. Henley, B. de Jong Boers and H. Knapen
1996 'Man and environment in Indonesia, 1500-1950; An international workshop, KITLV, Leiden, 27-29 June 1996', *Indonesian Environmental History Newsletter* 7:1-9.

Dampier, W.
1931 *Voyages and discoveries*. With introduction and notes by C. Wilkinson. London: Argonaut Press.

Dempster, J.P.
1975 *Animal population ecology*. London: Academic Press.

Dobbin, C.
1983 *Islamic revivalism in a changing peasant economy; Central Sumatra, 1784-1847*. London/Malmö: Curzon Press.

Donner, W.
1987 *Land use and environment in Indonesia*. London: Hurst.

Dove, M.R.
1985a 'The agroecological mythology of the Javanese and the political economy of Indonesia', *Indonesia* 39:1-36.
1985b *Swidden agriculture in Indonesia; The subsistence strategies of the Kalimantan Kantu'*. Berlin: Mouton.
1986 'The practical reason of weeds in Indonesia; Peasant vs. state views of *Imperata* and *Chromolaena*', *Human Ecology* 14:163-90.

Ellen, R.F.
1979 'Sago subsistence and the trade in spices; A provisional model of ecological succession and imbalance in Moluccan history', in: P.C. Burnham and R.F. Ellen (eds), *Social and ecological systems*, pp. 43-74. London: Academic Press. [ASA Monograph 18.]

Eng, P.J. van der
1993 *Agricultural growth in Indonesia since 1880; Productivity change and the impact of government policy*. Groningen: Universiteitsdrukkerij. [PhD thesis, Rijksuniversiteit Groningen.]

Forbes, H.O.
1885 *A naturalist's wanderings in the Eastern archipelago; A narrative of travel and exploration from 1878 to 1883*. London: Sampson Low.

Fox, J.J.
1977 *Harvest of the palm; Ecological change in Eastern Indonesia*. Cambridge, Mass.: Harvard University Press.
1991 'The heritage of traditional agriculture in Eastern Indonesia; Lexical evidence and the indications of rituals from the outer arc of the Lesser Sundas', in: P. Bellwood (ed.), *Indo-Pacific history 1990; Proceedings of the 14th congress of the Indo-Pacific Prehistory Association, Yogyakarta, Indonesia*, vol. 1, pp. 248-62. Canberra/Jakarta: n.n. [Bulletin of the Indo-Pacific Prehistory Association 10.]

Freeman, J.D.
1970 *Report on the Iban*. London: Athlone Press.

Geertz, C.
1963 *Agricultural involution; The process of ecological change in Indonesia*. Berkeley/Los Angeles: University of California Press.

Generale Missiven
1960-88 Generale missiven van Gouverneurs-Generaal en Raden aan Heren XVII der Verenigde Oostindische Compagnie. Edited by W.P. Coolhaas [vol. 1-8] and J. van Goor [vol. 9]. 's-Gravenhage: Nijhoff. [RGP, Grote Serie, 9 vols.]

Halhed, G., T. Church and J.D. Lewis
1820 'Report on the population of the District of Lumba Selapan', *Proceedings of the Agricultural Society Established in Sumatra* 1-Appendix C:1-15.

Hasselt, A.L. van
1882 *Volksbeschrijving van Midden-Sumatra*. Leiden: Brill. [Vol. 3 of P.J. Veth (ed.), *Midden-Sumatra; Reizen en onderzoekingen der Sumatra-Expeditie 1877-1879*.]
Heersink, C.G.
1995 'Interactie tussen mens en landschap in Zuid-Sulawesi tijdens de Vroeg-Moderne Periode'. Internal paper EDEN Project, KITLV.
Hefner, R.W.
1990 *The political economy of mountain Java; An interpretive history*. Berkeley: University of California Press.
Helfrich, O.L.
1889 'Bijdrage tot de geographische, geologische en ethnographische kennis der Afdeeling Kroë', *Bijdragen tot de Taal-, Land- en Volkenkunde* 38:517-632.
Heyne, B.
1814 *Tracts, historical and statistical, on India [...]*. London: Baldwin.
Jacobs, J.
1883 *Eenigen tijd onder de Baliërs; Eene reisbeschrijving met aanteekeningen betreffende hygiène, land- en volkenkunde van de eilanden Bali en Lombok*. Batavia: Kolff.
1894 *Het familie- en kampongleven op Groot-Atjeh; Eene bijdrage tot de ethnographie van Noord-Sumatra*. Leiden: Brill. 2 vols.
Knaap, G.J.
1987 *Kruidnagelen en Christenen; De Verenigde Oost-Indische Compagnie en de bevolking van Ambon 1656-1696*. Dordrecht/Providence: Foris. [KITLV, Verhandelingen 125.]
1995 'Slavery and the Dutch in Southeast Asia', in: G. Oostindie (ed.), *Fifty years later; Antislavery, capitalism and modernity in the Dutch orbit*, pp. 193-206. Leiden: KITLV Press. [Caribbean Series 15.]
Lamb, H.H.
1982 *Climate, history and the modern world*. London/New York: Methuen.
Lee, R.B. and I. DeVore (eds)
1968 *Man the hunter*. Hawthorne: Aldine.
Livi-Bacci, M.
1992 *A concise history of world population*. Cambridge, Mass./Oxford: Blackwell.
Lombard, D.
1974 'La vision de la forêt à Java (Indonésie)', *Études rurales* 53/56:473-85.
Loth, V.
1996 'Ecological history of the Banda Islands', *Indonesian Environmental History Newsletter* 7:10-2.
MacKenzie, W.G., W.T. Lewis and R. Bogle
1820 'Report on the population of the district of Dua-Blas', *Proceedings of the Agricultural Society Established in Sumatra* 1-Appendix B:1-20.

Maloney, B.K.
1995a 'Dry periods and forest fires in Indonesia', *Indonesian Environmental History Newsletter* 5:9-12.
1995b 'New perspectives on possible early rice cultivation in highland North Sumatra'. Paper presented at the EUROSEAS conference 'Keys to Southeast Asia', Leiden.

Marsden, W.
1975 *The history of Sumatra*. Kuala Lumpur: Oxford University Press. [Reprint of the third edition (1811), introduced by John Bastin.]

Mary, F. and G. Michon
1987 'When agroforests drive back natural forests; A socio-economic analysis of a rice-agroforest system in Sumatra', *Agroforestry Systems* 5:27-55.

Mohr, E.C.J.
1938 'The relation between soil and population density in the Netherlands East Indies', in: *Comptes rendus du congrès international de géographie, Amsterdam 1938; Tome deuxième, travaux de la section IIIc, géographie coloniale*, pp. 478-93. Leiden: Brill. [Union Géographique Internationale.]

Moyer, D.S.
1983 'Cultural constraints on marriage; Anti-exchange behaviour in nineteenth-century South Sumatra', *Bijdragen tot de Taal-, Land- en Volkenkunde* 139:247-59.

Murphy, J.O. and P.H. Whetton
1989 'A re-analysis of a tree-ring chronology from Java', *Proceedings of the KNAW, Series B* 92/3:241-57.

Nagtegaal, L.
1995 'Urban pollution in Java, 1600-1850', in: P.J.M. Nas (ed.), *Issues in urban development; Case-studies from Indonesia*, pp. 9-30. Leiden: Research School CNWS.

Owen, N.G. (ed.)
1987 *Death and disease in Southeast Asia; Explorations in social, medical and demographic history*. Singapore: Oxford University Press. [ASAA Southeast Asia Publication Series 14.]

Padoch, C.
1982 *Migration and its alternatives among the Iban of Sarawak*. The Hague: Nijhoff. [KITLV, Verhandelingen 98.]

Palmer, J.G. and J.O. Murphy
1993 'An extended tree-ring chronology (teak) from Java', *Proceedings of the KNAW, Series B* 96/1:27-41.

Pelzer, K.J.
1945 *Pioneer settlements in the Asiatic tropics; Studies in land utilization and agricultural colonization in Southeastern Asia*. New York: International Secretariat Institute of Pacific Relations.

Persoon, G.A.
1994 *Vluchten of veranderen; Processen van verandering en ontwikkeling bij tribale groepen in Indonesië*. [PhD thesis, Rijksuniversiteit Leiden.]

Pigafetta, A.
1969 *Magellan's voyage; A narrative account of the first circumnavigation.* Translated and edited by R.A. Skelton. New Haven/London: Yale University Press.

Presgrave, E.
1822 'Account of a journey from Manna to Pasummah Lebar and the Ascent of Gunung Dempo [...] in the year 1817', *Malayan Miscellanies* 2-2:1-93.

Raffles, S. (ed.)
1830 *Memoir of the life and public services of Sir Thomas Stamford Raffles [...].* London: Murray.

Reid, A.
1983 (ed.) *Slavery, bondage and dependency in Southeast Asia.* St. Lucia: University of Queensland Press.
1987 'Low population growth and its causes in pre-colonial Southeast Asia', in: N.G. Owen (ed.), *Death and disease in Southeast Asia; Explorations in social, medical and demographic history*, pp. 33-47. Singapore: Oxford University Press. [ASAA Southeast Asia Publication Series 14.]
1988 *Southeast Asia in the age of commerce 1450-1680. Vol. 1. The lands below the winds.* New Haven/London: Yale University Press.
1993 *Southeast Asia in the age of commerce 1450-1680. Vol. 2. Expansion and crisis.* New Haven/London: Yale University Press.
1995 'Humans and forests in pre-colonial Southeast Asia', *Environment and History* 1:93-110.

Rousseau, J.
1990 *Central Borneo; Ethnic identity and social life in a stratified society.* Oxford: Clarendon Press.

Schefold, R.
1988 'De wildernis als cultuur van gene zijde; Tribale concepten van "natuur" in Indonesië', *Antropologische Verkenningen* 7-3:5-22.
1990 *Harmonie en rivaliteit; Verbeelding van botsende principes in Indonesië.* Leiden: Rijksuniversiteit. [Inaugural lecture.]

Schoorl, J.W.
1994 'Power, ideology and change in the early state of Buton', in: G.J. Schutte (ed.), *State and trade in the Indonesian archipelago*, pp. 17-60. Leiden: KITLV Press. [Working Papers 13.]

Schulte Nordholt, H.
1996 *The spell of power; A history of Balinese politics 1650-1940.* Leiden: KITLV Press. [Verhandelingen 170.]

Seavoy, R.E.
1973 'The transition to continuous rice cultivation in Kalimantan', *Annals of the Association of American Geographers* 63:218-25.

Snouck Hurgronje, C.
1893-94 *De Atjéhers.* Batavia: Landsdrukkerij, Leiden: Brill. 2 vols.

Substance
1820 'Substance of the report on the condition of society', *Proceedings of the Agricultural Society Established in Sumatra* 1, II:1-52.

Teensma, B.N.
1989 'An unknown Portuguese text on Sumatra from 1582', *Bijdragen tot de Taal-, Land- en Volkenkunde* 145:308-23.

Umans, L.
1993 *Analysis and typology of indigenous forest management in the humid tropics of Asia*. Wageningen: Stichting BOS.

Valentijn, F.
1724-26 *Oud en nieuw Oost-Indiën*. Dordrecht: Van Braam, Amsterdam: Onder de Linden. 5 vols.

Volz, W.
1909 *Nord-Sumatra; Bericht über eine im Auftrage der Humboldt-Stiftung der Königlich Preussischen Akademie der Wissenschaften zu Berlin in den Jahren 1904-1906 ausgeführte Forschungsreise. Band I: Die Batakländer*. Berlin: Reimer.

Wallace, A.R.
1869 *The Malay archipelago; The land of the orang-utan, and the bird of paradise; A narrative of travel, with studies of man and nature*. New York: Harper.

Wiersum, K.F.
1996 'Domestication of valuable tree species in agroforestry systems; Evolutionary stages from gathering to breeding'. Paper presented at international conference on domestication and commercialization of non-timber forest products in agroforestry systems, ICRAF, Nairobi.

Wuisman, J.J.J.M.
1985 *Sociale verandering in Bengkulu; Een cultuur-sociologische analyse*. Dordrecht/Cinnaminson: Foris. [KITLV, Verhandelingen 109.]

HAROLD BROOKFIELD

Landscape history
Land degradation in the Indonesian region

Argument

In a volume on the environmental history of Indonesia, this chapter adopts the distinctive approach of writing about landscape history, a field wider than the range of problems and periods usually embraced within 'environmental history'. The history of the landscape is more inclusive. It embraces physical and ecological processes affecting the land surface, and necessarily draws substantially on evidence that is not historically documented. In the felicitous words of Rackham (1986:xiii) it brings together 'the inanimate world of climate, soil and landforms; the world of plants and animals; the world of archaeology; and the world of historical documents'. Land degradation is here viewed as a part of landscape history, permitting access to a multidisciplinary range of sources, one of which is informed observation of the landscape itself.[1] Use of a landscape history framework also permits access to evidence extending over much longer than documented time. On the interpretation of land degradation at any place, Blaikie and Brookfield (1987:9) suggested that at any point in time it is the product of an 'equation' in which natural degrading processes plus human interference are offset by natural restorative processes plus the aid given these by human artifice. It follows that degradation has to be viewed in terms of natural causes and processes, whatever their time span, as well as human causes and preconditions. A distinctive set of interpretations can follow.

In Southeast Asia, 'degraded land' usually refers to land converted from forest into various patterns of felled or burned forest, savannah, grassland, and cultivation, and thereby exposed to accelerated soil erosion.[2] Land in

* This chapter owes a great deal to valuable information supplied by Lesley Potter, and also by Jack Golson and Kurt Lambeck. The many useful suggestions offered by the editorial committee in Leiden have, so far as possible, been followed. The help of Helen Parsons in assembling the references, and in many other ways, has been indispensable.

1 Nibbering (this volume) is also writing landscape history, though he does not describe it as such. He draws on comparative evidence from the landscape itself in different regions as fundamental evidence.

2 Usually the grassland is described as *Imperata cylindrica (alang-alang)*, although this is by no means the only common grass, or even the principal one in many areas.

which the main problems are soil-nutrient depletion alone or soil compaction is seldom included, and salinization associated with land water is widely recognized as a problem only in certain areas of strong seasonal drought, particularly in northern Thailand and the inland basins of Timor (Ormeling 1956). Although the popular description of all grassland as 'degraded' land will be questioned below, this chapter will follow the regional fashion, and I will discuss only expanding grasslands and accelerated erosion due to gross vegetational change. Although natural erosive forces and a potentially natural origin for certain large grassland and savannah areas have been discussed from time to time, the prevailing view is that degradation in equatorial Southeast Asia is overwhelmingly the work of people. Miners, loggers, and planters do receive a share of the blame, but the principal nominated villains are the upland farmers, both for becoming too numerous, and for practising shifting cultivation.[3] There is only somewhat less agreement about the time period involved, or the degree of subsidiary relationship to local natural conditions.[4] Land deterioration and its accompanying erosion are seen as fundamentally modern. Donner (1987:137) cites from Klinkert (1937) an account of a catastrophic flood in East Java in 1861 for which he implies deforestation as an underlying cause, but this is in support of his general, and mainly contemporary, argument.[5] Indirectly he does call attention to the significance of rare, high-magnitude rainfall events in producing severe erosion; a high proportion of total damage takes place during such events. Before the EDEN project was initiated, only a few writers on the environmental history

3 Two modern quotations, both stressing ongoing process and both either stating or implying wholly human causation of degradation, may stand as examples 1. 'Currently, land use throughout Kalimantan is on an unstable trajectory of net conversion of rain forest to *alang-alang*, without yet the development of sustainable systems of productive agriculture, of productive managed forestry and of protection of nature reserves' (Leighton and Peart, 1988); and 2. 'In Indonesia and in Southeast Asia in general, there is a strong tendency for critical land to be wasteland occupied by *alang-alang* [*Imperata cylindrica*] [...] Outside Java critical land is covered with *alang-alang* and shrubs as a result of improper shifting cultivation practices' (Soerjani 1980:9). In Indonesia and Malaysia alone, shifting cultivation already has a huge literature. Notwithstanding the classic nature of some descriptions (Dove 1981; Chin 1985), there is no one single system of shifting cultivation, and probably never has been (Spencer 1966; Pelzer 1978a; Brookfield, Potter and Byron 1995). It has now been shown, especially in West Kalimantan and Sarawak, that from quite minor to almost total changes in land-use system have been achieved from a relatively simple 'shifting-cultivation' base over periods of from about a hundred to three hundred years (Padoch 1982; Padoch and Peters 1993; Sather 1990). It also seems possible that some of the areas in which there has been the least elaboration of farming systems are areas which have experienced significant decline in population within the past 150 years. None of this information makes much impact on those who write 'crisis scenarios'.

4 For example, Lane-Poole 1925; Van Steenis 1937; Endert 1946; Richards 1952; Ormeling 1956; Robbins 1960; Metzner 1977, 1982; Henty 1982.

5 Here the term 'underlying cause' is used to describe the cause of severity of damage, not of the incident itself of which in this case the 'proximate cause' was a 1,000 mm rainfall over a few days.

of Indonesia had drawn systematically on historical evidence.[6] There is also important nineteenth-century historical material from neighbouring Malaysia (Jackson 1968).

This chapter sets out to raise questions as much as to provide answers. Two principal questions are asked: first, how much of the degradation seen today has taken place only in modern times, or within the period since AD 1500?; second, how large is the contribution of natural forces and conditions relative to that of human beings? Two other questions logically arise. They are, third: how much of former degradation has been repaired?; and, fourth – in words similar to the second question – how much of the repair is due to natural processes and how much to human beings?[7] The questions cannot readily be separated, but it will help to try to do this. Except in the next section and towards the end of the chapter, less is written here about the third and fourth questions than about the first and second, since most of the discussion is about origins rather than repair. In Indonesia, research of the type most relevant to these questions is limited in scope, though we shall find some important material in the islands of Java, Timor, Irian/New Guinea and Borneo. Nothing is written in this chapter about degradation due to the timber industry, principally a post-1950 phenomenon, or about degradation of forest biodiversity due to extraction of non-timber products. This is discussed by Potter (this volume, and 1996b).

Degradation in Java, nineteenth and twentieth centuries

We begin with Java, where the sediment loads carried by rivers are among the highest in the world (Douglas and Spencer 1985:60-3), but where great areas that were grassland a century ago are now productive, terraced fields. This seeming contradiction ought to be resolvable, yet the literature is of limited help. Nibbering (1991b:32-3) and some other writers have noted that there is extremely little firm information on the actual derivation of the high sediment loads of Javanese rivers.[8] Although

6 They include Pelzer 1978b; Potter 1987, 1993, 1996a, 1996b, 1997; Brookfield, Potter and Byron 1995; Nibbering 1991a, 1991b; Nibbering this volume.

7 Blaikie and Brookfield (1987) and others have proposed: 1. that degradation has had its major effect on sites where the natural sensitivity of ecosystems to the forces of change is high, often because of conditions inherited from a much earlier past and 2. that repair of temporary damage has been more common where the resilience of ecosystems – their natural ability to recover from interference – is high.

8 There are serious questions about some of the erosion estimates derived from sediment-load measurements. Seckler (1987:86) cites an example of two sets of 'measurements' in the Solo basin, one yielding an erosion rate of 50 tonnes/ha/year, the other of between 1,800 and 4,800 tonnes/ha/year. The higher figure would mean that 38 cm of soil would be lost annually to erosion. These estimates were derived from sediment-load measurements, some of them possibly taken during short, unusually wet periods. Nibbering (1991b:32) recalls that in an East Java watershed improved measuring techniques led to reduction of the 'average erosion rate' from 3 mm per year to 1 mm. Moreover, a disproportionate share of this came from roads and

erosion remains widespread in some deforested uplands a lot of it is bank erosion by fast-flowing streams. Landslides are not uncommon on steep, young volcanic mountains, even under forest (Diemont, Smiet and Nurdin 1991).[9] There are wide differences in the erodibility of soils. A significant part of the sediment load may derive from very young and unconsolidated volcanic material in the headwaters, where large ash falls, *lahar* and rock avalanches have occurred several times even in the period of modern record (Dam 1994). Diemont, Smiet, and Nurdin (1991) place stress on these events and other natural causes of erosion. In addition, they draw attention to the common practice of cutting into hillsides to establish or extend irrigated ricefields; the marginal channels are used to dispose of the excavated material. Other material later slumps into these channels, and all this load is promptly carried into the rivers.

In regard to the upland cultivated land, there are large differences between the interpreters. Nibbering (1991b:34) comments on Hardjono's (1986:4) view that upland farmers in Java still rarely practise terracing, by stating that in the 1980s it was 'increasingly difficult to find areas where dry-farmed land is not terraced in one way or another'. Palte (1989), on the other hand, recounts failure to adopt adequate terracing in Central Java. Drawing on subsequent work on an area of commercial farming in West Java, Hardjono (1993:239) insists that both subsistence and commercial farming on the uplands 'lead to land degradation when they are practised on steep slopes'. She was drawing on data from an area where terracing was still not practised. By contrast, it could be said mainly of Java in the 1940s that, because of its sound slope management, Indonesia suffered far less erosion than did many other tropical countries (Naval Intelligence Division 1944, I:369). This was written before the worldwide surge of concern about soil erosion in the developing countries which, though it has its origins in the 1930s 'dust-bowl' period in the United States, arose mainly after World War II.

It may be that some of the modern 'crisis narratives' about erosion in Java reflect what Nibbering (1991b:34) describes as a 'historical lag' in which substantial human repair and generally improved management tend to be ignored. This applies especially in his own principal field area in the limestone Gunung Sewu southeast of Yogyakarta. The land was

---

villages, rather than agricultural land. Erosion estimates for Java differ by several orders of magnitude, and it has to be concluded that little is, in fact, known.

9 Major rainfall events will always produce not only flooding but also erosion, not only on cleared land but even under thick forest. The particular problem in the latter is that when deeply weathered ground is saturated in depth, the weight of the trees on slopes can overcome the binding action of their root systems, leading to very large landslides which can strip whole hillsides, and bring great quantities of debris into the valleys. In another part of the world, I recall in 1980 the forested central mountain range of St Vincent in the West Indies looking, from a few kilometres away, like the black and white keys of a piano after such a rare event. I have seen similar but less dramatic scars on forested mountainsides in several parts of Southeast Asia.

already damaged by grazing and burning before close agricultural settlement began in the first half of the nineteenth century. It was disastrously deforested and eroded in the early part of the twentieth century, but was largely under conservationist management and yielding an improved livelihood to its closely settled people in the 1980s. Nibbering's comparative account of three Javanese upland areas in this volume demonstrates the scale of the changes that have taken place in management practice – largely spontaneously by the farmers themselves – as palpable symptoms of degradation have been seen to threaten the livelihood of growing numbers of people. He also carries the story into the outer islands, where different stages of the same process of management intensification can be recognized.

The background to modern change

Population growth, major expansion of cultivation especially into the uplands, adoption of new crops, and development of new management technologies have all impinged on the natural landscape since 1500, and especially since 1850, in ways not experienced in earlier centuries.[10] Although a part of the expansion of cultivation has been based on the use of upland rice, the introduction of New World crops around the period 1520-1650, and their subsequent adoption and spread, provided the principal means for expansion of dryland agriculture of higher productivity than that based on the regional upland domesticates, taro, yams, millet, and Job's tears (*Coix lacrymae jobi*).[11] The New World crops widened the range of ecological niches that could be occupied, and new crops were often substituted for old because of their higher production performance and lower demand on soil nutrients. The pre-1500 range of crops in the region included both seed-propagated and vegecultural plants, and the sixteenth century brought more of each.

Among the new root crops one, the sweet potato (*Ipomoea batatas*) had an outstanding impact when it reached Irian/New Guinea, as we shall see below, but it was also adopted in other parts of Indonesia, in the Philippines and on the Asian mainland. In southern China, sweet potato was sufficiently well known by 1594 for planting material to be deliberately imported in order to alleviate famine (Yen 1974:327). Sweet potato itself probably first came to the region via India after regular Portuguese

10 In a discussion of the environmental problems of the uplands of Southeast Asia, Allen (1993) distinguished the period 1850-1950 from all that went before, and then the period since 1950 as one in which wholly new forces came strongly into play. As I do in this chapter, Allen noted that a good deal of former degradation has been repaired, especially by adoption of new farming technologies. However, he treated all that happened before 1850 as one.

11 The present millet in Java is the pan-Asian *Setaria italica*, but the species anciently grown in large parts of the archipelago was probably a cultivar of *Panicum crus galli*, or farmyard millet (Nibbering 1991b:27).

voyaging began early in the sixteenth century (Yen 1974). Maize seems to have come somewhat later, and the sweet potato was already present in the Philippines by the time that maize, and further sweet potato varieties, were successfully introduced by the Spaniards around 1560 (Spencer 1975).

Ultimately, maize and cassava, together with tobacco, have become the most important of the new introductions. In the Southeast Asian region, maize was first adopted on a large scale in southern China, where the Chinese already had available milling technology that could be used (Spencer 1975:10). In most of Indonesia, as in the Philippines, adoption of this annual crop seems to have been a more gradual process, though it is mentioned in the Moluccas as early as 1540 in a source used by Reid (1988:19).[12] It became a staple food in eastern Indonesia, especially Timor, before the late seventeenth century (Ormeling 1956; Fox 1977), but its adoption as a crop of upland areas in Java and Madura seems to have been delayed until the eighteenth century, even late in that century.[13]

Maize is less demanding of soil nutrients than upland rice, but its cultivation exposes more of the ground to rain impact when grown as an annual crop, rather than in a mixed or relay pattern. The same applies to the more recently adopted cassava. As soil fertility on the Java uplands has been depleted, a rotation of rice, maize, and cassava has quite widely been adopted, but it can lead to further decline in soil quality so that ultimately only maize and cassava become feasible crops (Palte 1989). Where these and other crops are grown in a mixed or relay pattern, in association with tree crops, or if good use is also made of livestock manure and of short-term fallowing, there is no necessary degradation, and the cropping pattern can be sustainable (Palte 1989; Nibbering 1991a, 1991b).

On erosion-sensitive soils without terracing the effect of annual cropping for food crops can be comparable to that of the cash crops pepper, tobacco, and gambier.[14] The effects of some of the early commercial agriculture, growing these and other field crops by very simple methods, are more fully described for Peninsular Malaysia than for Indonesia (Jackson 1968). Chinese cassava and gambier farmers created something like a 'hollow frontier' of farmland, backed by *Imperata cylindrica* grassland, advancing up interfluves between the Malay-occupied valleys. Later, all this became covered by rubber. In the same way, Dutch tobacco planters in Sumatra believed for some time that only a single crop of cigar-wrapper

12 As with some other early references to maize, the crop actually seen may have been sorghum (*Sorghum vulgare*).

13 In the Philippines, similarly late adoption is interpreted as related to importation of stone mills by Chinese traders (Spencer 1975:14).

14 In the case of the woody sprawler gambier (*Uncaria gambir*), damage arose not so much from its cultivation on open ground – as was the main cause of erosion associated with pepper and cassava – as from the large demand for wood to boil down its leaves, to derive the catechu-tannic acid which was the commercial product.

tobacco could be obtained from the cleared forests, and so left grassland behind them. In time, however, they did control the use of fire to encourage secondary forest (Pelzer 1978b). Farmers have subsequently reclaimed much of this land (Pelzer 1982). When rubber planting began in Peninsular Malaysia, growers following European practice adopted clean weeding, with disastrous results in the great 'red floods' that took place during a major rainfall event in 1926 (Winstedt 1927). These floods remained on record as the highest in Kuala Lumpur until 1971. While there is less material from Indonesia, the pattern was much the same (Brookfield, Potter and Byron 1995).

The forces at work in this historical period were complex. Population growth at a high rate has been important, especially since 1800, leading to the occupation of new areas. War and taxation have also driven people into the uplands. While clearance inevitably led to erosion in the short term, growth of labour numbers has increasingly facilitated subsequent conservationist management.[15] In terms of land degradation, cattle grazing and the development of commercial agriculture may have done much more dramatic damage than food-crop farming. The question of cattle grazing is important in the explanation of nineteenth-century land degradation in Java, as Nibbering shows (this volume). It leads us also to Timor, and to a critical point in development of the argument.

Grassland, savannah and the possible inheritance from the Pleistocene in Timor

Although present in small numbers earlier, cattle have become important in Timor only since their deliberate introduction in 1912, and they now share with shifting cultivators an important part of the blame for persistence of grassland, deforestation, and accelerated erosion in that island. Doubts about the modern and continuing anthropogenic creation of these problems in Timor have scarcely arisen in the literature since their very cautious presentation by Ormeling (1956). As in many other areas, pyromaniac or 'match-happy' indigenes seem to most writers to provide sufficient cause. Yet Timor, with its limited areas of forest, long dry season, and rain falling on relatively few days of the year, ought to be an area *par excellence* within Indonesia where, under the hypothesis that human activity would have had a supplementary rather than a fundamental role, the effect of climate should be given first consideration.

The severity of the dry season in Timor, especially in the north and in inland valleys, derives from the fact that for half the year, during the

[15] However, Palte (1989) explains a partial halting of efforts to improve soil and water management on steep slopes in Central Java as due to a major increase in opportunities for off-farm employment in modern times, making labour less available despite continued population growth. This may be an increasingly widespread problem, particularly in Java.

Downpour over Singkalong, Central Sulawesi; photograph by A.A. Cense, ca 1937 (KITLV photo collection 29461).

southeast monsoon season, the island is in the lee of Australia.[16] The important May to July rainfall on the southern side, which shortens the dry season, is mainly picked up over the Timor Sea. These climatic controls become much more significant in interpreting land-cover patterns if climatic history is taken into account alongside human history, and if this is done over a much longer period than we have considered up to this point.

During the late Pleistocene sea levels were very low, and the Timor Sea shrank to a width of 100-200 km. Well-evaluated sedimentological and limnological evidence from near the edge of the Sahul shelf, closer to Timor than to Australia, suggests that its late-Pleistocene rainfall was only from a half to a third of that falling today on the North Australian coast (Van Andel et al. 1967). Sea temperatures were also cooler, and the Sahul plain would have been an arid steppe considerably drier than the Australian coast between Darwin and Broome is today. Its coast would have been close enough to be within easy sight from Timor's mountains. Until sea levels rose sufficiently to widen the Timor Sea, Timor can have had only a short and weak wet season and most of the island would have been almost wholly arid for the greater part of every year. Forest could have persisted only on the highest ground.

Sea levels worldwide were lowest around 18,000 years ago. The ice sheets began to shrink rapidly at dates that varied around 16,000 years ago, and early work suggested that there would have been an immediate effect on sea level. Van Andel et al. (1967), for example, suggested that flooding of the Sahul shelf would already have begun about that time. More recent work, taking account of the isostatic effects of glacial melting and deeper seas, have revised this to the extent that change is seen as very small until after 14,000 years ago. The sea worldwide then rose rapidly by about 120 m until about 9,000 years ago, when it slowed to reach a peak about 6,000 years ago (Lambeck 1995). Subsequently it has declined slightly. For the Sahul shelf, unpublished reconstructions by Lambeck (personal communication) show that as late as 12,000 years ago much of the shelf was still land, and the Timor Sea was not yet more than 300 km wide. To the east there remained a wide, climatically dry, land barrier extending from south of the Aru Islands to beyond Torres Strait, not finally opened until after 8,000 years ago.

There were certainly people in Timor by 13,500 years ago, and they were foragers, or hunter-gatherers. Their activities would have interfered with the slow post-Pleistocene recolonization of the island by monsoon forest, though probably less so than the activities of the farming population that became firmly established only about 4,000 years ago (Glover 1986). Natural and human fires in combination could quite possibly

16 Although several months of every year are wholly or almost wholly without rain at a number of recording stations in Timor, the dry spell is more complete and longer during major regional droughts, associated with the periodic El Niño phenomenon which is discussed below.

have prevented any sustained establishment of monsoon forest over quite large tracts that today remain grassland and savannah. Similar arguments calling on the Pleistocene inheritance are advanced for grasslands and savannah on the south coast of Papua east of Torres Strait, where climatic conditions today are more favourable to forest than they are in parts of Timor. There, an agricultural Austronesian population seems to have become established more recently than in Timor, only some 2,000 years ago (Eden 1974).[17] The significance of inheritance in this argument is not to discount the consequences of human interference, but rather to open the possibility that in a seasonally very dry environment only a light human interference has been necessary to prevent recolonization by forest, after relaxation of the severe climatic conditions that had earlier created semi-arid vegetation.

The erosion that is rampant in Timor raises another dimension. Rapid uplift in the outer Sunda arc, and to a lesser degree in some other areas of Indonesia, has continued into modern times. The study of raised coral reefs on the north coast of East Timor suggests an uplift rate of half a metre per 1000 years through the Pleistocene and Holocene (Chappell and Veeh 1978). Steepening of slopes by uplift leads to entirely natural accelerated erosion, quite irrespective of any anthropogenic component. This is especially significant in the 'ridge-and-ravine' topography of inland Timor. Deep incision of short, fast-flowing streams, accompanied by slumping and sliding of the soil around gullyheads in the headwaters, are noteworthy consequences. They are to be expected on *prima-facie* grounds as a consequence of recent and ongoing steepening of gradients, though they are still often attributed solely to human deforestation (Bierenbroodspot 1986). Moreover, except for the limestones and limited areas of metamorphics, much of Timor is composed of highly erodible material, in particular the widespread Bobonaro scaly clay which has low permeability in depth, leading to lateral water flow within the soil and hence frequent slope failure.[18]

17 For the grass and savanna region south of the Fly and Digul rivers in southern Irian/New Guinea, both sides of the international boundary east of Merauke, Swadling (1983) brings together a number of suggestions from earlier literature, and legend, to raise the possibility that the mound-raising agricultural people of this zone are an even more recent arrival, coming from the north within perhaps only the last few hundred years, though there were other people there before them. This area, opposite Torres Strait, today borders a shallow sea that was exposed and under a very dry savanna until the early Holocene. There is little speculation in any of the literature on the origins of the modern savannas of this seasonally-dry southern 'bulge' of Irian/New Guinea, but it seems clear that the possibility of inheritance from Pleistocene and even early-Holocene aridity cannot be excluded.

18 In a much more stable environment far to the west in Malaysia, there is a major contrast in lithology between the old core areas of the Sundaland complex in western Borneo and Peninsular Malaysia, and the comparatively recent additions in eastern and northern Borneo. The natural suspended sediment yields of rivers draining wholly forested areas are half an order of magnitude greater among the mudstones of eastern Sabah than on the main range of the Malay Peninsula (Douglas 1994). Before these measurements were made, Douglas

Grassland and degradation in the mountains of Irian/New Guinea

Early origins for present grassland or other apparently degraded vegetation formations are suggested in several parts of the tropics. One such change, with no direct evidence of human interference being involved, is said to have taken place around 3,500 years ago in western India (Caratini et al. 1991), though this conclusion has been questioned (Meher-Homji 1996). Continuous evolution of grassland ecosystems is suggested from even before the beginning of the Pleistocene ice age in the stable landforms of Africa (Retallack 1992). On the island of Lakeba in Fiji, work with which the writer was closely associated placed the origin of a supposedly modern degraded formation well back in the Pleistocene. It also demonstrated by measurement that erosion during the 3,000 years of human occupation can only have been a fraction of the total erosion and redistribution of soil material that took place in association with degradation (Brookfield et al. 1979; Latham 1983). Mainly from African sedimentological evidence, it has been suggested that quite abrupt climatic change from drier to wetter conditions at the end of the Pleistocene led to periods of fairly intense erosion. Over one or more thousand years the surface on which substantially increased rain fell remained only lightly vegetated. The effect may have been at least as great as that produced by deforestation under human hands, and 'the current phase of man-induced accelerated geomorphic activity may be far from unique in recent geological history' (Street-Perrott, Roberts and Metcalfe 1985:183).[19]

The central mountains of Irian/New Guinea constitute a single region and had, until 1930, a single human history. Although there are many differences among the people, both agriculture and society of the larger population groups are more alike than unlike across the border that divides the ranges almost exactly in half. Grasslands occur in two altitudinal bands. Alpine and sub-alpine grasslands occupy areas above about 3,800 m; seemingly anthropogenic grasslands are widespread between 2,500 m and about 1,200 m where, in the valleys which exit the highlands, lowland rain forest takes their place. During the Pleistocene period the mountains were affected in ways even more dramatic than was Timor. When the glaciers were at their most extensive about 18,000 years ago they occupied an area of some 2,000 km^2 along the whole mountain spine, compared with 8 km^2 in the 1970s and even less today. Until about 12,000 years ago, as Hope and Hope (1976) determined, there was a semi-

(1990:215) remarked that in the Borneo mudstones and shales 'landslipping, channel-bed and bank erosion often supply large quantities of mud, sand and gravel to rivers in dense tropical forest devoid of any human disturbance'.

19 In southeastern Thailand it has been shown that late-prehistoric deforestation (4,500-2,500 years ago), associated with early copper mining and smelting, caused considerable erosion and re-deposition. It did so on a landscape already significantly transformed by rapid natural processes at the end of the Pleistocene (Cremaschi, Ciarla and Pigott 1992).

continuous swath of alpine and sub-alpine grassland and shrub extending over more than 1,500 km from just east of the Paniai lakes in Irian Jaya to the eastern end of the high ranges in Papua New Guinea. This was due mainly to lower temperature, but probably also to much lower precipitation. The treeline in the main ranges fell to around 2,300 m, then climbed again rather rapidly to reach 4,000 m by about 8,500 years ago. It subsequently declined some 200 m, partly for climatic reasons as the warm post-glacial 'hypsithermal' (9-8,000 to 5-4,000 years ago) came to an end, and partly because of human interference due to hunting with fire in the alpine grasslands.[20]

People were already present, as foragers and hunters, during and well before the period of maximum glaciation. There is palaeobotanical evidence indicating that they already used fire, creating or enlarging grasslands. Today, grasslands interspersed by cultivation occupy most of the intermontane valleys between about 1,500 and 2,500 m. These valleys are populated by some two million agricultural people. The cultivation and grassland are much more extensive in total area than the ancient alpine grasslands, which would have only just reached down to their present upper limits. Above and around these valleys is post-glacially established lower montane, upper montane and sub-alpine rain forest, forming distinct complexes within altitude bands that do not differ greatly from one end of the ranges to the other. Reading upward, each exhibits successively less biodiversity than the formation below it (Grubb and Stevens 1985). The differences are very marked, and transition zones between the three types are narrow. Interpretation suggests that formations have not recolonized upward as a whole, but that individual species have colonized according to tolerance of temperature, persistent cloud and fog, and soil limitations, and that the process of recolonization is still continuing.[21]

The origin of the mountain-valley grasslands has been discussed since the early 1960s.[22] At that time, little was firmly established concerning either the prehistory or palaeobotany of the region, but subsequent work has made the highland region of Papua New Guinea, and to a much lesser degree Irian Jaya, one of the most closely researched regions of the intertropical world from these points of view. The following summary draws

20 Hope and Hope 1976; Walker and Flenley 1979; Walker and Hope 1982; Haberle 1994.

21 Bringing together all available data from a wide region, Grubb and Stevens (1985:192-5) show that for trees and shrubs, ground-dwelling herbs, climbers and scramblers, and vascular epiphytes alike, the numbers of families and genera decline upwards, and so also do the numbers of genera within families, and of species within genera.

22 A view that they are wholly anthropogenic and, in the Papua New Guinea portion, represent a continuing history of human migration, clearing new forest and leaving a 'hollow frontier' behind, was strongly presented by Robbins (1960, 1963). It was challenged by Brookfield (1964) on the basis that it disregarded complex internal variations in highland climate, as well as what was then known about human migration in the preceding one or two centuries. Moreover, it could not readily be extrapolated to Irian Jaya.

mainly on the most recent statements (Golson 1989, 1991; Haberle 1994), disregarding a large number in between. Although most of the researched data comes from Papua New Guinea its conclusions apply almost without change to Irian Jaya. The small amount of palaeobotanical work done in the Irian Jaya mountains indicates this beyond doubt.

Equally on both sides of the border, and with almost exact equivalence in dates, it is now firmly established that widespread agricultural occupation including wetland management by ditch-drainage had taken place by about 6,000 years ago. A significant change in agricultural occupation, almost certainly involving the tillage of grasslands, began around 2,000 years ago. About 1,400 years ago the pollen record indicates that the agroforestry practice of planting nitrogen-fixing *Casuarina oligodon* in dryland cultivated areas had begun at a few sites, becoming more extensive in the course of the next few hundred years.[23] There was then a major increase in occupation about 350 years ago, involving the first agricultural settlement of land above about 2,300 m. This is associated with the introduction, via Indonesia in the early-to-mid seventeenth century, of the South American sweet potato (*Ipomoea batatas*) which soon became the principal staple in all but a few areas. The most recent period, perhaps less than 250 years, has involved the evolution and spread of a cluster of intensive soil-management practices, some of them elaborate, and also some large human migrations accompanied by warfare. Important also has been a set of systems for sustaining social relations all based primarily on use of the domesticated pig in ritual and compensation payments, and in competitive social exchange. Cultivation of food for the pigs about doubled the human land requirement.

Before the firm recognition of cultivation around 6,000 years ago there is evidence of what may have been clearance for agriculture at valley sites in both Irian Jaya and Papua New Guinea, in the latter involving some rather simple management of a wetland dated at 9,000 years ago. The main evidence for agriculture in this period is clay sedimentation dated before 7,000 years ago in the Baliem valley of Irian Jaya, and in the Papua New Guinea site overlying the 9,000-year-old ditches. In terms of the history of vegetation this sedimentation presents the greatest problems because, while a 6,000-year-old beginning for agriculture lies well in the warm hypsithermal, at 9,000 years ago the valleys at 1,500 m would only recently have become warm enough for farming, and it is unlikely that the lower montane rain forest would yet have been fully established. Indeed, Walker and Hope (1982) very germanely felt it was possible that the intermontane valleys at this level had carried no type of forest in the coldest millennia, because of cold-air drainage from the extensive open

[23] There has been a very large increase in the areal extent of *Casuarina oligodon* planting since the 1930s, and especially since 1950.

lands above.[24] Whether the clay deposition in the early Holocene represents very early clearance for cultivation, or the effect of increased rain on a landscape still only lightly vegetated, following the African speculation mentioned above (Street-Perrott, Roberts and Metcalfe 1985), remains unknown although the balance of opinion favours cultivation. In any case, it seems that although there has been major extension of grasslands, we have here another instance in which inheritance from an earlier period of drier and cooler climate has quite possibly contributed to the formation of a modern landscape element.[25]

Finally, in regard to the mountain grasslands, what of the patchy but locally quite extensive erosion that has taken place? Some of it, for example in the Baliem valley of Irian Jaya, seems fairly clearly located in the sweet-potato period since the seventeenth century, if the pattern of stone-walling of gardens on limestone slopes belongs wholly to this time. This would be a reasonable supposition given its wide modern use around the valley and in the outfall gorge. There are some areas, including one on a limestone hill in the central valley near Wamena, where the remains of stone walls enclose plots which are now almost totally bare limestone pavement, denuded of soil.[26] This is an extreme though not unique case, and

24 The question of possible aridity is also important. In a detailed study of the summit climate on Mt Kinabalu in Sabah, based on observations during a recent drought year, Kitayama (1996) noted that air subsidence became common after sunset, when the day-time convective uplift of air ceased. This subsidence, sufficient to overcome the wetting effects of lowered saturation-vapour pressures with low night temperatures and cause very dry conditions, was probably intensified by the climatic conditions accompanying the drought (El Niño, see below). It could arise spasmodically under normal conditions if the temperature inversion associated with the tropical Hadley cell circulation falls sufficiently low to reach the high mountains, which it always does on the eastern side of the oceans. Kitayama noted the finding of daily irregular fluctuations, probably due to inversions, near the summit of Mt Wilhelm in Papua New Guinea (Hnatiuk, Smith and McVean 1976). He further related this to the widespread presence of sclerophyllous vegetation and small leaf size on most species on Mt Kinabalu, and hypothesized that this might be a response to repeated moisture stress. In their comparative study, Grubb and Stevens (1985:118) also noted a sustained drop in mean leaf size between the lower montane, upper montane and especially the sub-alpine forest in Papua New Guinea. The tree line before about 12,000 years ago was down to about 2,500 m, only 1,000 m above the main valley floors. On the basis of this evidence the Walker and Hope (1982) cold-air drainage hypothesis gains significantly in persuasion.

25 It is an obvious question to ask how far the reasoning from the mountains of Irian/New Guinea may also be applicable in Java, where frost is not uncommon in the mountains, even at fairly low altitudes, during ENSO events, and where a distinctive *Casuarina junghuhniana* forest in the highlands of seasonally dry East Java is regarded as fire maintained, probably by anthropogenic fires (Van Steenis 1972). The effect of ENSO events could be important, and there may also be a longer history of distinctive environmental conditions.

26 There are other areas, for example in the Chimbu region of Papua New Guinea, where the surface soil on some interfluves is simply the denuded, underlying mineral shale which, because of its rapid weathering and high mineral fertility, continues to produce crops of the undemanding sweet potato. Early fieldwork in 1959 established that these quite widespread soils have a name which means simply 'shale soil', as distinct from other and deeper soils. Some such areas were so in 1959 and were still so in 1984 (Brookfield and Brown 1963; Brookfield 1996). There is no present evidence that the current raised-bed system in this area generates significant erosion. This is one area where we know that the present people have been there

there are many other grassland areas which remain capable of cultivation but lie unused because the people have better land elsewhere (Allen and Crittenden 1987).

The subsequent fate of this land, whether to be reclaimed or remain abandoned, has much to do with the degree of sensitivity to further degradation and erosion for while the management methods that have been evolved might be successful on some, they would not be successful on all. Even this is not to say that all but the most totally denuded areas remain incapable of any management. The skills and large input of labour required to keep them in service have not been available, or applied. Although intensification of agronomic practices continued into the 1930s, the more recent history is of spread of agrotechnologies to new areas and of improvements of a secondary order, not of major new innovation in land management. This is in contrast with the modern history of the Javanese uplands, discussed above, and it suggests that the pressures that were felt by Javanese farmers have not been experienced by Irian/New Guinea highland farmers in modern times.

Introducing ENSO, El Niño and La Niña

Still within the Irian/New Guinea case, but with a wider reference, it becomes relevant to look more closely than has been normal practice at climatic variation within the Holocene period, a question which introduces the history of ENSO (El Niño-Southern Oscillation) and its effect in the West Pacific region. With a periodicity that ranges, in the limited historical record, from one or two years to two or more decades, it brings severe drought (El Niño) and also some very heavy rain (La Niña) to Indonesia, New Guinea and Australia.[27] Elsewhere in this volume, Henley details the large impact on production, health, and population numbers of dry and wet events, separately and in combination, during a part of the nineteenth century in North Sulawesi.[28] Although the record from the tropical West Pacific region is reasonably complete only as far back as the last quarter of the nineteenth century, there is abundant evidence of earlier events, perhaps best calibrated against the record since the 1790s from eastern Australia and India (Nicholls 1992, personal communication).[29]

---

for not much more than one century, so that the possibility of degradation under a different land-management system at an earlier time cannot be discounted, though it cannot be established either.

27 The extreme rainfall events of 1926 and 1971 in Malaysia, and probably that of 1861 in Java, referred to above, took place in La Niña events.

28 Brookfield and Allen (1991) and Brookfield (1993) advance historical evidence from Java and New Guinea to suggest that abrupt change from exceptional rain to severe drought, or vice versa, is the most deadly event for lost production and even famine, more severe in its impact than flood or drought alone. Henley (this volume) presents evidence which seems to support this contention.

29 The record of drought and frost in the New Guinea mountains is reviewed in depth by

Remoter evidence suggests a waxing and waning in the strength of ENSO over much longer time, with strong indication that the modern pattern has been present continuously only during the past 3,000 years.[30] This is the later Holocene period, when agriculture was already firmly established in Irian/New Guinea, and the montane climate was much more variable than in the early Holocene; there were at least four small re-advances of mountain ice. It was during this later period that dryland farming evolved both tillage and agroforestry. If, as is commonly believed, taro was dominant among a possibly wider suite of crops before very recent centuries, its dryland cultivation would have necessitated use of bush fallow. Sedimentation either increased or was resumed in certain swamps, and there is growing palaeobotanical evidence both of grasses and of fire. Whatever the nature of the more recent sweet-potato revolution, substantial environmental change under a variable climate had already preceded it at an accelerating rate.

Climatic variability has continuing importance. In severe droughts accompanying recent ENSO events, in 1941, 1972, and 1982, a high-altitude pattern in which high daytime temperatures were followed by severe night frosts arose in valleys above 2,000 m, causing near-total crop loss both in Papua New Guinea and Irian Jaya. There were light frosts even in the mid-montane valleys as low as 1,500 m. As we shall see below in regard to areas further west in Indonesia, the modern and historical fire record shows close association with drought arising from ENSO incidents. We have, from Corlett (1984), clear evidence that fire-penetration of sub-alpine forest took place during what is implied as an exceptional drought in the mid-seventeenth century, around or soon after the time of sweet-potato introduction and before the subsequent agricultural revolution, if such it was. Brookfield (1989) linked this evidence to a possible region-wide drought about this time, and compared it with other ENSO-related

---

Allen (1989) and in a regional context which includes Indonesia by Allen, Brookfield and Byron (1989).

30 Brookfield (1989) noted suggestions in some of the literature from the eastern Pacific that the phenomenon can be traced only after about 5,000 years ago. That would have meant that its onset would have coincided with the end of the hypsithermal in the Irian/New Guinea highlands. It now seems more likely that ENSO has been a continuous feature of Pacific climate, but that its vigour has varied substantially through time. In a review of evidence, largely pollen evidence, from both sides of the Pacific, McGlone, Kershaw and Markgraf (1992) remark of the early Holocene period that the evidence for the western Pacific is principally of climatic equability during which ENSO fluctuations were dampened. Unfortunately, they did not include the substantial pollen record from Irian/New Guinea in their comparative analysis, but they do point out that as far north as northern Queensland the period before about 8,000 years ago was not only one without evidence of vegetation adaptation to a drought(and fire)-flood sequence, but was also somewhat drier and cooler than it later became. Thereafter, but still without any major increase in variability, the western Pacific seems to have become wetter, and only after 5,000 years ago and then much more strongly after 3,000 years ago does the present high level of variability appear in the palaeobotanical record, implying that the modern pattern of ENSO events had fully emerged, or re-emerged.

fire events in the mountains.[31] Modern analogy would suggest that episodic grassland extension in the valleys might have been related to fires spreading and becoming hotter during such droughts, entering forest or secondary woodland, when exceptional dryness due to air subsidence may also have been significant at abnormally low altitudes. On a historical time scale, the incidence of such conditions can be inferred to have been quite frequent over the last 3,000 years.

Conclusions from eastern Indonesia

In both Timor and Irian/New Guinea, we seem to have a long and complex story. It includes creation of grassland that may in large part derive from human maintenance of a grassland never reforested after the Pleistocene. It also includes later creation or extension of grassland under dryland farming using fire that may have escaped during periods of drought, especially periods of major drought associated with the larger El Niño events. What seems fairly certain is that even though the human element has been of major importance, it has operated within conditions defined by environment, climatic history, and climatic variability. In Timor, continuing semi-aridity may give as strong a place to the natural as to the human causes.

In Irian/New Guinea the mid-montane grasslands may or may not be in significant part an inheritance from the Pleistocene. Here a proportion of the observable losses of topsoil probably does belong to the past 300 years of agricultural expansion during which conservationist soil management has widely been developed, but much of it may belong to an earlier time, within the past three to five thousand years. On the basis of contemporary evidence it is quite likely that seventeenth-century adoption of the sweet potato was so rapid and extensive in the mid-montane valleys because it could thrive on previously degraded grassland soils, whereas other crops could not and had to be grown only in more restricted environments.

The grasslands of western Indonesia: Kalimantan

In western Indonesia the effect of the Pleistocene was more strongly on sea level than on climate. Although a powerful case has been made for climatic conditions significantly drier than those of the present day (Ashton 1972), it derives from the interpretation of soils and vegetation, and does not rest on firmly established palaeobotanical evidence. The strongest support is inferential: most of the shallow seas around western Indonesia were dry land during the Pleistocene maximum and, from reconstructions by Lambeck (personal communication), Bali, Java, Borneo,

[31] Knapen (this volume) refers to 'one of the most serious droughts of the seventeenth century' in 1660.

Sumatra, and the mainland remained linked by continuous land until later than 10,000 years ago. Climatic conditions more continental than those of the present would have followed. Evidence derived from study of the mammals, discussed below, strengthens the probability that Pleistocene climates were drier than those which have prevailed during the Holocene.

The earliest clear record of conversion from forest to grassland in western Indonesia is in the Batak region of Sumatra, dated by pollen analysis at 1,500 to 3,000 years ago (Maloney 1980, 1985). Some of the most extensive lowland grasslands in modern Indonesia are in Kalimantan, especially but not only in the southeastern part of the island. Most Kalimantan grasslands are *Imperata cylindrica* (*alang-alang*), although other species are also present. Their history and its climatic context have been reviewed, principally by Potter in Brookfield, Potter and Byron (1995:158-203). Potter (1987, 1996a, 1996b, 1997) has elsewhere reviewed the evidence in greater detail. This section of the chapter relies heavily on Potter's work.[32]

There has been substantial modern expansion of the grasslands, and one whole area in West Kalimantan (the Ela Hulu, in the Melawi basin) seems to have come into existence only since World War II, although it was already an area of secondary forest with shifting cultivation at the end of the nineteenth century (Enthoven 1903). Modern grazing of cattle, associated with annual burning off to provide fresh pasture, has been blamed for the present extent of grassland in this area. Potter's more detailed studies in areas of seasonal drought in the lee of the Meratus mountains, and elsewhere in the east, provide much greater historical depth. It is known that big areas in southeastern Kalimantan were cleared of forest for pepper cultivation in the sixteenth and seventeenth centuries, but it is not known how much of this has become grassland and how much has reverted to forest.

In the Riam Kiwa area studied in greatest depth, Potter distinguishes between areas on the forest margins where the grassland-forest boundary has fluctuated during this century, though with a net extension of grassland, and more persistent core areas of 'sheet *alang-alang*' which were already grassland before the end of the nineteenth century, and probably have a much longer history. It is clear that fire has been a major means by which large areas of grassland have been maintained. Fires have spread most extensively, and sometimes burned into the forest, during the frequent and severe El Niño droughts which affect this part of Borneo, and to a lesser extent a much larger part of the island.[33] Fire is no new phenomenon:

[32] Potter has seen the present text, and kindly supplied a copy of her (unpublished) recent updating paper on the topic (Potter 1996b). Additional information kindly provided is acknowledged in the text.

[33] There is good documentary record of fire in most (but not all) of the major El Niño events

the occurrence of charcoal in forest soils has now been reported several times, with an earliest radiocarbon date of 17,700 years ago (Goldammer and Siebert 1989). Reading from the record established further east, we may presume that the whole of the later Holocene has been a period of occasional fire risk, of greater or lesser magnitude.

Southeastern Borneo seems to have become the most closely occupied part of the island in medieval times and fires, deliberately set for the hunting of deer and pasturing of cattle, seem likely to have a history of at least three or four hundred years. Dayak hunters may for many centuries have set dry-season fires to assist capture of a range of animal life, and wildfires may have escaped from clearings even in the forest in exceptional dry spells. There is at present no direct evidence from which earlier vegetation history can be reconstructed in Borneo. Indirect evidence is derived from the habits of wild mammals, especially ungulates; some papers included in this volume, or presented at the Workshop on which it is based, have considerable relevance. The chapters by Boomgaard and Kathirithamby-Wells in this volume and a paper presented to the workshop by Reub contain significant discussion.

The larger ungulates all require relatively open conditions for browsing. During the Pleistocene they were able to migrate on land as far as the Wallace Line and, although human interference in the vegetation probably remained light for most of this period, a more open forest could have provided all that these larger mammals needed. Some of them, such as the one-horned Javan rhinoceros, became extinct from Borneo after denser vegetation was re-established in the Holocene (Kathirithamby-Wells, this volume). Among the ungulates, the most significant in regard to the Borneo grasslands is perhaps the *banteng* (*Bos sondaicus* or *javanicus*). This animal, only very distantly related to the progenitors of European or Indian cattle (Kikkawa, Amano and Suzuki 1995), presumably migrated as far as eastern Java and Bali on the edge of the Wallace Line during the Pleistocene. Like other ruminants, it has a strong preference for disturbed forest and open savannah, and has been found unable to survive in close forest (Hoogerwerf 1938). The *banteng* became concentrated in areas periodically cleared by people, and particularly in areas where seasonal drought created the conditions conducive to long persistence of grassland. It was domesticated at the southeastern limit of its migrations, in eastern

---

since the 1870s, and there is also undated record which could relate to El Niño years such as 1891 and 1914 (Hooze 1893; Endert 1927), though it may relate simply to fires in normal dry seasons. The 1914 event is well recorded from other sources which also indicate fire. There is also record of the severity of drought in Borneo during the major global event of 1877-78. With one exception, probably 1846 (Pijnappel 1860), research has not yet uncovered records for Borneo that might relate to major events in the early and middle nineteenth century. Henley's record from Gorontalo in North Sulawesi (this volume) suggests that the 1846 drought, which also occurred there, was only one of several between 1821 and 1865.

Java or Bali, in an unknown period to become the distinctive Bali breed of cattle (sometimes *Bos javanicus javanicus*).[34]

There are now few true *banteng* in Borneo, mainly in Sabah, in a reserve shared with elephants, Sumatran rhinoceros and other species. They are also found in Kayan Mentarang on the Upper Bahau, where grassland is deliberately maintained for them and deer in an area recently abandoned by its former inhabitants (Potter 1996b). Elsewhere, even in Sarawak where they may now be extinct, they have often been described in the past, especially on river-plain savannahs and in secondary forest in several parts of the island.[35] The '*banteng*' which have in recent times lived in and around the grasslands of southeastern Kalimantan may be the descendants of feral Bali cattle. In the early twentieth century they were often described as 'cattle', not confined and merely rounded up when required for consumption. In more recent times these animals are privately owned and under control, making it very unlikely that they were wild *banteng*, or any wilder than crosses between remaining wild *banteng* and feral cattle (Potter, personal communication). It is very probable that they have been there for several centuries, and that burning of grassland has had to do with their maintenance as well as that of deer. There is an area for inquiry here, opened up by several chapters in this book, that calls for closer examination.

Repair and restoration; changes in grassland and the forest

An important question now arises. Grasslands may, as Potter (1996b:8) suggests, be 'a kind of land reserve, able fairly easily to be converted to more intensive agricultural systems when this is required'. This was certainly true of grassland in nineteenth-century Java and it is proving to be true of important tracts of grassland in both Borneo and Irian/New Guinea. But it is not necessarily true of all grasslands. Some are underlain by thin soils that, through erosion and compaction, have lost most of

34 Wharton 1968; Payne and Rollinson 1973; Ad Hoc Panel 1983; Rollinson 1984; Groves 1995. Madura cattle and Java cattle are both cross-bred between Bali cattle and Zebu cattle imported from India, probably before the end of the first millennium AD. Though smaller in size, and with narrower heads than wild *banteng*, Bali cattle are otherwise physically very similar to their progenitors. Cross-bred Madura and Java cattle, and Zebu cattle proper, are quite different in appearance.

35 Wharton (1968:150-1) provides maps of the location of these reports. There are few in the southeast. The preferred habitat of *banteng* was first closely described by Beccari (1904). The managed ecology of a reserve in southern Java in which *banteng* (or *banteng* crossed with Bali cattle) are conserved is analysed in depth by Sumardja and Kartawinata (1977). Although the main effect of management is to provide them with tracts of *Imperata cylindrica* (and *Chromolaena odorata)*, both of which they eat, their preference is for other and scarcer grasses. Without management, forest would return to the reserve and the *banteng* would be unable to survive. It may have been population decline in the interior of Borneo in the later nineteenth century (discussed below, and by Knapen this volume) that reduced the habitat of *banteng*, leading to partial or total extinction.

whatever agricultural capability they had. Some grassland areas are on soils that are naturally very infertile; once stripped of a forest cover slowly established over centuries, they are unlikely to carry either farms or forest in the foreseeable future. The podzolic soils of the *kerangas* lands in Borneo carry only a light forest that will not readily return after clearance. A case in point is the Sook plain in Sabah, denuded of its forest by fire in the 1914 drought, and still grassland; *per contra,* charcoal in depth, dated 350-500 years ago, has been found in *kerangas* soils now under forest in East Kalimantan (Shimokawa 1988).

Much also depends on the seed bank remaining in the soil. At least on most Borneo soils, though not all, some introduced weeds may compete successfully with *Imperata,* and allow apparently vanished pioneer tree species such as *Trema* and *Macaranga* to be re-established (Eussen and De Groot 1974). It seems improbable that this would happen readily in the 'sheet *alang-alang*' areas, far from living trees, and with a long history of grass cover. It is much more likely in the areas of patchy grassland, interspersed among secondary forest, which are the more common condition in Borneo. In the region as a whole some quite large areas have in fact returned to forest. On the fringes of the central mountains in Irian/New Guinea are areas now under dense forest where mortars and pestles can quite widely be found. In Borneo, Chin's (1985:29-33) unique ethno-historical account of the migrations of Dayak now on the Upper Baram, in Sarawak, reports that they occupied an area in the Apo Kayan highland in East Kalimantan until about the end of the 1870s, and that when they left it was heavily infested with *Imperata cylindrica.* Recent survey work shows no modern grassland in this area and it now forms part of a nature reserve, devoid of people.

Grassland can also be managed out of existence wherever the quality of the soil permits. It can be managed by tillage, to yield crops on a long-term basis, and it can be totally displaced under both arable and agroforestry systems. Grassland can also be shaded out by tree plantations, and in some locations crowded out by competitive weeds of greater agricultural value such as the recently-introduced *Chromolaena odorata* and some other *Compositae.*[36] This is not new. Grasslands have been tilled in Central Irian/New Guinea since about 2,000 years ago and since unknown, but certainly not recent, dates in Java and Timor. In the latter island some areas of better soil are permanently managed by terracing, as in western Indonesia (Metzner 1977). Tillage of the grasslands in Kalimantan is much more recent, not being reported before the end of the 1920s. Parts of the grassland in Kalimantan were shaded out under rubber a decade or more before this

36 *Chromolaena odorata* was formerly termed *Eupatorium odoratum.* It is generally thought to have been introduced during World War II. A related plant of similar aggressive habit, *Eupatorium pallescens,* was introduced in the late nineteenth century, both to control *Imperata cylindrica* and, in fallow, to serve as a nitrogen-rich green manure (Stoutjesdijk 1935).

(Brookfield, Potter and Byron 1995).

When we examine grasslands as degradation more closely, therefore, we find a picture that is much less simple than it seemed to be at first sight. Only some grassland is truly a 'green desert'; a very large part of it is reclaimable for agriculture, or is naturally reclaimable by the forest. Moreover, except where climatic conditions are such as to inhibit recovery of forest, as in much of Timor, withdrawal of human interference would lead quite soon to such recovery, and there is abundant evidence of this from inland Borneo. The Dayak on the Upper Baram in Sarawak, studied by Chin, are among a great many whose ancestors suffered very heavy mortality in the second half of the nineteenth century. Quite widely in eastern Sarawak, and in other areas, the first literate observers recounted much larger villages, and many more than exist today. Depopulation is interwoven into ethnic history (Lian 1987:23-4).[37] Declines in numbers have certainly been very large, though they cannot be quantified.[38]

The people of these now sparsely peopled areas have made far fewer changes in their shifting-cultivation farming system in the direction of greater permanence than have the people of more densely occupied areas in western Sarawak and West Kalimantan. Yet their activities have not destroyed the forest, and it has survived or even returned under lighter human pressure, albeit in a condition different from much older forest. As people, however small in number, have worked over the region for cultivation, extraction of produce, hunting and gathering, there have been repeated changes in the condition of the forest. Probably, a high proportion of contemporary forest is several generations in age, and retains the product of human use in the form of biodiversity which may be reduced in species numbers, but contains a higher content of plants of value to people (Spencer 1966:127). It must be concluded from the Borneo evidence that many areas which are described as 'primary forest' in most of the modern literature are more probably secondary forest of no very great age, perhaps only from one to three hundred years. This observation is not new even in Southeast Asia, and it has also been made – and has been demonstrated historically – in other parts of the tropical world.[39]

37 See also Knapen (this volume).

38 The only exception is in the ethnohistorical evidence recounted by Chin (1985:29-33). The Lepo Ga' Kenyah numbered about 1,000 men and women when living on the Apo Kayan in Kalimantan, but only 700 men and women when they arrived on the Baram in Sarawak in the early 1880s. After a visit by some of them to the river town of Marudi on the Lower Baram about 1885 a major epidemic, which Chin believes to have been of cholera or smallpox, killed most of the people leaving only 30 men. The story was written down from an old man's recollections in 1966, initially in Kenyah. In 1958 the population of the longhouse numbered 68, but had increased to 148 (47% below 15 years of age) at the time of Chin's field work in 1977-78.

39 It was made of the forests of West Kalimantan, in the days when observers used their eyes rather than their prejudices, by Von Gaffron (1858:224), cited by Potter (this volume, and 1996b). Von Gaffron wrote that most of the old forests through which he passed were not more than 130 to 150 years old.

Conclusion

Finally to return to the four questions posed at the outset, it is clear that while a substantial part of Indonesia's degraded land came into this condition only in the last two or three centuries there has also been significant repair of degradation in the same period, both through the adoption of improved land management and through natural recovery of forest on land on which human interference has become lighter. What is noteworthy about this conclusion, and in opposition to a good deal of the general literature, is that both increase and reduction of population can create conditions for repair of degradation, as well as for its creation. Explaining restoration involves exploration of a complex interplay between repair by human artifice, either aiding or aided by nature, and natural processes as such. On the improvements in upland Java, Nibbering (this volume) notes the way in which farmers have used natural processes on the terraces to complement their own intensification efforts. Over a long time span, both short-term and secular change in climatic conditions have been powerful forces in the creation and persistence of grasslands and savannah, but in all cases human interference has also been involved. This emerges most clearly in Irian/New Guinea, second in Timor and third in Kalimantan.

To go beyond this gloss would require a great deal more research among material which certainly exists but has not yet been adequately employed. A history of Indonesian agriculture which incorporates the question of land management remains to be written. For Java there is not yet even a comprehensive history of irrigation. Perhaps such histories of the landscape, as well as of economy and society, will emerge from the EDEN project? From any point of view, the key area must be Java with its enormous increase in population and cultivation since the end of the eighteenth century, and total transformation of both uplands and lowlands during that period.

What remains important in writing Indonesian landscape history is that change in the landscape and its qualities takes full account both of historical antecedents and of natural processes, and also examine both natural and induced repair of damage done in the past. Such history cannot take account only of documentary evidence, but must also rely on evidence obtained by other disciplines, and in the field. Nor can it be limited to any particular historical period since the forces that create the landscape have varied within historical time, and have changed greatly through prehistoric time. There is inheritance from the past in the modern landscape, and it needs always to be taken into account.

A landscape cannot tell the observer its own history, though it may contain much evidence from which that history can be reconstructed. Even at a general level, a view of landscape that is informed by comparison with other landscapes can prompt questions, and help avoid error. There has been so much written about land degradation in modern times that is erroneous. This large literature is based on supposition more than observation, and on an almost total want of proper historical research. It fails to take account of what a range of disciplines have to offer if their literature is searched. This chapter has set out a few of the dimensions that ought to be involved in reconstructing the history of the Indonesian landscape. It has indicated where a few answers might lie, but mainly has asked questions which, in addition to their historical and scientific interest, are of practical significance in terms of modern management of the land.

References

Ad Hoc panel of the Advisory Committee on Technology Innovation, Board on Science and Technology for International Development, Office of International Affairs, National Research Council

1983 *Little-known Asian animals with a promising economic future.* Washington, DC: National Academy Press.

Allen, B.

1989 'Frost and drought through time and space, part I: The climatological record', *Mountain Research and Development* 9:252-78.

1993 'The problems of upland land management', in: H. Brookfield and Y. Byron (eds), *South-East Asia's environmental future; The search for sustainability*, pp. 225-37. Tokyo: United Nations University Press.

Allen, B., H. Brookfield and Y. Byron

1989 'Frost and drought through time and space, part II: The written, oral and proxy records and their meaning', *Mountain Research and Development* 9:279-305.

Allen, B. and R. Crittenden

1987 'Degradation and a pre-capitalist political economy; The case of the New Guinea highlands', in: P. Blaikie and H. Brookfield (eds), *Land degradation and society*, pp. 145-56. London: Methuen.

Ashton, P.S.

1972 'The Quaternary geomorphological history of western Malesia and lowland forest phytogeography', in: P.S. Ashton and M. Ashton (eds), *The Quaternary era in Malesia; Transactions of the second Aberdeen-Hull Symposium on Malesian Ecology, 1971, Hull (UK)*, pp. 35-62. Hull: Department of Geography, University of Hull. [Miscellaneous Series 13.]

Beccari, O.

1904 *Wanderings in the great forests of Borneo; Travels and researches of a naturalist in Sarawak.* London: Constable. [Reprinted Singapore: Oxford University Press, 1986.]

Bierenbroodspot, J.

1986 *Soil erosion and its control in West Timor.* Canberra: NTT Integrated Area Development Project, Australian International Development Assistance Bureau.

Blaikie, P. and H. Brookfield (eds)
1987 *Land degradation and society*. London: Methuen.

Brookfield, H.
1964 'The ecology of highland settlement; Some suggestions', *American Anthropologist* 66-4:20-38.
1989 'Frost and drought through time and space, part III: What were conditions like when the high valleys were first settled?', *Mountain Research and Development* 9:306-21.
1993 '"The failure of just one crop"; Reflections on the social and environmental history of famine, with special reference to south-eastern Asia', in: H.-G. Bohle, T.E. Downing, J.O. Field and F.N. Ibrahim (eds), *Coping with vulnerability and criticality; Case studies on food insecure people and places*, pp. 183-201. Saarbrücken/Fort Lauderdale: Breitenbach. [Freiburger Studien zur geographischen Entwicklungsforschung 1.]
1996 'Untying the Chimbu circle; An essay in and on hindsight', in: H. Levine and A. Ploeg (eds), *Work in progress; Essays in New Guinea Highlands ethnography in honour of Paula Brown Glick*, pp. 63-84. Frankfurt am Main: Lang.

Brookfield, H. and B. Allen
1991 'Environmental and human responses to climatic variability in the west Pacific "maritime continent" region', in: J.E. Hay (ed.), *South Pacific environments; Interactions with weather and climate; Proceedings of a conference organized by Environmental Science, University of Auckland, 2-7 September 1991*, pp. 71-82. Auckland: University of Auckland.

Brookfield, H. and P. Brown
1963 *Struggle for land; Agriculture and group territories among the Chimbu of the New Guinea highlands*. Melbourne: Oxford University Press.

Brookfield, H., M. Latham, M. Brookfield, B. Salvat, R.F. McLean, R.D. Bedford, P.J. Hughes and G.S. Hope
1979 *Lakeba; Environmental change, population dynamics and resource use*. Canberra: SOCPAC Printery for UNESCO. [UNESCO/UNFPA Population and Environment Project in the Eastern Islands of Fiji; Island Report 5.]

Brookfield, H., L. Potter and Y. Byron
1995 *In place of the forest; Environmental and socio-economic transformation in Borneo and the eastern Malay Peninsula*. Tokyo: United Nations University Press.

Caratini, C., M. Fontugne, J.P. Pascal, C. Tissot and I. Bentaleb
1991 'A major change at ca. 3500 years BP in vegetation of the Western Ghats in north Kanara, Karnataka', *Current Science* 61-9/10:669-72.

Chappell, J. and H.H. Veeh
1978 'Late Quaternary tectonic movements and sea-level changes at Timor and Atauro Island', *Geological Society of America Bulletin* 89:356-68.

Chin, S.C.
1985 *Agriculture and resource utilization in a lowland rainforest Kenyah community*. Kuching: Sarawak Museum. [Sarawak Museum Journal 35-56 (new series); Special Monograph 4.]

Corlett, R.T.
1984 'Human impact on the subalpine vegetation of Mt Wilhelm, Papua New Guinea', *Journal of Ecology* 7:841-54.

Cremaschi, M., R. Ciarla and V.C. Pigott
1992 'Paleoenvironment and late-prehistoric sites in the Lopburi region of Central Thailand', in: I. Glover (ed.), *Southeast Asian archaeology 1990; Proceedings of the third conference of the European Association of Southeast Asian Archaeologists*, pp. 167-77. Hull: Centre for South-East Asian Studies, University of Hull.

Dam, M.A.C.
1994 *The late Quaternary evolution of the Bandung Basin, West-Java, Indonesia*. [PhD thesis, Vrije Universiteit Amsterdam.]

Diemont, W.H., A.C. Smiet and Nurdin
1991 'Re-thinking erosion on Java', *Netherlands Journal of Agricultural Science* 39:213-24.

Donner, W.
1987 *Land use and environment in Indonesia*. London: Hurst.

Douglas, I.
1990 'Sediment transfer and siltation', in: B.L. Turner II, W.C. Clark, R.W. Kates, J.F. Richards, J.T. Mathews and W.B. Meyer (eds), *The earth as transformed by human action; Global and regional changes in the biosphere over the past 300 years*, pp. 215-34. Cambridge: Cambridge University Press with Clark University.
1994 'Land degradation in the humid tropics', in: N. Roberts (ed.), *The changing global environment*, pp. 332-50. Oxford: Blackwell.

Douglas, I. and T. Spencer
1985 'Present-day processes as a key to the effects of environmental change', in: I. Douglas and T. Spencer (eds), *Environmental change and tropical geomorphology*, pp. 39-73. London: Allen and Unwin.

Dove, M.R.
1981 *Subsistence strategies in rain forest swidden agriculture; The Kantu' at Tikul Batu*. [PhD thesis, Stanford University.]

Eden, M.J.
1974 'The origin and status of savanna and grassland in the southern Papua', *Transactions of the Institute of British Geographers* 63:97-110.

Endert, F.H.
1927 'Floristisch verslag', in: D.W. Buijs, H. Witkamp, F.H. Endert, H.C. Siebers and D.F.K. Bosch (eds), *Midden-Oost-Borneo expeditie 1925*, pp. 200-83. Weltevreden: Kolff.
1946 'Droogte resistente gewassen', *Tectona* 36:165-214.

Enthoven, J.J.K.
1903 *Bijdragen tot de geographie van Borneo's Wester-afdeeling*. Leiden: Brill. 2 vols.

Eussen, J.H.H. and W. de Groot
1974 'Control of *Imperata cylindrica* (L.) Beauv. in Indonesia', *Mededelingen Fakulteit Landbouw-Wetenschappen Gent* 39-2:451-64.

Fox, J.J.
1977 *Harvest of the palm; Ecological change in Eastern Indonesia*. Cambridge, Mass.: Harvard University Press.

Gaffron, W.G. von
1858 'Nota omtrent den getah-pertja-boom', *Natuurkundig Tijdschrift voor Nederlandsch-Indië* 16:224-8.

Glover, I.C.
1986 *Archaeology in eastern Timor, 1966-67*. Canberra: Department of Prehistory, Research School of Pacific Studies, Australian National University.

Goldammer, J. and B. Siebert
1989 'Natural rain forest fires in eastern Borneo during the Pleistocene and Holocene', *Die Naturwissenschaften* 76:518-20.

Golson, J.
1989 'The origins and development of New Guinea agriculture', in: D.R. Harris and G.C. Hillman (eds), *Foraging and farming; The evolution of plant exploitation*, pp. 678-87. London: Unwin Hyman.
1991 'Bulmer Phase II; Early agriculture in the New Guinea Highlands', in: A. Pawley (ed.), *Man and a half; Essays in Pacific anthropology and ethnobiology in honour of Ralph Bulmer*, pp. 484-91. Auckland: The Polynesian Society.

Groves, C.P.
1995 'Domesticated and commensal mammals of Austronesia and their histories', in: P. Bellwood, J.J. Fox and D. Tryon (eds), *The Austronesians; Historical and comparative perspectives*, pp. 152-63. Canberra: Department of Anthropology, Research School of Pacific and Asian Studies, Australian National University.

Grubb, P.J. and P.F. Stevens
1985 *The forests of the Fatima Basin and Mt Kerigomna, Papua New Guinea with a review of montane and subalpine rainforests in Papuasia*. Canberra, Australia: The Australian National University. [Department of Biogeography and Geomorphology, Publication BG/5.]

Haberle, S.
1994 'Anthropogenic indicators in pollen diagrams; Problems and prospects for late-Quaternary palynology in New Guinea', in: J.G. Hather (ed.), *Tropical archaeobotany; Applications and new developments*, pp. 172-201. London/New York: Routledge.

Hardjono, J.
1986 'Environmental crisis in Java', *Prisma; The Indonesian Indicator* 39:3-13.

1993 'Immediate needs in upland Java', in: H. Brookfield and Y. Byron (eds), *Southeast Asia's environmental future; The search for sustainability*, pp. 239-41. Tokyo: United Nations University Press.

Henty, E.E.
1982 'Grasslands and grassland succession in New Guinea', in: J.L. Gressit (ed.), *Biogeography and ecology of New Guinea*, pp. 459-73. The Hague: Junk. [Monographiae Biologicae 42.]

Hnatiuk, R.J., J.M.B. Smith and D.N. McVean
1976 *The climate of Mt Wilhelm*. Canberra: Department of Biogeography and Geomorphology, Research School of Pacific Studies, Australian National University. [Mt Wilhelm Studies 2.]

Hoogerwerf, A.
1938 'Honderd uren de "ongeziene" gast der bantengs in Oedjoeng-Koelon', *Tropische Natuur* 27:25-35, 39-47, 64-72, 77-86.

Hooze, J.A.
1893 'Topografische, geografische, mineralogische en mijnbouwkundige beschrijving van een gedeelte der afdeling Martapoera in de residentie Zuider en Oosterafdeeling van Borneo', *Jaarboek van het Mijnwezen in Nederlandsch-Indie (Batavia)* 2:1-431.

Hope, J.H. and G.S. Hope
1976 'Palaeoenvironments for man in New Guinea', in: R.L. Kirk and A.G. Thorne (eds), *The origin of the Australians*, pp. 29-54. Canberra: Australian Institute of Aboriginal Studies.

Jackson, J.C.
1968 *Planters and speculators; Chinese and European agricultural enterprise in Malaya, 1786-1921*. Kuala Lumpur: University of Malaya Press.

Kikkawa, Y., T. Amano and H. Suzuki
1995 'Analysis of genetic diversity of domestic cattle in East and Southeast Asia in terms of variations in restriction sites and sequences of mitochondrial DNA', *Biochemical Genetics* 33:51-60.

Kitayama, K.
1996 'Climate of the summit of Mount Kinabalu (Borneo) in 1992, an El Niño year', *Mountain Research and Development* 16:65-75.

Klinkert, G.P.
1937 'De catastrofale overstroming op Java in het jaar 1861', *Tectona* 30:731-6.

Lambeck, K.
1995 'Late-Pleistocene and Holocene sea-level change in Greece and southwestern Turkey; A separation of eustatic, isostatic and tectonic contributions', *Geophysical Journal International* 122:1022-44.

Lane-Poole, C.E.
1925 *The forest resources of the territories of New Guinea and Papua*. N.p.: n.n. [Australian Parliamentary Paper 73-F.12389.]

Latham, M.
1983 'Origine de la formation à talasiga', in: M. Latham and H.C. Brookfield (eds), *Iles Fidji orientales; Etude du milieu naturel, de son utilisation et de son évolution sous l'influence humaine*, pp. 129-41. Paris: ORSTOM. [Travaux et Documents de l'ORSTOM 162.]

Leighton, M. and D. Peart
1988 Ecological research for sustainable land management and yield of forest products on the poor soil of West Kalimantan. [Mimeo.]

Lian, F.J.
1987 *Farmers' perceptions and economic change; The case of Kenyah swidden farmers in Sarawak*. [PhD thesis, Australian National University, Canberra.]

Maloney, B.K.
1980 'Pollen analytical evidence for early forest clearance in North Sumatra', *Nature* 287:324-6.
1985 'Man's impact on the rain forest of West Malesia; The palynological records', *Journal of Biogeography* 1:537-8.

McGlone, M., A.P. Kershaw and V. Markgraf
1992 'El Niño/Southern Oscillation climatic variability in Australasian and South American palaeoenvironmental records', in: H.F. Diaz and V. Markgraf (eds), *El Niño; Historical and paleoclimatic aspects of the Southern Oscillation*, pp. 435-62. Cambridge: Cambridge University Press.

Meher-Homji, V.M.
1996 'Past environments through palynology; A short appraisal with reference to the Western Ghats', *Environment and History* 2:249-52.

Metzner, J.K.
1977 *Man and environment in eastern Timor; A geoecological analysis of the Baucau-Viqueque area as a possible basis for regional planning*. Canberra: Australian National University. [Development Studies Centre Monograph 8.]
1982 *Agriculture and population pressure in Sikka, Isle of Flores; A contribution to the study of the stability of agricultural systems in the wet and dry tropics*. Canberra: Australian National University. [Development Studies Centre Monograph 28.]

Naval Intelligence Division
1944 *Netherlands East Indies*. London: Naval Intelligence Division of the Admiralty. 2 vols. [Geographical Handbook Series B.R. 518.]

Nibbering, J.W.
1991a 'Crisis and resilience in upland Central Java; A case-study in Gunung-Kidul District', in: J. Hardjono (ed.), *Indonesia; Resources, ecology, and environment*, pp. 104-32. Singapore: Oxford University Press.
1991b *Hoeing in the hills; Stress and resilience in an upland farming system in Java*. [PhD thesis, Australian National University, Canberra.]

Nicholls, N.
1992 'Historical El Niño/Southern Oscillation variability in the Australasian region', in: H.F. Diaz and V. Markgraf (eds), *El Niño; Historical and palaeoclimatic aspects of the Southern Oscillation*, pp. 151-73. Cambridge: Cambridge University Press.

Ormeling, F.J.
1956 *The Timor problem; A geographical interpretation of an underdeveloped island*. 's-Gravenhage: Nijhoff, Groningen/Djakarta: Wolters.

Padoch, C.
1982 *Migration and its alternatives among the Iban of Sarawak*. The Hague: Nijhoff. [KITLV, Verhandelingen 98.]

Padoch, C. and C. Peters
1993 'Managed forest gardens in West Kalimantan, Indonesia', in: C.S. Potter, J.I. Cohen and D. Janczewski (eds), *Perspectives on biodiversity; Case studies of genetic resource conservation and development*, pp. 167-76. Washington, DC: American Association for the Advancement of Science Press.

Palte, J.G.L.
1989 *Upland farming on Java, Indonesia; A socio-economic study of upland agriculture and subsistence under population pressure*. Amsterdam: Koninklijk Nederlands Aardrijkskundig Genootschap, Utrecht: Geografisch Instituut Rijksuniversiteit Utrecht. [Nederlandse Geografische Studies 97.]

Payne, W.J.A. and D.H.L. Rollinson
1973 'Bali cattle', *World Animal Review* 7:13-21.

Pelzer, K.J.
1978a 'Swidden cultivation in Southeast Asia; Historical, ecological, and economic perspectives', in: P. Kunstadter, E.C. Chapman and S. Sabhasri (eds), *Farmers in the forest; Economic development and marginal agriculture in northern Thailand*, pp. 271-86. Honolulu: University Press of Hawaii.
1978b *Planter and peasant; Colonial policy and the agrarian struggle in East Sumatra, 1863-1947*. 's-Gravenhage: Nijhoff. [KITLV, Verhandelingen 84.]
1982 *Planters against peasants; The agrarian struggle in East Sumatra, 1947-1958*. The Hague: Nijhoff. [KITLV, Verhandelingen 97.]

Potter, L.M.
1987 'Degradation, innovation and social welfare in the Riam Kiwa valley, Kalimantan, Indonesia', in: P. Blaikie and H. Brookfield (eds), *Land degradation and society*, pp. 164-76. London: Methuen.
1993 'Banjarese in and beyond the Hulu Sungai, South Kalimantan; A study of cultural independence, economic opportunity and mobility', in: J.T. Lindblad (ed.), *New challenges in the modern economic history of Indonesia; Proceedings of the first conference on Indonesia's modern economic history*, pp. 264-98. Leiden: Programme of Indonesian Studies.
1996a 'Forest degradation, deforestation and reforestation in Kalimantan; Towards a sustainable land use?', in: C. Padoch and N. Peluso (eds),

Borneo in transition; People, forests, conservation and development, pp. 13-40. Kuala Lumpur: Oxford University Press.

1996b 'Forests and grasslands, drought and fire; The nineteenth- and twentieth-century environmental history of Borneo'. Paper prepared for the fourth biennial conference of the Borneo Research Council, Bandar Seri Begawan, Brunei Darussalam, 10-15 June 1996.

1997 'The dynamics of *Imperata*; Historical overview and modern farmer perspectives, with an example from South Kalimantan, Indonesia', *Agroforestry Systems* 36:31-51.

Pijnappel, J.

1860 'Beschrijving van het westelijke gedeelte van de Zuider- en Ooster-afdeeling van Borneo (De Afdeeling Sampit en de Zuidkust)', *Bijdragen tot de Taal-, Land- en Volkenkunde* 3:243-346.

Rackham, O.

1986 *The history of the countryside*. London: Dent.

Reid, A.

1988 *Southeast Asia in the age of commerce 1450-1680. Vol. 1: The lands below the winds*. New Haven/London: Yale University Press.

Retallack, G.J.

1992 'Middle-Miocene fossil plants from Fort Ternan (Kenya) and evolution of African grasslands', *Paleobiology* 18-4:383-400.

Richards, P.W.

1952 *The tropical rain forest*. Cambridge: Cambridge University Press.

Robbins, R.G.

1960 'The anthropogenic grasslands of Papua and New Guinea', *Symposium on the impact of man on the humid tropics vegetation*, pp. 313-29. Goroka, Territory of Papua and New Guinea: Administration of the Territory of Papua and New Guinea and UNESCO Science Cooperation Office for South East Asia.

1963 'Correlations of plant patterns and population migrations into the Australian New Guinea Highlands', in: *Plants and the migrations of Pacific peoples; A symposium*, pp. 313-29. Honolulu: Bishop Museum Press.

Rollinson, D.H.L.

1984 'Bali cattle', in: I.L. Mason (ed.), *Evolution of domesticated animals*, pp. 28-38. London: Longman.

Sather, C.

1990 'Trees and tenure in Paku Iban society; The management of secondary forest resources in a long-established Iban community', *Borneo Review* 1:16-40.

Seckler, D.

1987 'Issues in the economic evaluation of soil and water conservation programmes', in: P. Blaikie and H. Brookfield (eds), *Land degradation and society*, pp. 84-96. London: Methuen.

Shimokawa, E.

1988 'Effects of a fire of tropical rain forest on soil erosion', in: H. Tagawa and N. Wirawan (eds), *A research on the process of earlier recovery of tropical rain forest after a large-scale fire in Kalimantan Timur, Indonesia*, pp. 2-11. Kagoshima: Research Centre for the South Pacific, Kagoshima University.

Soerjani, M.
1980 'Symposium on the prevention and rehabilitation of critical land in an area development', *Proceedings of Biotrop Workshop on Alang-Alang*, pp. 9-14. Bogor, Indonesia: SEAMEO-Biotrop. [Biotrop Special Publication 5.]

Spencer, J.E.
1966 *Shifting cultivation in southeastern Asia*. Berkeley/Los Angeles: University of California Press. [University of California Publications in Geography 19.]
1975 'The rise of maize as a major crop plant in the Philippines', *Journal of Historical Geography* 1:1-16.

Steenis, C.G.G.J. van
1937 'De invloed van den mensch op het bosch', *Tectona* 30:634-53.
1972 *The mountain flora of Java*. Leiden: Brill.

Stoutjesdijk, J.A.J.H.
1935 '*Eupatorium pallescens* D.C. op Sumatra's Westkust', *Tectona* 28:919-26.

Street-Perrott, F.A., N. Roberts and S. Metcalfe
1985 'Geomorphic implications of late-Quaternary hydrological and climatic changes in the Northern Hemisphere tropics', in: I. Douglas and T. Spencer (eds), *Environmental change and tropical geomorphology*, pp. 165-83. London: Allen and Unwin.

Sumardja, E.A. and K. Kartawinata
1977 *Vegetation analysis of the habitat of banteng (Bos javanicus) at the Pananjung-Pangandaran nature reserve, West Java*. Bogor: SEAMEO-BIOTROP. [Biotrop Bulletin 13.]

Swadling, P.
1983 *How long have people been in the Ok Tedi impact region?* Boroko: Papua New Guinea National Museum. [PNG National Museum Record 8.]

Van Andel, T.H.V., G.R. Heath, T.C. Moore and D.F.R. McGeary
1967 'Late Quaternary history, climate and oceanography of the Timor Sea, northwestern Australia', *American Journal of Science* 265:737-58.

Walker, D. and J.R. Flenley
1979 'Late-Quaternary vegetational history of the Enga Province of upland Papua New Guinea', *Philosophical Transactions of the Royal Society of London* 286:265-344.

Walker, D. and G.S. Hope
1982 'Late-Quaternary vegetation history', in: J.L. Gressit (ed.), *Biogeography and ecology of New Guinea*, pp. 263-85. The Hague: Junk. [Monographiae Biologicae 42.]

Wharton, C.H.
1968 'Man, fire and wild cattle in Southeast Asia', in: *Proceedings; Tall timbers fire ecology annual conference,* pp. 107-67. Tallahassee, Florida: Tall Timbers Research Station.

Winstedt, R. O.
1927 'The great flood, 1926', *Journal of the Malayan Branch of the Royal Asiatic Society* 5:295-309.

Yen, D. E.
1974 *The sweet potato and Oceania; An essay in ethnobotany.* Honolulu: Bishop Museum Press. [Bernice P. Bishop Museum Bulletin 236.]

ANTHONY REID

Inside out
The colonial displacement of Sumatra's population

'The traveller of yesterday was almost entirely confined to the plains, the gardens, the dazzling shores and teeming life of the sea. To tell the truth, the historian is not unlike the traveller. He tends to linger over the plain, which is the setting for the leading actors of the day, and does not seem eager to approach the high mountains nearby.' (Braudel 1972, I:29.)

The present occupants of the Indonesian archipelago do not appear to have a very long history there. Linguistic and archaeological evidence suggests the movement of Austronesian languages southwards from Taiwan within the last five thousand years, carried by peoples also noted for rice agriculture and pottery (Bellwood 1985:102-15). Many important maritime migrations took place much more recently. Should we not then expect these people to colonize first the fertile lowland valleys and coastal plains we now see covered in ricefields, close also to fish-stocks? Only as population pressures or internal conflicts drove some pioneers out of the coastal lowlands would we expect to find them penetrating the mountainous cores of the islands.

Yet the origin myths of many Indonesian peoples speak not of the sea, the coast, and gradual migration inland, but of a kind of ethnogenesis in the mountains.[1] In islands like Bali and Nias there seems to be a positive aversion to the sea, and a habit of building villages on ridgetops rather than near waterways. Even small islands which have played intensely maritime roles in historic times often hark back to an older tradition of settlement high up in the interior. Tidore (North Moluccas), for example, sent *kora-kora* expeditions as far as New Guinea and received tribute from dozens of islands, but its first capital was not at the coast where Soa is now but six kilometres up the mountain at Gurabanga.

1 See, for instance: Barnes 1974:28-37; Lebar 1972:126; Reid and Reid 1988:84; Coté 1980.

During the rapid population growth of the twentieth century, the most significant migrations in Indonesia have not been of lowlanders into the sparsely settled highlands, as has been a marked feature in China and Vietnam (and to some extent the goal of successive Indonesian governments through the transmigration scheme), but the reverse. Bataks, Minangkabaus, Banjarese, Minahasans, Toraja, and Atoni moved from their relatively crowded and impoverished highland valleys into the coastal cities and lowland plans.

How should we explain this paradoxical feature of what is usually considered Asia's most 'maritime' region? Were pre-colonial conditions more congenial to highland settlement, with the lowland cities becoming attractive only under high-colonial and modern conditions? Did colonialism distort a more 'natural' highland development we would have expected without it? Or is this a false paradox, based on unreliable origin myths associated with sacred mountains? Before considering possible explanations, I will very briefly review the pattern in the larger Indonesian islands, and then assemble the evidence on Sumatra.

Broader Indonesian patterns

Borneo as a whole has been exceptionally sparsely populated even by Southeast Asian standards throughout recorded history. Its vast coastal, swamp, and deltaic regions were largely uncultivated until about a century ago, and many of the people who now populate these lowlands have come from outside the island – notably from South Sulawesi, South China, the southern Philippines, and Java.[2] Even the Iban and related groups which now form a large proportion of the 'Dayak' or indigenous non-Muslim population of Sarawak and West Kalimantan have a language very close to Malay and traditions which suggest immigration to Borneo from the Malacca Straits area within the last millennium (Lebar 1972:180-1; Sandin 1967:1-2).

Though Borneo has a relatively small area of true highlands favourable to the early stages of agriculture, these appear to have played a disproportionate role in the development of the oldest settled agriculture. The Kelabit of North-Central Borneo in particular have apparently practised wet rice agriculture for centuries in alluvial valleys above 1,000 metres around the northeastern end of the current Sarawak-Kalimantan border, and developed a complex culture there of metalworking and pottery, as well as creating carved stone graves and monuments perhaps

2 The immigration process goes on. The 1991 census of Sabah (Malaysia 1995) reported 139,403 Malaysian citizens of Indonesian origin, and 425,175 non-Malaysian citizens in Sabah, of whom I have estimated 70% to be Indonesians (with most of the remainder Filipino). A quarter of Sabah's population are therefore recent arrivals, and these form the majority in the eastern lowlands.

thousands of years old (Lebar 1972:159; King 1993:47-8, 100-2, 116-7, 265). The more numerous Kayan and Kenyah groups in adjacent areas near the headwaters of the east and northwestern Borneo rivers are swidden cultivators, but also developed a high level of craftsmanship, stone monuments, and hierarchic social structure. According to Raymond Kennedy all Kayan groups trace their genealogy back to the ancestral homeland in the Apo Kayan. He believed this 'lofty rolling plateau country' was the centre of diffusion for Borneo's more 'civilized' peoples (cited by Lebar 1972:168-9).

The more numerous Kadazans (or Dusuns), who dominated the fertile west coast of today's Sabah at the beginning of Chartered Company rule, trace their origins to Nunuk Ragang (literally 'red banyan tree'), which is today understood to be at the junction of the Liwagu and Gelibang Rivers well to the east of the Crocker range which forms the backbone of Sabah. According to the legend, population pressure caused a dispersal from these uplands towards the valleys of the west coast (Topin n.d.:44; Evans 1953: 187-8).

By far the most dense population of pre-colonial Borneo was in the Hulu Sungai, some hundred kilometres up the Barito above Banjarmasin. Exploiting the alluvial basins of the eastern tributaries of the Barito, flowing down from the Meratus Mountains, or around the lakes and swampy areas where retreating floodwaters provided fine conditions for wet rice growing, the Banjarese of this small area probably represented almost half of Borneo's population in the early eighteenth century (Bock 1881:234; Schophuys 1936:84-126). Banjar outmigration began during the 1850s, first around southeastern Borneo and in the 1870s into the middle Mahakam River of East Borneo. By 1890 they had begun to colonize East Sumatra and the Malayan Peninsula. At the 1930 census there were still 551,571 inhabitants of the Hulu Sungai at a density of 47 per km^2 (as well as another 200,000 downstream in the Banjarmasin District), as against a population density for the whole of Dutch Borneo of only four persons per km^2. But by then there were nearly 80,000 Banjarese migrants in Sumatra and 45,000 in Malaya (Potter 1993:270-9).

In Sulawesi one must distinguish between the Bugis, Makassar, and Mandar peoples of the southwestern peninsula, who appear to have developed rice agriculture in the coastal lowlands by the sixteenth century, and most of the rest of the island where agriculture was primarily a highland activity. The heartland of Minahasa rice growing and demographic development appears to have been in the high intermontane basin which contains Lake Tondano (Henley 1994, and personal communication). Other highland lakes such as Lake Poso and Lake Lindu were important centres of sedentarization, although the main centre of the mysterious megalith remains of Central Sulawesi is in an upland valley mid-way between these two lakes. The Toraja of the central massif also developed a complex stateless civilization at elevations above 1,000 metres.

Even for the cultures which cannot be considered either highland or interior, lakes merit attention as early population centres and foci for some of the first states. Lake Limboto appears to have played a role in the development of the fish and rice culture of Gorontalo, while Pelras (1981) argues that it was Lake Tempe in Wajo (South Sulawesi) which enabled the Bugis to develop as a distinct culture. The iron and nickel deposits around Lake Matano appear to have had a role in the early development of Bugis kingship at Luwu (Reid 1983; Caldwell 1988).

Timor has a relatively narrow coastal belt below 300 metres in height, and until recently it remained little cultivated. The dominant agricultural pattern was a highland one, producing rice and maize through shifting cultivation which became increasingly difficult to sustain in the dry conditions of most areas. The Atoni Pa Meto ('people of the dry land') who comprise three-quarters of the population of West Timor considered themselves inherently agricultural highlanders. Like many peoples of the eastern Indonesian islands, they have resisted government efforts in this century to bring them down to the coastal lowlands where they can be controlled and transformed (Schulte Nordholt 1971; Lebar 1972:103).

Java is the large island most difficult to fit into the 'Sumatran' pattern I am proposing. West Java does have physical and climatic similarities with Sumatra, and its extensive upland valleys probably have more agricultural potential than the lowlands. Moreover in the sixteenth century a coastal Islamic kingdom, Javanese-speaking Banten, created a sharp cultural dichotomy with the quasi-stateless Sundanese of these highland valleys. But unlike the Sumatran uplands, the Priangan struck the earliest eighteenth-century Dutch visitors as surprisingly unpopulated, which De Haan (1910:381-8) attributed to epidemics, notably smallpox, and an uncertain food supply. Since Dutch control generally preceded adequate Dutch reporting in this area, however, it may be a case of a highland people valuing their independence and retreating eastwards before the Dutch coffee-sergeants.

The Sumatran evidence assembled below may suggest a need to look again at the understudied history of West Java, and perhaps to question older assumptions that Sundanese civilization began around the Bogor-Jakarta area, where the main evidence for some kind of state is focused. Did 'the westward expansion of Muslim Javanese along the north coast cut off the Sundanese from the sea, forcing them to settle in the Priangan highlands', as Pigeaud (1962:291) argued, or were they there all along, with the 'kingdom' of Pajajaran acting only as the mediator between this mountain civilization and the outside world, as Batak 'kings' did in the same period (Pinto 1989:20-9)?

Central and East Java is not the model for Indonesia so much as the exception, in that there, 'unlike neighbouring areas of Southeast Asia, there was [...] a remarkable degree of ethnic homogeneity from an early period' (Hefner 1990:35). Lowland and deltaic irrigated rice agriculture was practised since the fourteenth century and probably earlier, eventually driving out the malaria-carrying *Anopheles sundaicus* from the cultivated regions. Although the Javanese were very seldom politically united during their long history, the relatively easy movement along the Solo and Brantas Rivers appears to have created a greater degree of cultural and economic interaction between coast and interior than applied in the other large islands.

Java does however provide the model which enables us to understand how the 'age of commerce' (Reid 1990, 1993b) affected the balance of population in some other areas. The oldest Hindu-Javanese centres were far from the coast, around wet rice cultivating areas of the Mataram plain and the headwaters of the Solo and Brantas Rivers. From about 1300 to 1600, however, when the advantages of trade pulled population and rice production towards the coast, first Majapahit and then Demak, Cirebon, Gresik, and Surabaya all flourished while we hear almost nothing of the interior areas. This phase is comparable to the change brought about in the same period in the northern (Aceh) coast of Sumatra and in the South Sulawesi peninsula, where lowland irrigated rice also expanded to feed the burgeoning port-cities.

The complex crisis of the seventeenth century, in which the forced monopolies of the VOC were the most obvious if not the most fundamental element, appears to have restored the comparative advantage of the upland interior (Reid 1993b:270-303). The rulers of the new Mataram unified the fertile interior plains of Surakarta and Yogyakarta, and Sultan Agung assaulted the *pasisir* (northern coastal belt) and removed much of its remaining population inland. When the Mataram court explained the number of its arms-bearing subjects to Rijklof van Goens in the 1650s, 630,000 were inscribed for the Mataram heartland and other interior areas as against only 290,000 (30%) from the *pasisir* districts (including Cirebon) (Van Goens 1656:225). When *cacah* (households) were counted in 1755 for the division of the realm, there were still more than twice as many in the interior heartland as in the then VOC-controlled *pasisir* – though Ricklefs has shown that these were probably conventional figures which ignored losses of population in the war-torn interior (Hageman 1860:267; Ricklefs 1986). When the first Dutch count took place in 1795 the *pasisir* and Madura together already equalled the (possibly underestimated) *cacah* counts of interior Central and East Java (Hageman 1860:267). More confidently we may use the first Dutch population count for the whole island, in 1831, when the interior which Peter Boomgaard calls the *kejawen* and the *pasisir* each held about 2.2

Map 1. 'Schetskaart van het rijk Atjih' (Veth 1873:136).

million people, with the predominantly lowland and coastal Oosthoek a further 0.8 million (Boomgaard 1989:166).

We can reasonably conclude that while Dutch intervention on the coast in the seventeenth century contributed to its instability and unattractiveness for habitation, the reverse was the case after 1755. Dutch control of the *pasisir* between 1755 and 1830 encouraged a second shift of population outwards from the Mataram heartland, some of it no doubt resulting from the savage Java War of 1825-30. The modern outmigration to the lowlands had much lower barriers of geography and culture to overcome in Java than it did in the other islands, and it began 50-100 years earlier.

Sumatra

Sumatra represents the most striking example of upland agriculture and lowland emptiness before the late nineteenth century. Agronomists and prehistorians have recognized this phenomenon for some time, but historians continue to be misled by the glitter of states such as Sriwijaya, Pasai, Aceh, Barus, Palembang, Jambi, Riau, and Siak in written sources both external and internal, and the silence of these same sources about the populous interior. Intrepid nineteenth-century travellers who penetrated into the high valleys of the Bukit Barisan range, from Raffles and Burton and Ward onwards, were repeatedly astonished by the large populations and productive agriculture they encountered. But since these highlanders did not leave written records or build centralized states, we continue to write histories in which they are extremely marginal actors, at best the people of the *ulu* who occasionally sent their goods, tribute, and slaves down to the rulers of the river-ports. The map of northern Sumatra included in an 1874 book by Europe's foremost geographical specialist on Indonesia, P.J. Veth (1873), is indicative of the problem – everything beyond the narrow coastal plain is marked as 'very mountainous but wholly unknown interior' and Lake Toba's very existence seems doubted in a shrunken scrawl one-tenth its true size in the vicinity of Tarutung (see Map 1).

The evidence is accumulating that the earliest agriculture in Sumatra was not in the river deltas or coastal plains where we might have expected it, but in the high basins of the Bukit Barisan, generally above 500 metres. Pollen analysis of cores taken to the south of Lake Toba and around Lake Kerinci have shown evidence of rice agriculture as much as five thousand years ago in the first case and three thousand in the second (Maloney 1989; Bellwood 1985:230-1). Japanese agronomic historians have made clear that the intramontane basins of highlands in Sumatra, as in Thailand, are particularly suitable for the early stages of rice cultivation, and one has judged the Besemah (Pasemah) basin as exhibiting features of low-technology wet rice agriculture (no plough, mini-*sawah* fed by

Lake Toba, North Sumatra; photograph by P.J. van Winter, 1931 (KITLV photo collection 20000).

streams, harvesting by *ani-ani*) also found two thousand years ago in Japan and Yunnan (Ishii 1975:164-91; Tanabe 1994:23-46; Sakurai Yumio, personal communication). Most of Sumatra's pre-Islamic monuments are found in highland plateaux many days' journey from a port – the extraordinary stone statues, cists, and menhirs of Besemah (0-500 AD), the Tantric Buddhist temple remains of Padang Lawas (tenth-thirteenth centuries), and the inscriptions and statuary of Adityavarman in Pagarruyung (fourteenth century). Dongson-like bronzes have been found near the mountain lakes of Kerinci and Ranau in highland South Sumatra, while one of the Besemah stone statues at Pagaralam includes the representation of a Dongson drum of Heger I type (Van Heekeren 1958:20-1, 63-78).

The coasts of Sumatra, on the other hand, were inhospitable to early agriculture, with the single exception of the narrow northern littoral of Aceh, where since about the sixteenth century permanent ricefields were built to feed an urban and coastal population grown large in the age of commerce. The east coast south of the Asahan River was ringed with tidal peat swamps, and even behind these the permanence of rainfall (with no significant dry season), made shifting agriculture very difficult to practice. Cities such as Palembang and Jambi were fed almost entirely with rice from the headwaters of the rivers of which they controlled the outlet (Andaya 1993:18-9; Oki 1986:21, 24-35). The west coast had relatively little flat land, much of it also abandoned to swamps in the delta areas. Christine Dobbin described the settlements of the coast north of Padang in the early nineteenth century as:

> 'small, drab-looking coastal villages which differed markedly from those of the uplands. They were small in terms of both size and population; even the leading ones, such as Ulakan [...] presented "a wretched appearance" (Nahuijs). Behind these villages stretched sparsely peopled, marshy plains [...] Even where the plain was broadest, large tracts of marshy swamp and *alang-alang* grass separated one village from another.' (Dobbin 1983:42-3.)

In eighteenth-century Bengkulu the British constantly worried why the area under their domination was so underpopulated. They blamed poor soil, disease, frequent abortion, a discouragingly difficult marriage system, and the flight of people from British labour demands, but some argued that the coastal strip could still support fifty times its population (Kathirithamby-Wells 1977:116-20).

By contrast the upland valleys occupied by (south to north) the Ranau, Komering, Rawas, Besemah, Rejang, Kerinci, Minangkabau, Mandailing, Angkola, Toba Batak, and Karo Batak were all reportedly heavily populated at the point at which they were first seen by Europeans in the nineteenth century. Much of the population that did exist on the coast had migrated there in recent times from relatively heavily populated highlands – like the Besemah and Rejang in parts of the Bengkulu coast, the Minangkabau who went down to the west-coast ports to trade, the Toba

Bataks who had settled in the ports of Barus and Tapanuli (Sibolga) on the west coast, and in agricultural areas of Asahan on the east coast (where genealogies suggest immigration from the highlands in the eighteenth century), or the Karo Bataks who were growing pepper in the Deli-Langkat area of East Sumatra by the early nineteenth century. Some of these peoples' origin myths include reference to a contact of the founding ancestors with Java (sometimes as Majapahit) or India, but all believe that they became distinctive peoples in their own highlands, at some sacred origin place from which all have subsequently dispersed.

Population figures are scarce and unreliable before the modern transformation, but tend to confirm this picture. Thomas Stamford Raffles's visit to Pagarruyung in the Minangkabau highlands in 1818 had convinced him that the population within a radius of 50 miles 'cannot be estimated at less than a million', while Burton and Ward in 1824 reckoned the 'whole Batak country' to contain 1.5 million people (Reid 1995:174, 180). In 1817 an English visitor had claimed 100,000 people in Besemah Lebar alone (Andaya 1993:230), and the following year Raffles thought the Pagaralam area of Besemah which he visited 'one of the finest countries in the world', with many villages having more than 500 inhabitants and rice costing one-fifth what it did on the coast at Bengkulu (Raffles 1835, I:347-8).

Population pressure in the Besemah ricelands at the end of the eighteenth century was beginning to push Besemah to colonize other upland basins in the headwaters of the Ogan, Komering and Musi Rivers on the eastern side of the divide, and the Manna and Alas Rivers on the west side, as well as further south into Ranau and Lampung. But in general the Besemah avoided moving below 500 metres because their system of *sawah* irrigation required a drop of water off the hills. 'We can only cultivate at the foot of mountains with running water for our *sawah*', as one told a Dutch official (cited in Hoven 1927:13; also Andaya 1993:229-31).

By comparison with the seemingly abundant nineteenth-century population of the highlands, the coastal strip of Bengkulu had only 60,000 inhabitants when the British surveyed it in 1819 (Kathirithamby-Wells 1977:116), though by the middle of this century between two-thirds and three-quarters of Bengkulu Province's population has lived in that strip. The 1920 census still listed only 61,000 *orang Palembang*, but classified 212,000 South Sumatrans as *orang Ulu*, which included Ogan, Komering, Semendo, Rejang, Besemah, and Serawai.[3]

[3] *Statistisch jaaroverzicht 1927*:36. These ethnic totals must be treated with great caution since almost 2.8 million inhabitants (20%) of the Outer Islands were given as of 'unknown' ethnicity. The 1930 census is more careful to tie up such ends, but at the expense of creating larger groupings. The 'Palembangers' category jumped to 733,000 in that census (presumably by greatly expanding the definition of the term), while there were 83,000 Besemah, 44,000 Semendo, 67,000 Serawai and 89,000 Rejang-Lebong, but no mention of Ogan and Kemering (*Volkstelling 1930*, VIII:91-2).

For West Sumatra it is possible to construct a table (Table 1) of highland and lowland population over a somewhat longer term, though with the usual caveats about the earliest figures. Highland population was concentrated in the three intramontane valleys of Tanah Datar, Limapuluh Kota, and Agam, together with the Solok basin to the south of Lake Singkarak, which was less important an agricultural area but has been the site of much modern development.

Table 1. Changing population balance in West Sumatra.

	ca 1830[a]	1852[b]	1920[c]	1990[d]
Tanah Datar	between 200,000	153,604	260,835	380,709
Solok	and 600,000	70,752	221,316	782,551
Limapuluh Kota	300,000	103,567	182,672	387,847
Agam	100,000	197,217	246,890	491,520
Total highlands	ca 800,000 (91%)	556,980 (80%)	958,784 (64%)	2,042,627 (51%)
Coastal lowlands	ca 80,000 (9%)	138,937 (20%)	546,777 (36%)	1,956,050 (49%)

[a] Dutch figures of 1824 and 1833 cited by Dobbin (1983:16, 43). The lowland figure is my estimate based on Dutch figures of 13,000 for Padang and 50,000 for the coastal plain to its north.

[b] 1852 Dutch census cited by Graves (1971:24). The 'Total highlands' figure is greater than the sum of the four key districts because of the addition of Ophir and Rau, minor highland areas to the north.

[c] Figures derived from Graves (1971:24-5). Again the total highlands figure is higher for the same reason.

[d] Biro Pusat Statistik 1990:22. *Kabupaten* boundaries are somewhat different from those of the Dutch *onderafdelingen*, and city populations now administratively separate have been added to *kabupaten* totals.

The earliest figures are such rough estimates that they might be discarded were it not that they predate the devastating Padri War, which may well have been a factor in beginning the remarkable modern phase of outmigration from the Minangkabau highlands where it was fought. Even if we consider the pattern only from the more reliable 1852 figures, the change in the balance of population in the colonial period is evident. Between 1852 and 1990 the highland population increased only fourfold (and that in the original three rice-bowls only threefold), while the lowland population increased 14-fold. By comparison the population of Java increased 11-fold in the same period. Even more highland Minangkabau migrated to lowlands on the eastern coast of Sumatra or in Java or Malaya than to the west-coast lowlands which remained within the provincial boundaries of West Sumatra. By the time of the 1930 census 211,000 Minangkabaus or 11% of their population already lived outside their homeland – the largest groups being in adjacent Jambi (58,000), Riau (51,000), East Sumatra (51,000), and Malaya (14,000 – not including the Negri Sembilan migrants of the eighteenth century) (*Volkstelling 1930*, VIII:91-2; Naim 1984:31). After independence Jakarta and the other major cities of the Republic became the greatest magnet for Minangkabau migrants. Lance Castles already estimated 60,000 Minangkabaus there in 1962 and Minangkabau leaders claimed as many as 10% of the Jakarta population in the 1970s (though undoubtedly with exaggeration) (Naim 1984:116). Modern censuses show not ethnicity but place of birth, which of course catches only the first generation of migrants. By that measure, 15% of those born in West Sumatra were no longer there at the 1990 census, with the largest groups of migrants being in Jakarta (154,000), Riau (146,000) and West Java (88,170) (Biro Pusat Statistik 1990:67-9).

The Batak highlands

Toba Batak oral tradition, and that of many *marga* of the other Batak subgroups, traces their origins back to the divine ancestor Si Raja Batak, who descended on the slopes of the mountain (Pusuk Buhit) dominating the western shore of Lake Toba. There he built the village and wet ricefields of Sianjur Mula-Mula, and gave birth to two sons who became ancestors of the Batak moieties – Lontung and Sumba. Today, no doubt influenced by the elaborate *marga* lineages published in the 1920s (Hoeta Galoeng 1926; Ypes 1932), Toba Bataks relate to each other by tracing common ancestry back to these founding figures. The longest genealogies do not extend beyond 25 to 30 generations, however, whereas it seems likely that peoples related to the Batak were occupying the shores of Lake Toba for more than two thousand years (Vergouwen 1964:21).

The concentration of Batak peoples around Lake Toba, at an elevation of around a thousand metres, cannot simply be explained by the unsuitability of the lowlands of North Sumatra for settlement. Both coasts of Sumatra have rich soils and safe harbours at the latitude of Lake Toba, and the East Coast in particular has become in the twentieth century one of the outstanding agricultural regions of Indonesia. Nevertheless these lowlands too remained a scarcely populated jungle until the 1860s, while the Bataks developed complex agricultural civilizations in their highland valleys around the lake, gradually spreading southwards and eastwards to other highland locations, but seldom cultivating near the coast.

The estimate of Burton and Ward in 1824 of 1.5 million Bataks in total is generally regarded as much too high, since even today the population of the central Batak highlands has barely crawled to such a figure. Lance Castles's careful estimates are now the most useful for the nineteenth century, based on an extrapolation of Junghuhn's detailed figures of 1840. He thereby arrives at a figure of 352,000 for the 1840 population of

Tapanuli Residency, about two-thirds of it in the present Toba Batak *kabupaten* of North Tapanuli. Junghuhn himself believed, however, that the population may have been twice as large before the terrible destruction wrought by the militant Islamic Padri movement which invaded the Batak area between Burton and Ward's visit and his own (Castles 1975:198-9).

On any reckoning, the small area comprising the island of Samosir and the highland valleys to the south of Lake Toba contained several times the population of the lowlands eventually incorporated into the Residency of East Sumatra. When the latter's population was first counted in 1880 there were only 119,000 people there (Cunningham 1958:11), of whom about 30,000 were recent migrants and contract labourers from outside Sumatra, while some (probably too few) must have been Karo and Toba Bataks already attracted down to the lowlands by the growing opportunities represented by the estates. Even allowing for undercounting of the interior, in the mid-nineteenth century there are unlikely to have been more than 100,000 people in the 90,000 km^2 of the East Sumatran lowlands. The small island of Samosir (670 km^2) within Lake Toba alone had almost as many people. Its population when first counted in 1907 was 74,000 at a density of 110 per km^2 (the highest anywhere outside Java and Bali), despite soils and rainfall much inferior to those of eastern Sumatra (Sherman 1990:23-4). Given its high outmigration we can assume it was already extremely thickly populated by Sumatran standards in the nineteenth century.

Once the East Sumatran plantation district began to import labour on a massive scale, the change in the balance of North Sumatran population was even more striking than that of West Sumatra (Table 2). The biggest influx in 1870-1890 was of Chinese and in 1890-1930 of Javanese contract labourers, but there was also, at first as a trickle and since 1940 as the overwhelmingly dominant factor, an internal flow of Bataks coming down to the lowlands.

The Bataks were slower than the Minangkabaus to leave their mountain fastnesses. The early directions of outmigration from the crowded Toba Batak heartland around Lake Toba were predominately to other highland valleys, in Dairi and Alas to the north and Silindung, Angkola, and Asahan in the south (Tichelman 1936; Cunningham 1958:82). The first large-scale Toba migration to the lower slopes was facilitated by the colonial government, anxious to develop wet rice agriculture in the Simalungun area when it became a new frontier for plantation development. From 1915 onwards Toba Batak cultivators opened *sawah* in lower Simalungun where irrigation facilities were provided by the government. By 1930, 12% of the Toba Batak population lived in East Sumatra's lowlands, 49,000 of them in Simalungun and 18,000 in Asahan (Naim 1984:49; Cunningham 1958:85-7). In total, the 1930 census showed 'Bataks' already as the leading outmigrants of Indonesia, with 15.3% of their

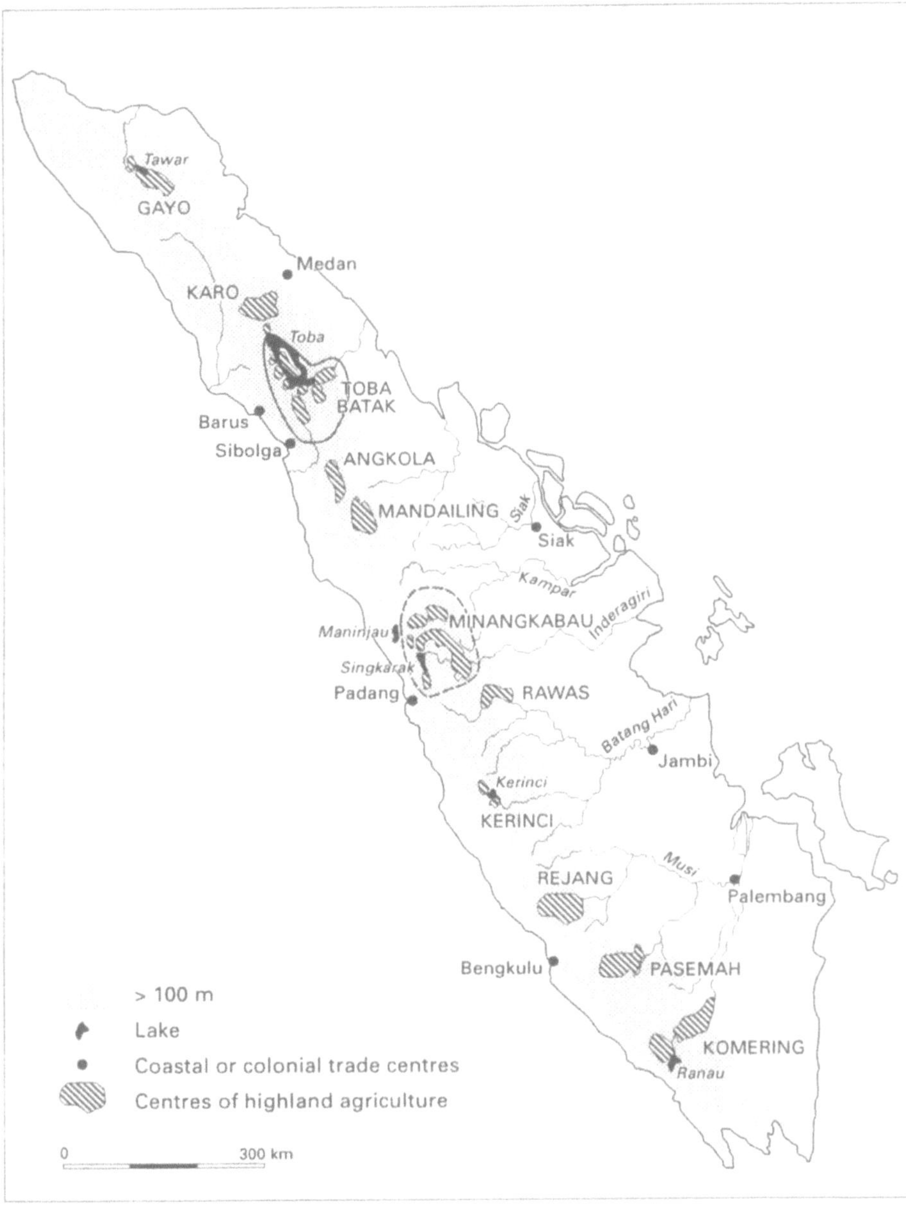

Map 2. Highland rice centres in Sumatra.

Table 2. Change of population balance in North Sumatra (excluding Nias).

	mid-19th century[a]	1930[b]	1990[c]
North Tapanuli highlands	ca 200,000	418,736	695,777
South Tapanuli	ca 140,000	276,681	954,245
Karo highlands	ca 60,000	86,411	257,981
Dairi	ca 20,000	54,053	276,980
Total highlands	ca 420,000	835,881	2,184,983
	(79%)	(33%)	(23%)
East Sumatra minus Karo	ca 100,000	1,587,272	7,192,773
Sibolga and Barus	ca 10,000	92,036	285,912
Total lowlands	ca 110,000	1,679,308	7,478,685
	(21%)	(67%)	(77%)

[a] The Tapanuli figures are based on Castles (1975:196-8), who estimated populations roughly half what they were in 1930, which was the scale of increase suggested by Junghuhn's partial figures. I have however assumed a rate of population increase slightly above this norm for Dairi, because it was a frontier for Toba Batak immigration well before 1930; and somewhat below it for Karo, which was throughout the twentieth century distinguished from other Batak areas by its low rate of population increase, and which suffered more than other areas from the Dutch conquest. Joustra's population estimates in 1906-8 were 84,000 for the Karo highlands (thought to be relatively well documented), 25,000 for Dairi, 75,000 for Samosir and 248,000 for the remainder of the North Tapanuli highlands (Joustra 1910:53, 66, 78).

[b] *Volkstelling 1930*, IV.

[c] Biro Pusat Statistik 1990:19.

number living outside the original Batak Province of Tapanuli, as against comparable figures of 14% for the Banjar, 11% for the Minangkabau, 10.5% for the Bugis, and 9.5% for the Menadonese – and at the other end of the scale only 3.4% for Javanese, despite all the official encouragement for them to move, and 0.1% for Balinese and Sasak (Naim 1977:51).[4]

At the same time Mandailings from South Tapanuli, quicker to profit from colonial education opportunities and as Muslims more readily fitting into the Malay establishment of the East Coast, began to occupy clerical and commercial positions in the eastern lowlands. There were 45,000 of them in the east coast in 1930, more than half as many as the Mandailings remaining at home (Naim 1984:49).

The revolution of 1945-1949 with its destruction of the authority of not only the Dutch colonial state, but also the Malay sultanates and the Western plantations, opened the floodgates to more extensive Batak migration to the lowlands. Although there are again no census figures by

4 Boundaries of course affect this ranking. When Minangkabau and Minahasans descended to their local coastal port it was still within the province dominated by their ethnic group, while Toba and Mandailing Bataks moving to East Sumatra were counted (in 1930, but not thereafter) as cross-province migrants.

ethnicity, Cunningham (1958:95) has shown that an average of seventeen Toba Batak churches a year were opened in East Sumatra in the period 1950-55, as against barely two a year in the fifty preceding years. The strength of the flow of outmigration is illustrated by his study of one village of the Meat valley running down to Lake Toba. Of the 59 people in the village in 1950, no less than 34 (58%) had left for the East Coast by 1955, though the high Batak fertility had brought the remaining population up to 42 (Cunningham 1958:59, 64).

Although the Karo Bataks of the Karo plateau (at roughly 1,300 metres elevation, to the north of Lake Toba) were closest to the Medan area and may at one time have dominated it, their twentieth-century descent to the plains is the most recent of all. Only after the revolutionary experience of 1945-1949 did Karos massively embrace the opportunities of education and modernization, as well as joining the Toba in occupying former plantation land near Medan. By 1981 there were 52,000 Karo in Medan – much fewer than the absolute numbers of Toba (183,000), Mandailing or Minangkabau, but a bigger proportion (say 10%) of the whole Karo Batak people (Pelly 1983:103).

The trend of the last twenty years has been for migration not to the local coastal cities, where opportunities tend to be limited by the ethnic niche each occupies, but to Jakarta. The number of North Sumatra-born migrants in other provinces has increased dramatically at the three censuses: 188,326 in 1971, 417,659 in 1980, and 770,093 in 1990. Of the last figure, 200,000 were in Jakarta and 116,000 more in West Java, meaning that North Sumatrans (predominantly Bataks of various sorts) were the third largest group in the capital after Javanese and Sundanese (Biro Pusat Statistik 1994:35).

The most recent migration trends are better shown by the question asked in the last two censuses as to where people were five years ago. This showed that 177,289 people had left North Sumatra in the five years before the 1980 census and 277,647 before that of 1990. Both the absolute figures for 1990 and the pace of increase show that Bataks (in the broadest sense) are now the ethnic group most likely to emigrate outside their province, even though that province is now defined to include the original area to which they migrated (East Sumatra).

In 1995 I conducted similar surveys in Tiga Binanga (Karo) and Samosir (Toba) to those which Cunningham carried out in 1956. These showed that the pace of migration has accelerated further to the point where virtually all youth leave the village in their late teens, with a tiny fraction returning after marriage to care for the old people. But where Cunningham's outmigrants headed almost exclusively for the East Sumatran lowlands, they are now destined for Jakarta, Bandung, and the new industrial estates of Batam, near Singapore.

The nine households of the hamlet (*huta*) of Lumban Tonga-tonga

(Pangururan, Samosir) produced 69 children since the 1950s. Sixty-one of these are old enough to have migrated, of whom only four (7%) remain in the village. Forty-five people, or 73% of the whole generation, have left the province of North Sumatra, the largest concentrations being in Bandung (23) and West Sumatra (7). This group is not distinguished educationally, with only three appearing to have tertiary qualifications of any kind.

Toba Batak, especially in Samosir, continue to have large families, for whom outmigration is the only real option. By contrast Karo Bataks have the lowest birth rates in North Sumatra and the highest education rates. With Karos the motive for migration is their exceptional commitment to using education to push the next generation into a better life.

In the much larger Karo village of Gunung, 90% of the youngest generation born to parents now under 60 had left the highlands by the time they were 20 years of age, compared with 58% of those born to parents aged 60-75. In the youngest generation, the favourite destination was Jakarta (40%), followed by Medan, Batam, and Bandung. The educational record of these children would be high for a developed industrial country, and is astonishing for children born in an Indonesian village with no educational opportunities at all before 1950. 95% of the younger generation and 41% of the older one finished high school, while 45% and 22% had tertiary education.

In the twentieth century, and particularly after political independence removed their peculiar alienness under Dutch rule, the cities have become the great poles of attraction for ambitious highlanders. As one Toba Batak clergyman who lived through the transition pointed out: 'When I was a boy in Tapanuli and left to study in Singapore [...] the women in the family moaned and cried, fearing that I would fall off the earth [...]. Today the women will moan and cry if a son does not have ambition and does not want to leave the *huta*' (cited by Cunningham 1958:78). The urban Indonesian population of just 4 million or 7% in 1930 became 14 million (15%) in 1961 and 55 million (31%) in 1990 (Hugo et al. 1987:89; Biro Pusat Statistik 1990). With the interesting exception of Bandung in Java, all the urban growth centres were in the lowlands. Although the Sumatran highlands sustained the bulk of that island's population up to 1900, they now contain no major city, no airport (to the dismay of tour operators), and no state university. More than ever, the path to modernization is to leave the highlands.

Hypotheses

Why did pre-modern Sumatrans, and many other Indonesians, for the most part cluster in upland valleys? If we can establish these reasons, can we explain why they have been so reversed in the course of the twentieth century that highlanders are now rushing to the lowland cities and leaving their pleasant cool valleys to become depopulated museums? Are these changes a necessary part of the modernization process, or to some extent a distortion introduced by colonial control?

Five factors appear to have played some part in the attractiveness of the highlands in the past: agriculture, health, statelessness, security, and culture.

1. *Agriculture*

The lowland valleys and deltas where most of Southeast Asia's current rice production is concentrated was not in fact hospitable to earlier generations of cultivators. Floods were a constant problem, ruining crops, endangering cattle, and threatening households and their fresh water supplies. Only large-scale irrigation and drainage works could control the large volumes of water in the lowlands and direct it to maintaining the permanent, inundated ricefields which we know today to be highly efficient. As Odum has pointed out, the wet rice agriculture of lowland valleys and deltas is 'one of the most productive and dependable of agricultural systems yet devised by man', but it requires a complete alteration of the natural environment and very high inputs of labour (cited by Tanabe 1994:14). With relatively simple technology and small family units of labour, highland valleys with small but permanent streams of water are far easier to manage. While it is conventional to draw a sharp line between upland rain-fed shifting cultivation and lowland permanent wet ricefields, in reality both types have long operated together as part of a spectrum of different agricultural strategies. Small plots of irrigated *sawah* were relatively easy to construct even without ploughing if adjacent to natural streams, and highlanders used them in addition to shifting plots on higher slopes.

Secondly, the density of the rain forest in most of Sumatra, Borneo, West Java, and northern Moluccas presents particular problems for pioneer cultivators. In most of these regions (unlike eastern Java and Nusa Tenggara) there is a minimal dry season, little light penetrates the dense forest canopy, and soils obtain few nutrients from decaying vegetable matter. The forest is difficult to fell, and never really dries out enough to burn easily. Unsuccessful burning is the major cause of swidden failure in such areas, and is especially likely in thick primary forest (Dove 1985:131-56). Such dense forest also has little game for hunting. The areas with a more pronounced dry season are at a distinct advantage for the development of settled agriculture, both of dry and wet type. As it happens some of the mountain valleys of the Bukit Barisan do have such a dry season when the southwest monsoon drops its heavy rain on the high western range close to the coast, sparing the protected valleys behind them. No more than 50 mm falls each month on average in June, July, and August in the Lake Toba area and the valleys to the north and south of it

(Sherman 1990:19). To a lesser degree the same phenomenon occurs in the valleys of West Sumatra. This season is critical for shifting cultivators to burn off, and for the rice to ripen satisfactorily.

For these reasons agrarian historians are unsurprised that the high valleys produced the earliest concentrations of settled rice agriculture. These highland farmers moved to the coastal areas chiefly in the late nineteenth and twentieth centuries. Some, like the Banjarese at the end of the nineteenth century, appear to have extended their cultivation to the swampier delta areas when the pressure of humans on resources was such as to make the draining and cultivation of those swamps cost-effective. Elsewhere it was colonial intervention in the form of drainage and irrigation, often intended in the first instance to serve Western plantations, which was the stimulus for opening new lowland areas for rice. At the end of the colonial period some of the most heavily populated parts of Indonesia were the Brantas delta and the northern coast of Java from Cirebon to Demak, where Dutch engineers had regulated the flood-prone lowlands.

2. *Health*

It is too easy to point to malaria as the chief reason why highlands were relatively healthier than lowlands until modern times. Pierre Gourou is the most emphatic proponent of the view that all Southeast Asia's problems were caused by the anopheles mosquito.

> 'It attacks something like one-third of the human race, but in practice all the inhabitants of the hot, wet belt may be considered to be more or less infected [...]. [After citing horrific figures for deaths] the most serious fact perhaps is that one death from malaria corresponds to at least 2,000 days of illness. Undoubtedly, malaria is largely responsible for the poor health, small numbers, absence of enthusiasm for work, stationary demographic character, and backwardness of tropical peoples.' (Gourou 1960:6-8.)

A.W. Nieuwenhuis in 1930 also believed that centuries of exposure to malaria was the reason Indonesians were smaller, sicker, and less dynamic than Polynesians, with each generation becoming physically and intellectually weaker (cited by Van der Brug 1994:82-3).

But even Gourou conceded that among the peasants of the Tongking delta malaria was not a great problem, since most of the anopheles vectors in Southeast Asia did not breed in muddy ricefields. The same applied to heavily populated rice-bowls in Java. But the most fatal seemed indeed to be the lowland areas of the 'empty centre' of pre-colonial Southeast Asia in Malaya, eastern Sumatra and southern and western Borneo, where *Anopheles umbrosus* ruled the coastal swamps, and *Anopheles sundaicus* took over as soon as the swamp forests were cleared. In this region one was really safe from malaria only above 2,000 feet or 600 metres – where most of the Sumatrans made their homes. A recent thesis by Peter van der Brug has shown convincingly that the appalling mortality of Batavia between

1733 and 1776, when 50% of VOC soldiers died within a year of reaching the city, was caused by the depredations of *Anopheles sundaicus* once brackish fish-ponds were developed immediately adjacent to the city. Those élite Dutch who could afford to build their mansions only a few kilometres inland survived the epidemic (Van der Brug 1994:201-6).

There were other factors than malaria. Water-borne diseases such as typhoid and cholera must have killed far more in the lowlands, where clean water sources were rare and floods particularly dangerous, than in highlands where the clear water of mountain streams could often be piped or carried to settlements. Coastal and river-mouth cities were the worst death-traps of all. Certainly the historical sources for nineteenth-century Sumatra suggest that highlanders were healthier and better formed (Raffles 1835:350; Dobbin 1983:42-3). Even well into the twentieth century highlanders believed their homelands far healthier than the coasts. 'Coolness is associated [by the Rejang] with fertility, peace and sound health. The Malays, by contrast, live in a warm climate [...] there is much disease and the people are poor and miserable. These images conform in most respects to the facts' (Jaspan 1964:17; also Sherman 1990:22-3.)

These relative advantages were reversed in the twentieth century. The 'taming' of the lowlands through drainage of swamps, the felling of forest, and the extension of rice paddies, together with the much better provision of clean water and health facilities in cities and more accessible areas, has made the lowlands much healthier places to live today.

3. *Statelessness*

In Indonesia the state has always been essentially coastal and sustained by foreign resources, while the highlands have been miracles of statelessness, tenuously held together by kinship systems and ritual obligations rather than bureaucracy.

In the Mediterranean context Fernand Braudel sought to explain why mountains were the 'asylum of liberty' by reference to their relative emptiness: 'the population is so inadequate, thinly distributed and widely dispersed as to prevent the establishment of the state, dominant languages, and important civilizations' (Braudel 1972, I:38). This cannot be the reason in the pre-colonial archipelago, where I have shown population densities to have been relatively high in upland valleys and plateaux. The Batak, Minangkabau, and Rejang certainly did establish complex civilizations in these valleys, with their own writing systems, but without giving rise to a recognizable state. Minangkabau and Batak kings were charismatic or magical figures who projected their power outwards but did not *rule* their own people (Drakard 1993; Castles 1972).

So persistently has each step towards stronger states in the archipelago arisen from trading ports, with external aid and inspiration, that one is inclined to seek the indigenous political dynamic in a genius for managing without states. Indic, Chinese, Islamic, and European ideas and actors each inspired state forms more ambitious and effective than those that went before.[5] With some exceptions in Java and Bali, all states in the archipelago have been based at river mouths or strategic coastal locations. They absorbed some of the interior people and established loose relations of tribute with others, but most highland populations defended their autonomies by a mixture of guerrilla warfare, diplomatic flexibility, and deliberate exaggeration of myths about their savagery. Masashi Hirosue (1996) has recently revived persuasively an argument that the ferocity of Batak cannibalism was deliberately exaggerated by Bataks to ensure that they were left in peace by coastal rulers and alien adventurers. Some of the highlanders of South Sumatra consciously designated themselves as freemen (*orang mardika*), who paid no tax and owed no obligations to the *raja* of the river-mouths, nor to the Dutch and British (Collins 1979:90-2).

In the seventeenth and eighteenth centuries the establishment of European colonial power at fortified coastal strongpoints probably tended to drive Indonesian populations further into highlands beyond their reach. The VOC discouraged the local population from settling near Batavia and other forts for strategic reasons. Villagers sought to escape labour burdens imposed by the Dutch or British Companies, as well as periodic retribution, by retreating further inland. This factor was gradually reversed in the nineteenth century as the Pax Neerlandica spread in the lowlands. The low-level war and raiding endemic to the stateless highlands was an increasingly high price to pay for freedom once a novel kind of rule of law became available in the twentieth century. Although few surrendered their independence to the colonizer without a fight, the highlanders were fortunate to encounter him at a late period when he also offered positive opportunities for modernization.

4. *Security*

Those who followed the option of statelessness had to find other means to defend themselves. Mountain barriers and sheer distance were a major asset. State power, particularly in its colonial form, operated primarily by sea. In local warfare hilltop villages had particular defensive advantages.

Coastal communities with weak states or no states suffered terrible depredations from maritime raiding expeditions. Best documented (since the careful work of James Warren) is the Iranun/Balangingi raiding of the eighteenth and nineteenth centuries, estimated to have taken 200,000-300,000 captives from coastal areas of the Philippine and Indonesian archipelagos in the period 1770-1870 (Warren 1981:208). In the Straits of Malacca area *orang laut* also carried off many people in coastal com-

5 Kulke 1986; Wisseman 1977; Schulte Nordholt 1993; Reid 1993a.

munities for sale to the prosperous sultanates of Jambi, Palembang, and elsewhere (Andaya 1993:96). These raids undoubtedly discouraged people from settling in any areas accessible to raiding vessels (Castles 1975:199).

Lakes are associated with the ethnogenesis of many interior peoples: Toba for the Bataks; Kerinci for the Kerinci; the central Kapuas Lake system for the Ibanic peoples, Tondano for the Minahasa, Limboto for Gorontalo; Poso and Lindu in Central Sulawesi, Tempe and Matano for the origin of Bugis states, and Lanau for the Magindanao and Illanun of Mindanao. The lakes must have been a source of inexhaustible fresh water and of fish, but there may also have been a strategic factor – most apparent in Samosir Island in Lake Toba. When the enemy had to attack by water one could see him coming long beforehand and there was less fear of real surprise.

5. Culture

Those who participate in states and live in cities are inclined to categorize those who do not as uncivilized, barbarous, and pagan (in the literal as well as the religious sense). This savage/civilized dichotomy is an ancient one in Southeast Asia, and visitors like Marco Polo, Ibn Battuta and Ma Huan relayed it to wider audiences. In the sixteenth and seventeenth centuries Islam in Indonesia (like Christianity in the Philippines) became the major boundary-marker between the civilized and savage, urban and upland, and wars to extend this boundary made it a strongly felt one. The success of this first phase of Islamization among upland Minangkabau and Javanese, on the other hand, eroded the cultural boundary almost completely.

Where it remained strong this cultural boundary no doubt prolonged the highlanders' sense of distinctiveness and apartness from the modernizing state project. Until the twentieth century interior Bataks, Torajans, Alfurs, Dayaks, and Dusuns knew that they would be looked at with contempt when trading or migrating to the lowland towns unless they adapted Malayo-Muslim dress, language, diet, and behaviour. As long as that cultural barrier remained high, highlanders seldom migrated voluntarily to the lowlands, which they associated rather with the involuntary movement of enslavement or exile from their own community. For some the domination of lowlands by Europeans was a further reason to avoid them.

The conversion of highlanders to Islam or Christianity marked a crucial breakdown of this cultural barrier in the long term, even if in the short term it sometimes strengthened it. As Mandailing or Besemah became Muslims and Toba Batak, Torajan or Dusun turned to Christianity they embraced the modernizing project of the twentieth-century cities. They had an identity they could use effectively in the cities, whether or not a distinctive highland one. The Karo Batak example is instructive here. Although inhabiting the immediate hinterland of Medan, and growing vegetables for it and other lowland cities from virtually the moment a road was put through to the Karo plateau in 1909, they remained suspicious of the lowland and its power structures. Colonial attempts to expand education, notably through the Dutch Protestant Mission, were met with little enthusiasm and even hostility. Only with the Japanese occupation and the Indonesian Revolution in the 1940s did Karos on a large scale accept that their destiny too lay in the modern state, education, urbanization, and all the associated cultural changes. Having once made that decision, they could not leave their highlands soon enough. Karos are today probably the best-educated of Indonesia's ethnic groups, and one of the most urbanized (Kipp 1993).

Urban progress or colonial distortion?

The modern world of nation states has no place for mountain 'asylums of liberty' which were also pockets of insecurity of life and property. There was no question that the present century would see the highlands of Indonesia, as of other parts of the world, incorporated into a nation state with the usual demands for monopoly of the use of force within fixed borders. The colonial means by which this incorporation came about, for the most part at the turn of the last century, would however determine much of what followed.

The Dutch colonial state, the agent of this incorporation, was an extreme form of the sea-based coastal polity. It had been built by naval power, and until the end of the nineteenth century remained wary of venturing far from the reach of that power. On the one hand this made it possible for highlanders to keep their statelessness longer than their equivalents in other parts of the world. By the time they were forcibly incorporated into the colonial state it was shedding its predatory quality in favour of a modernization project. This undoubtedly enabled highlanders to perceive the positive opportunities embedded within the disaster of colonial conquest, and to seize them more enthusiastically than lowlanders embittered by centuries of conflict. Minahasans, Minangkabaus, Mandailings, and Toba Bataks became the best-educated Indonesians of the first half of this century, joined by Karos and Torajans in the second half.

On the other hand the incorporation of highlands by lowlands was as one-sided a process as could be imagined, with all centres of power, influence, capital, and knowledge being firmly on the coast. No interior or highland centres developed to provide foci of modernization like a Mexico City, Bogota or La Paz; a Delhi or Lahore; a Nairobi, Entebbe or Johannesburg. Even in areas such as Tapanuli (North Sumatra), Central Sumatra, South Sumatra, and Timor, where the bulk of population, agriculture, and economic opportunity was still in the highlands in 1900, colonial centres remained firmly in Sibolga, Padang, Palembang, and Kupang. The reasons were in part the realities of Dutch military power mediated over three centuries. When one considers the significantly different pattern of Spanish and British colonies, however, it is difficult to avoid a sense that the heritage of the Low Countries lay too heavy on Dutch statesmen for them to conceive a successful city not accessible to ships. Even the attraction of a base among Christian or potentially Christian highlanders could not outweigh the lure of the sea.

Independence did nothing to break this pattern. Officials living in their coastal capitals continued to distrust the highlands, where comforts were few and the reach of government tenuous. Many upland villages continued to be relocated to valley floors, where the people could be reached by government, schools, police, and religious institutions. Once the colonial communications networks were in place, linking highlands to the ports rather than to each other, it was never likely that the pattern could be broken. Aeroplanes and busses did not have to follow the same coastal and lowland routes as ships and railways, but airports and roads were built as if they did. Indonesia's highlands still enjoy a pleasant climate and scenery and the heritage of ancient civilizations. Their most thriving industry is now tourism, but poor communications continue to hold back this industry like all others. Even the low-cost expedient of flying tourists to the highland lakes by seaplane has so far been passed up by established urban interests.

Highlanders are convinced that the route to the modernization they all desire is in leaving the land of their ancestors. The peculiar shape of Indonesia's modern history, and the prominence of the colonial factor in it, helps to explain why the highlands are now assuming the peripheral quality in fact which they always had in the European imagination.

References

Andaya, B.W.
1993 *To live as brothers; Southeast Sumatra in the seventeenth and eighteenth centuries.* Honolulu: University of Hawaii Press.

Barnes, R.H.
1974 *Kédang; A study of the collective thought of an eastern Indonesian people.* Oxford: Clarendon Press.

Bellwood, P.
1985 *Prehistory of the Indo-Malaysian archipelago.* Sydney: Academic Press.

Biro Pusat Statistik
1990 *Penduduk Indonesia; Hasil sensus penduduk 1990.* Jakarta: Biro Pusat Statistik.
1994 *Indikator kesejahteraan rakyat 1993; Sumatera Utara.* Medan: Biro Pusat Statistik.

Bock, C.
1881 *The headhunters of Borneo; A narrative of travel up the Mahakkam and down the Barito [...].* London: Sampson Low, Marston, Searle and Rivington.

Boomgaard, P.
1989 *Children of the colonial state; Population growth and economic development in Java, 1795-1880.* Amsterdam: Free University Press. [CASA Monographs 1.]

Braudel, F.
1972 *The Mediterranean and the Mediterranean world in the age of Philip II.* Translated by Sian Reynolds. New York: Harper and Row. 2 vols.

Brug, P.H. van der
1994 *Malaria en malaise: De VOC in Batavia in de achttiende eeuw.* Amsterdam: De Bataafsche Leeuw. [PhD thesis, Rijksuniversiteit Leiden.]

Caldwell, I.
1988 *South Sulawesi A.D. 1300-1600; Ten Bugis texts.* [PhD thesis, Australian National University, Canberra.]

Castles, L.
1972 'Statelessness and stateforming tendencies among the Bataks before colonial rule', in: A. Reid and L. Castles (eds), *Pre-colonial state systems in Southeast Asia*, pp. 67-76. Kuala Lumpur: Malaysian Branch of the Royal Asiatic Society. [MBRAS Monographs 6.]
1975 'Sources for the population history of northern Sumatra', *Masyarakat Indonesia* 2-2:189-209.

Collins, W.
1979 *Besemah concepts; A study of the culture of a people of South Sumatra.* [PhD thesis, University of California.]

Coté, J.
1980 The colonisation and schooling of the To Pamona of Central Sulawesi, 1895-1925. [MA thesis, Monash University, Melbourne.]

Cunningham, C.
1958 *The postwar migration of the Toba Bataks to East Sumatra.* New Haven: Yale University Southeast Asia Studies.

Dobbin, C.
1983 *Islamic revivalism in a changing peasant economy; Central Sumatra, 1784-1847.* London/Malmö: Curzon.

Dove, M.
1985 *Swidden agriculture in Indonesia; The subsistence strategies of the Kalimantan Kantu'.* Berlin: Mouton.

Drakard, J.
1993 *A kingdom of words; Minangkabau sovereignty in Sumatran history.* [PhD thesis, Australian National University, Canberra.]

Evans, I.H.N.
1953 *The religion of the Tempasuk Dusuns of North Borneo.* Cambridge: Cambridge University Press.

Goens, R. van
1656 *De vijf gezantschapsreizen van Rijklof van Goens naar het hof van Mataram, 1648-1654.* Edited by H.J. de Graaf. 's-Gravenhage: Nijhoff. [Werken Linschoten-Vereeniging 59.]

Gourou, P.
1960 *The tropical world; Its social and economic condition and its future status.* Translated by E.D. Laborde. London: Longmans. [First edition 1953.]

Graves, E.
1971 *The ever-victorious buffalo; How the Minangkabau of Indonesia solved their 'colonial question'.* [PhD thesis, University of Wisconsin.]

Haan, F. de
1910 *Priangan; De Preanger-Regentschappen onder het Nederlandsch bestuur tot 1811.* Vol. 1. Batavia: Bataviaasch Genootschap van Kunsten en Wetenschappen.

Hageman, J.
1860 'Geschied- en aardrijkskundig overzigt van Java op het einde der achttiende eeuw', *Tijdschrift voor Indische Taal-, Land- en Volkenkunde (TBG)* 9:261-419.

Heekeren, H.R. van
1958 *The bronze-iron age of Indonesia.* 's-Gravenhage: Nijhoff. [KITLV, Verhandelingen 22.]

Hefner, R.W.
1990 *The political economy of mountain Java; An interpretive history.* Berkeley: University of California Press.

Henley, D.
1994 'Population and environment in precolonial northern Sulawesi'. Paper presented at the 13th Biennial Conference of the Asian Studies Association of Australia, Perth, 13-16 July 1994.

Hirosue, M.
1996 'Sumatran port cities and "cannibalism" in the hinterlands'. Paper presented at the 14th Conference of the International Association of Historians of Asia, Bangkok, 22 May 1996.

Hoeta Galoeng, W.
1926 *Poestaha taringot toe tarombo ni halak Batak.* Laguboti: Zendings-drukkerij.

Hoven, W.
1927 *De Pasemah en haar verwantschaps-, huwelijks- en erfrecht.* Wageningen: Veenman. [PhD thesis, Rijksuniversiteit Leiden.]

Hugo, G.J., T.H.Hull, V.J. Hull and G.W. Jones
1987 *The demographic dimension in Indonesian development.* Singapore: Oxford University Press.

Ishii, Y. (ed.)
1975 *Thailand; A rice-growing society.* Honolulu: University of Hawaii Press.

Jaspan, M.A.
1964 *From patriliny to matriliny; Structural change among the Redjang of Southwest Sumatra.* [PhD thesis, Australian National University, Canberra]

Joustra, M.
1910 *Batakspiegel.* Leiden: Van Doesburgh.

Kathirithamby-Wells, J.
1977 *The British West Sumatran Presidency, 1760-1785; Problems of early colonial enterprise.* Kuala Lumpur: University of Malaya Press.

King, V.T.
1993 *The peoples of Borneo.* Oxford: Blackwell.

Kipp, R.
1993 *Dissociated identities; Ethnicity, religion and class in an Indonesian society.* Ann Arbor: University of Michigan Press.

Kulke, H.
1986 The early and imperial kingdom in Southeast Asian history', in: D.G. Marr and A.C. Milner (eds), *Southeast Asia in the 9th to 14th centuries,* pp. 1-22. Singapore: Institute of Southeast Asian Studies.

Lebar, F. (ed.)
1972 *Ethnic groups of insular Southeast Asia. Vol. 1: Indonesia, Andaman Islands, and Madagascar.* New Haven: Human Relations Area Files.

Malaysia
1995 *Laporan am banci penduduk/General report of the population census, 1991.* Kuala Lumpur: Department of Statistics.

Maloney, B.
1989 'Grass pollen and the origins of rice agriculture in North Sumatra', *Modern Quaternary Research in Southeast Asia* 11:135-61.

Naim, M.
1977 'Besarnya migrasi sukubangsa Minangkabau dan sukubangsa-sukubangsa lain di Indonesia – beberapa perkiraan statistik', *Masyarakat Indonesia* 4-1/2:29-56.
1984 *Merantau; Pola migrasi suku Minangkabau.* Yogyakarta: Gadjah Mada University Press.

Oki, A.
1986 'The river trade in Central and South Sumatra in the 19th century' in: T. Kato, M. Lutfi and N. Maeda (eds), *Environment, agriculture and society in the Malay world,* pp. 3-48. Kyoto: Centre for Southeast Asian Studies, Kyoto University.

Pelly, U.
1983 *Urban migration and adaptation in Indonesia; A case-study of Minangkabau and Mandailing Batak migrants in Medan, North Sumatra.* [PhD thesis, University of Illinois.]

Pelras, C.
1981 'Célèbes-sud avant l'Islam, selon les premiers témoignages étrangers', *Archipel* 21:153-84.

Pigeaud, Th.G.Th.
1962 *Java in the 14th century; A study in cultural history*. Vol. IV. The Hague: Nijhoff. [KITLV, Translation Series 4, 4.]

Pinto, F. Mendes
1989 *The travels of Mendes Pinto*. Translated by Rebecca Catz. Chicago: University of Chicago Press.

Potter, L.M.
1993 'Banjarese in and beyond Hulu Sungai, South Kalimantan; A study of cultural independence, economic opportunity and mobility', in: J. T. Lindblad (ed.), *New challenges in the modern economic history of Indonesia; Proceedings of the first conference on Indonesia's modern economic history*, pp. 264-98. Leiden: Programme of Indonesian Studies.

Raffles, S. (ed.)
1835 *Memoir of the life and public services of Sir Thomas Stamford Raffles [...]*. New edition. London: Duncan. 2 vols.

Reid, A.
1983 'The rise of Makassar', *Review of Indonesian and Malaysian Affairs* 17:117-60.
1990 'An "age of commerce" in Southeast Asian history?', *Modern Asian Studies* 24-1:1-30.
1993a 'Kings, kadis and charisma in the 17th century archipelago', in: Anthony Reid (ed.), *The making of an Islamic political discourse in Southeast Asia*, pp. 83-107. Clayton: Centre of Southeast Asian Studies, Monash University. [Monash Papers on Southeast Asia 27.]
1993b *Southeast Asia in the age of commerce, 1450-1680. Vol. 2: Expansion and crisis*. New Haven/London: Yale University Press.
1995 *Witnesses to Sumatra; A travellers' anthology*. Kuala Lumpur: Oxford University Press.

Reid, H. and A. Reid
1988 *South Sulawesi*. Berkeley: Periplus.

Ricklefs, M.C.
1986 'Some statistical evidence on Javanese social, economic, and demographic history in the later seventeenth and eighteenth centuries', *Modern Asian Studies* 20-1:1-32.

Sandin, B.
1967 *The Sea Dayaks of Borneo before White Rajah rule*. London: Macmillan.

Schophuys, H.J.
1936 *Het stroomgebied van de Barito; Landbouwkundige kenschets en landbouwvoorlichting*. Wageningen: Veenman. [PhD thesis, Landbouwuniversiteit Wageningen.]

Schulte Nordholt, H.
1993 'Leadership and the limits of political control; A Balinese "response" to Clifford Geertz', *Social Anthropology* 1-3:291-307.

Schulte Nordholt, H.G.
1971 *The political system of the Atoni of Timor*. The Hague: Nijhoff. [KITLV, Verhandelingen 60.]

Sherman, G.
1990 *Rice, rupees and ritual; Economy and society among the Samosir Batak of Sumatra.* Stanford: Stanford University Press.

Statistisch jaaroverzicht 1927
1928 *Statistisch jaaroverzicht van Nederlandsch-Indië [...]; Jaargang 1927.* Weltevreden: Landsdrukkerij.

Tanabe, S.
1994 *Ecology and practical technology; Peasant farming systems in Thailand.* Bangkok: White Lotus.

Tichelman, G.L.
1936 'Bataksch arbeidsreservaat', *Koloniaal Tijdschrift* 25:399-403.

Topin, B.
1993 'The origin of the Kadazan/Dusun; Popular theories and legendary tales', in: *Pesta Kaamatan Peringkat Negeri 1993*, pp. 42-4. Kota Kinabalu: Kadazandusun Cultural Centre.

Vergouwen, J.C.
1964 *The social organisation and customary law of the Toba-Batak of northern Sumatra.* The Hague: Nijhoff. [KITLV, Translation Series 7.]

Veth, P.J.
1873 *Atchin en zijne betrekkingen tot Nederland; Topographisch-historische beschrijving.* Leiden: Kolff.

Volkstelling 1930
1933-36 *Volkstelling 1930.* Batavia: Departement van Economische Zaken/ Landsdrukkerij. 8 vols.

Warren, J.F.
1981 *The Sulu zone, 1768-1898; The dynamics of external trade, slavery, and ethnicity in the transformation of a Southeast Asian maritime state.* Singapore: Singapore University Press.

Wisseman, J.
1977 'Markets and trade in pre-Majapahit Java', in: K.L Hutterer (ed.), *Economic exchange and social interaction in Southeast Asia*, pp. 197-212. Ann Arbor: Center for South and Southeast Asian Studies, University of Michigan. [Michigan Papers on South and Southeast Asia 13.]

Ypes, W.K.H.
1932 *Bijdrage tot de kennis van de stamverwantschap, de inheemsche rechtsgemeenschappen en het grondrecht der Toba- en Dairibataks.* Leiden: Adatrechtstichting.

Intensive maize cultivation in Minahasa, North Sulawesi, November 1945
(KITLV photo collection 6643).

DAVID HENLEY

Carrying capacity, climatic variation, and the problem of low population growth among Indonesian swidden farmers
Evidence from North Sulawesi

Anthropologists studying the subsistence aspects of traditional swidden farming communities in island Southeast Asia generally agree that the population densities observed among such groups are well below the levels indicated by the carrying capacity of their environment even under their existing agricultural systems. Michael Dove, on the basis of his fieldwork among Kantu' swidden and swamp-rice farmers of West Kalimantan, estimated that a population increase of 38% above the observed density of 11.5 people per km^2 'theoretically could be accommodated under the extant pattern of land use' (Dove 1985:381). Harold Conklin's classic study of Hanunóo agriculture on Mindoro in the Philippines put the equivalent margin there as high as 60%, despite an already respectable observed density of 24 inhabitants per km^2 (Conklin 1957:146).[1] For the Iban shifting cultivators of Sarawak, Freeman estimated that land degradation would set in once the population density exceeded about 27 per km^2 (Freeman 1955:135). In the Baleh region which he studied, however, the actual average density was between 4 and 7 per km^2.[2] And when Roy Ellen attempted to calculate the hypothetical maximum carrying capacity of the tuber-based swidden farming system which he studied on Seram, where the observed density was just 1 person per km^2 of 'potentially cultivatable land', his results indicated 'the capability of the present locality to support at the very least a fivefold increase in population and in all probability a population over twenty times that existing at present' (Ellen 1978:181-2, 256).

* Research in the National Archives of Indonesia in 1994 and 1995 was made possible by a grant from the Netherlands Foundation for the Advancement of Tropical Research (WOTRO).

1 Excluding land unsuitable for cultivation for environmental or ritual reasons, the observed density was 30 inhabitants per km^2.

2 Freeman 1955:136. I have assumed the average size of a '*bilek* family' to be 5.75 members, as in Freeman's own calculation of the maximum sustainable population density.

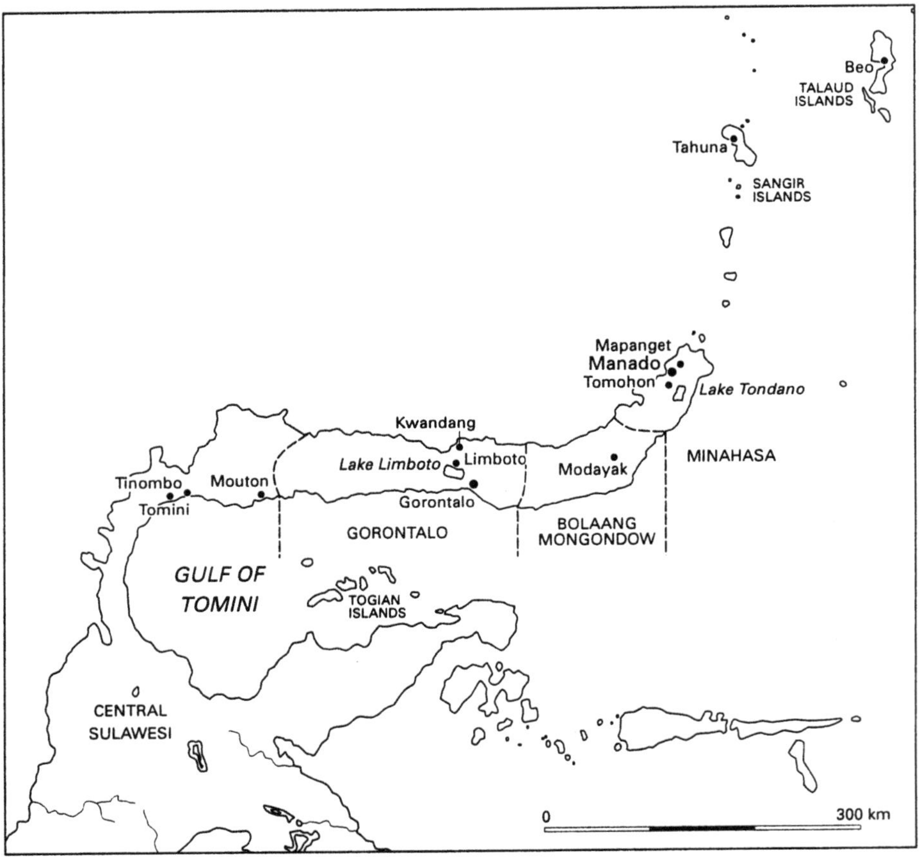

North Sulawesi

While such results offer eloquent support for the general opinion that pre-colonial Indonesia, and especially the swidden-farming outer islands, must have been sparsely populated, we should not be too quick to interpret them as indicating that low population densities had nothing to do with subsistence factors. That slow population growth among swidden farmers cannot be explained by limits to their food supply has primarily been a conclusion not of detailed empirical studies like those cited above, but of sideline work by scholars primarily concerned with other, non-economic and non-demographic aspects of the societies concerned. Missionary anthropologist Albert C. Kruyt, for instance, believed that only warfare, disease, and the frequent practice of abortion could account for the very low population to land ratio and the large tracts of uninhabited country which he observed in Central Sulawesi before Dutch rule (Kruyt 1903).[3] More recently cultural anthropologist Reimar Schefold, in his standard work on the Sakuddei of Siberut in the Mentawai Islands, also rejects the idea that the sparsity of their population can be explained in terms of subsistence limitations. Noting that 'the supply of basic vegetable foodstuffs, at least, regularly exceeds what can be consumed', Schefold concludes that only high mortality from endemic and epidemic disease offers a credible explanation for the observed average population density of about two persons per km^2 (Schefold 1988:68-70).[4] Historian Anthony Reid, on the other hand, quoting early European impressions of Southeast Asia as a relatively healthy environment, favours warfare and cultural factors affecting fertility as the main reasons for low population growth throughout the region in pre-colonial times (Reid 1987, 1988:15-8).

If we return to the existing quantitative studies of traditional subsistence systems, however, two things are immediately apparent. Firstly, the maximum carrying capacity is always calculated not directly in terms of the amount of food which could be produced sustainably per unit area of land, but rather by taking the amount of land actually cultivated by the existing population each year, multiplying it by an estimate of the minimum sustainable fallow period in years, and comparing the result with the total area of land available for cultivation. As Rappaport has emphasized, the limiting factor in this calculation is not food, but land (Rappaport 1984:72, 91). In other words, the result can only be relevant in demographic terms if it is assumed that the existing rate of land utilization per year is consistently sufficient to meet all the food needs of the population concerned. But the second striking thing about the available detailed studies on the economics of Indonesian subsistence farming is that they do not always support this assumption.

3 See also Adriani and Kruyt 1950-51, I:75-8.

4 The Sakuddei agricultural system combines temporary swiddens with permanent sago and taro plantations. Sago is the staple food (Schefold 1988:133-40).

Climatic variation and subsistence crises

Michael Dove's period of fieldwork among the Kantu' included one harvest which he describes as 'disastrous' (Dove 1985:149, 214). In 1974, ten out of the thirteen households included in his study did not harvest enough rice to feed them until the following harvest, and eight of the ten did not even have enough to sow as seed rice in the next planting season. The total shortfall was so great that the households without seed rice were forced to beg it from distant relatives in other longhouses up to a day's travel away (Dove 1985:170, 269, 293). In pre-colonial times, when hostile relations often prevailed between Kantu' settlements, such expeditions must have been much more precarious undertakings. The main reason for the bad harvest of 1974 was climatic: wet weather had preceded the burning of the swiddens prior to planting, so that the burn had been incomplete. In the following season, by contrast, a month of drought to dry out the slashed vegetation before the burn resulted in an excellent harvest for 1975. Noting also that total rainfall during the first five months of 1976 amounted to only 60% of what had fallen during the same period in 1975, Dove emphasizes that climatic variation and unpredictability is a key feature of the local environment and one which has far-reaching consequences for Kantu' agriculture and subsistence (Dove 1985:42-3, 53, 149, 162-3).

Another anthropologist to reach the same conclusion was J.D. Freeman, whose fieldwork among the Iban of Sarawak in 1949 and 1950 also happened to coincide with an unusually bad agricultural year (the worst in at least a decade) in which heavy rains interfered with the firing of the farms. Only one-third of the families included in his survey were able to meet their own basic rice requirements that year, against an estimated 70 to 80% in 'normal' years and 'virtually all' families in favourable years. Not surprisingly, Freeman characterizes Iban shifting cultivation as 'a highly uncertain undertaking' and concludes that 'the subsistence economy of the Iban is one of scarcity rather than plenty' (Freeman 1955:96-7, 104). And while his own fieldwork account implies that most families can cope with a shortfall in rice by selling off property or by supplementing their diet with sago, he also quotes with approval a nineteenth-century colonial administrator who observed that it was not unusual for poor harvests 'to put the inhabitants to great straits to obtain means of maintaining life' (Freeman 1955:104-5, 97).

Dove and Freeman were able to acknowledge the variability and sometimes inadequacy of the yields from swidden farming partly thanks to their relatively extended periods of fieldwork, which gave them more chance of witnessing a year of bad weather and the resulting food crisis.[5] In

5 Dove's stay in Borneo covered three rice harvests (Dove 1985:6) and Freeman's apparently more than one. By contrast the data presented by Conklin who does not discuss the impact of climatic variation refer essentially to a single agricultural year (Conklin 1957:4).

order to obtain a balanced overview, however, a genuinely historical approach is called for. One of the few authors who has combined field observation with quantitative evidence for climatic variability over long periods is geographer Joachim Metzner, whose studies of the relationship between man and environment in the Lesser Sundas offer the clearest evidence so far for the importance of drought and harvest failure in Indonesian subsistence agriculture. Analysing rainfall data from meteorological stations in the Sikka area of Flores from 1905 to 1977, Metzner concludes that one of the most important features of the rainfall regime from an agricultural point of view is its unpredictability (Metzner 1982: 41). In his complementary study of the Baucau-Viqueque area of East Timor, also using long series of meteorological records, he goes further by declaring that 'a pronounced variability of the rainfall regime' is nothing less than 'Timor's dominant climatic factor'. For subsistence farmers, moreover, 'temporary setbacks caused by climate tend to be synonymous with hunger' since there is seldom sufficient seed to permit a second sowing in the event of crop failure (Metzner 1977:64, 72, 292). Food shortages resulting from climatic variation, then, were common in traditional farming communities in Timor and contributed to the generally poor nutritional state of the population (Metzner 1977:257, 259). Another geographer, F.J. Ormeling, agreed in his 1955 study of man and environment in West Timor that famine had been a recurrent feature of Timorese history (Ormeling 1955:26, 237, 240). Both Ormeling and Metzner also include crop failure and famine in their lists of the factors which restrained population growth during pre-colonial times (Metzner 1982:90; Ormeling 1955:180, 237).

North Sulawesi in the nineteenth century

In this contribution I want to present some comparable evidence for an area much less well known for the inclemency of its natural environment, North Sulawesi. Most of my data come from nineteenth-century Dutch administrative sources, principally from the period between 1820 and 1870, supported by some later meteorological observations. My initial inclination to use the word 'pre-colonial' in the title, in line with Reid's classic formulation of the problem of low population growth in pre-colonial Southeast Asia (Reid 1987), was therefore too ambitious. The two areas on which I focus, Minahasa and Gorontalo, had been the site of VOC outposts since the seventeenth and the eighteenth century respectively. By the early nineteenth century they had both been significantly affected by the Dutch presence in both political and economic terms: Minahasa as a supplier of rice and coffee to the colonial state, and Gorontalo as a supplier of gold from the mines and alluvial deposits scattered along the peninsula. Towards the end of the period covered by my data, subsistence activities in both areas were also affected by the introduction of new forms of taxation and new types of compulsory labour service, mainly in the form of roadbuilding. It will not be possible to describe the changing political context in detail, and this weakness is acknowledged.

Especially at the beginning of the period in question, however, both areas still approximated to a 'traditional' or 'premodern' situation in three respects. Subsistence agriculture, first of all, continued to form the backbone of the economy and to be performed by traditional methods. Secondly, except for deliveries of Minahasan rice to the government the trade in foodstuffs remained limited, mainly due to factors connected with the indigenous political context (Henley 1995). Thirdly, health conditions and hence demographic patterns remained little affected by Western medicine – although this was no longer true of Minahasa after 1850, a point on which more will be said in the section on Minahasa below. More fundamentally, the sources which I will be quoting also afford sufficient glimpse of the areas outside Dutch control – the lands around the Gulf of Tomini, Bolaang Mongondow, and the Sangir and Talaud Islands – to show that the events described were not unique to Minahasa and Gorontalo.

A further disclaimer is also necessary at the outset. In what follows I will not be claiming that climatically induced subsistence crises were the only, or even the most important, factor retarding population growth. As will be evident from the information presented, disease was always more significant in terms of the number of direct victims which it claimed. With the exception of smallpox, however, most outbreaks of epidemic disease were strongly correlated with extreme climatic events and with food shortages, so that their respective contributions to increased mortality often cannot be disentangled. The typical mortality crisis in nineteenth-century North Sulawesi involved a combination of food scarcity and epidemic disease. The best place to observe this pattern over a long period is in the Dutch correspondence and other records from Gorontalo, a subregion particularly prone to natural disasters.

A drought-prone subregion: Gorontalo

In the middle of the nineteenth century the Gorontalo area as a whole (about 12,000 km^2) was sparsely populated, with an average density of not more than ten and possibly as low as six persons per km^2 (Henley 1994:16). There was a pocket of denser population in the central Limboto depression, which was unusual in regional terms in that it included a substantial area of irrigated rice cultivation. The staple crop, however, was maize, while much rice was also grown on dry fields in the normal swidden-farming manner. Located partly in the extended rainshadow cast by the highlands of Central Sulawesi on the other side of the Gulf of Tomini during the southeast monsoon, Gorontalo is also significantly drier than the more easterly parts of North Sulawesi. In Gorontalo town itself, partly due to unusual orographic effects, the average annual precipitation is a very low 1,211 mm (Braak 1946:75). Many upland areas in the same drainage area, however, have much higher rainfall, and at Kwandang on the north coast the annual average is a healthy 2,725 mm (ASEAN 1982:137). But perhaps the most striking thing about the rainfall pattern in Gorontalo is its unpredictability, both in annual and in seasonal terms (Van Hoëvell 1891:34-6). The impact of this unpredictability is clearly evident in the following chronicle of extreme climatic events, epidemics, and food shortages afflicting Gorontalo over a period of half a century (see Appendix 1). The record begins in 1820, one year after the Dutch re-established their outpost in Gorontalo after an absence of more than twenty years during which the only extreme event remembered in Dutch sources was an episode of serious flooding in 1805 or 1806.[6] My examination of the earlier VOC record is not yet complete and is therefore left out of consideration here.

After 1869 the record continues, but the impact of climatic extremes is progressively weakened by better communications, greater food imports (a factor already significant in the 1860s), improved medical care, and even the beginnings of organized famine relief. During the drought crisis of 1877 and 1878, for instance, the European authorities 'felt themselves obliged to offer help by setting up a brick factory in Gorontalo at which everybody who so desired could work for a daily wage of *f* 0.40, to be paid either in cash or in maize imported for that purpose from Minahasa'.[7]

Even allowing for a certain amount of hyperbole in Dutch reporting of the various crises (Allen, Brookfield and Byron 1989:294), the chronicle presents a rather convincing picture of a society close to the limits of its food-producing capacity, and with evident demographic consequences. In some cases, such as the smallpox epidemic of 1820, food shortages were admittedly the result rather than the cause of disease outbreaks which rendered people unable to perform agricultural work. More often, however, epidemics followed (rather than preceded) food shortages clearly brought on by unfavourable weather (1839, 1846, 1854, 1855). Some of the Dutch sources are explicit regarding the direction of causality here. During the disastrously dry 'wet' season of 1845-1846, for example, the assistant resident of Gorontalo described the 'tremendous number of deaths to stomach disease' as 'a result of lack of proper food over a period of several months due to the intense heat and absence of rain'.[8] It is possible that such reports underestimated more direct connections between drought and disease via poorer than usual hygiene or perhaps an enhanced malaria environment. However, there were also years in which drought was not

6 Floods in 1805 or 1806 are mentioned in: Ass. Res. Gorontalo to Res. Manado, 10 July 1842 (ANRI Manado 117).

7 KV Manado 1878 (ANRI Manado 53).

8 Quoted in: Res. Manado to Gov. Moluccas, 15 April 1846 (ANRI Manado 83).

Table 1. Climatic variability at six locations in North Sulawesi.[1]

	Annual precipitation						Precipitation days per year			
	years on record	mean (mm)	S.D. (mm)	C.V.	in driest year (mm)	in wettest year (mm)	years on record	mean	min.	max.
Tinombo (NW coast Tomini Gulf)	32	1287	275	0.214	844	1695	32	84	50	149
Kwandang (Gorontalo)	57	2725	821	0.301	1260	3568	–	n.a.	n.a.	n.a.
Modayak (Bolaang Mongondow)	31	2462	523	0.212	1327	3366	20	202	151	236
Mapanget (Minahasa)	26	3278	571	0.174	2009	4420	26	197	135	244
Tahuna (Sangir)	50	3897	782	0.201	2435	6093	23	188	141	236
Beo (Talaud)	25	3118	618	0.198	4164	1333	18	161	117	210

S.D. = standard deviation; C.V. = coefficient of variation (S.D./mean).
Note: all records are more or less discontinuous.
Source: ASEAN 1982.

1 According to present-day administrative boundaries, Tinombo is technically located in Central Sulawesi.

followed by epidemic disease, yet people (and implicitly adults, a point to which I will return) were nevertheless reported to have died from pure hunger (1821, 1826).

Intraregional comparisons

Unfortunately, nineteenth-century sources on the independent parts of North Sulawesi outside Gorontalo and the other area controlled by the Dutch, Minahasa (of which more below), are sparse and generally say little about subsistence or health conditions. Nevertheless, there is enough evidence to show that food shortages and epidemics associated with drought also occurred in genuinely 'pre-colonial' areas. During the drought of 1853, for instance, the assistant resident of Gorontalo reported that 'in the lands around the Gulf of Tomini, famine reigns everywhere as a result of the long drought and the resulting crop failure'.[9] In Mouton on the north side of the gulf the food shortage was alleviated by sago imported from the Togian Islands, although on Togian itself there was so little water in the streams that instead of being processed in the normal manner using a settling tank, the sago trunks had to be chopped up and shipped in dry form like timber.[10] In the course of 1854 the epidemic which followed the drought was reported from the Tomini lands, Bolaang Mongondow, and the Sangir archipelago as well as from Minahasa and Gorontalo.[11] During the 1855 drought there were also reports of food shortages on Sangir, where the lack of rain caused 'great damage to trees and crops'.[12]

The degree of climatic variability in Gorontalo, as already noted, was somewhat exceptional. Table 1, based on twentieth-century (and in some cases late nineteenth-century) precipitation records from selected meteorological stations for which more than 20 years of data are available, gives some idea of the relative levels of inter-annual variation in different parts of the region.

Any conclusions drawn from this table can only be very preliminary, since rainfall patterns within a single subregion, as we have seen for Gorontalo, varied widely according to altitude and other local factors. Nevertheless, the information it contains suggests some significant patterns. While Kwandang in Gorontalo is not the driest of the stations listed, it has easily the highest coefficient of rainfall variation. The station with the lowest average precipitation, Tinombo, has a considerably more reliable climate. Interestingly, Tinombo in the nineteenth century was an important centre of tobacco production and export, its markets including Gorontalo and Minahasa. Presumably the low but

9 MV Manado, January 1854 (ANRI Manado 54).
10 KV Manado 1853 (ANRI Manado 51); MV Manado, January 1854 (ANRI Manado 54).
11 MV Manado, May, July and September 1854 (ANRI Manado 54).
12 MV Manado, September 1855 (ANRI Manado 54).

dependable rainfall there was particularly favourable for the cultivation of this crop. Sangir, Talaud, and Modayak on the plateau of Bolaang Mongondow combine moderate to high mean annual rainfall with coefficients of variation much lower than that of Kwandang. All three of these areas seem to have had a relatively high population density in pre- and early-colonial times, at least compared to most parts of Gorontalo (Henley 1994), which would seem to confirm the demographic significance of climatic variability. The station with the lowest variability of all, however, is Mapanget in Minahasa, and it is to Minahasa that I now want to turn.

A subregion of 'moderate' climatic variability: Minahasa

In late colonial days Minahasa, and especially the high, cool plateau around Lake Tondano, was widely known for its equable climate as well as the productivity of its rich volcanic soil (Smits 1909:18). Dry rice and maize, the latter planted on the newly cleared rice swiddens at the height of the usually mild dry season, complemented each other in the agricultural cycle and provided what Dutch observers regarded as an exceptional level of food security. Indeed, in all the historical sources for Minahasa – in sharp contrast to those for Gorontalo – I have never come across an explicit reference to *hongersnood* or famine, still less to actual deaths from starvation. Yet Minahasa offers unique evidence that even without visible famine, variations in the food supply still had a significant impact on the mortality rate. Whereas the available population statistics for Gorontalo are too scanty to allow any sort of quantitative reconstruction of how the subsistence crises there affected demographic change, for Minahasa we are fortunate enough to have an annual record of births and deaths from 1849 onwards (see Appendix 2). For 1853 and most subsequent years, moreover, there are also statistics on the size of the annual rice and maize harvests. Only for climatic variation and disease is it still necessary to rely on qualitative observations, and even these are generally of high quality. This combination of data makes it possible to gain an overview of how demographic change related not only to climatic anomalies and epidemics, but also to the production of the most important subsistence crops from year to year. The rates of mortality and population growth are given in absolute terms; for reference, the total recorded population of Minahasa was 96,815 in 1850, 100,308 in 1860 ,and 115,007 in 1870.[13]

The quantitative data presented in Appendix 2 are problematic for several reasons. A first reservation concerns their accuracy. The Dutch authorities themselves did not regard their population statistics as

13 KV Manado 1850, 1860, 1870 (ANRI Manado 50, 52, 53).

accurate until 1859, and never developed much faith in their maize production figures despite the fact that they insisted on quoting them to the single cob. Secondly, the data refer to a period of accelerated political changes, including an expansion of compulsory coffee cultivation and corvée labour services and the introduction in 1852 of a monetary head tax to replace the earlier obligatory rice sales (which had averaged perhaps 30% of the total rice crop in the first half of the century). One reason for the relatively poor rice crop in 1861 was the heavy workload imposed upon the population in the first part of that year by compulsory roadbuilding duties in connection with an impending visit by the Governor-General. This interfered with agricultural activities to an unusual extent and caused some crops to be planted behind schedule.[14] On the positive side, the fact that epidemic smallpox had already been eliminated by inoculation by this stage and that quinine was being used on a considerable scale to combat malaria – factors which, between them, probably account in large part for the strong net population growth over the period covered – should make it easier to identify the purely 'nutritional' component of the mortality rate.

Whatever its shortcomings, the uniqueness of this information surely makes an attempt at analysis worthwhile. The link between climatic anomalies and poor harvests, first of all, is clear enough. The dry events of 1850, 1853, 1855, and 1865 all correspond to more or less poor rice harvests, and those of 1853 and 1855 appear to have affected the maize harvest too, although the maize production figures for those years are particularly incomplete.[15] The long rains of 1861 and 1862 also had a marked effect on both rice and maize harvests, although as I have mentioned there were also political reasons for the depressed harvests of those years, and 1867 presents an unexplained exception in this respect. Regarding the more important question of mortality, disease is clearly still the pre-eminent killer, most obviously in the case of the great dysentery epidemic of 1853 and 1854, but also during the 'fever' (predominantly malaria?) epidemics of 1850-1852 and the malaria outbreak of 1860.

Upon close inspection, however, a relationship is also discernible between food production and mortality. This was first identified as long ago as 1875 by the Dutch government investigator A.C.J. Edeling, who pointed out that a delay of one year typically separated each poor harvest from its demographic impact in the statistics.[16] Unfortunately the effect of the first drought in 1850 is hidden by the inaccurate mortality statistics for 1851, and that of the second in 1853 by the great number of deaths caused by the associated epidemic. The high mortality and

14 KV Manado 1861 (ANRI Manado 95).

15 That the maize harvest of 1855 was affected by the drought is confirmed by a qualitative source (KV Manado 1855, in ANRI Manado 51).

16 A.C.J. Edeling, Memorie omtrent de Minahasa 1875 (ARA MvK, Verbaal 17-4-1877/20).

negative population growth during 1856, however, clearly reflect the impact of the 1855 drought, since this was not accompanied by epidemic disease and there had been a clear recovery of population growth in 1855 following the relatively good year of 1854 between the droughts. In 1861 the story was the same, with dramatically increased mortality in 1862 following the poor harvests of the previous year despite the absence of epidemic disease. With the eye of faith it is also possible to see the decelerating growth of 1864 to 1866 as a result of steadily falling rice yields from 1863 to 1865, and the higher growth rates towards the end of the decade as a reflection of the good harvests of both rice and maize in that period.

There remain, admittedly, some awkward facts which are difficult to account for. Why, for instance, should the poor rice and maize harvests of 1862 – even poorer than those of 1861 – not have resulted in as high a mortality rate in 1863 as in 1862? On the other hand, the rate of population growth in 1863 was still lower than in any subsequent year of the record, and lower too than in the plentiful years of 1857-1859. To make full sense of the figures I will probably have to look in more detail not only at the political and cash crop (coffee) context, but also at the harvest and population statistics at district level, a task on which I have not yet begun. Nevertheless, I think the existence of a relationship between extreme climatic events, reduced food production, and increased mortality, even in the absence of epidemic illness, has already been sufficiently demonstrated. And the fact that it has been demonstrated for Minahasa, the area with the most reliable climate, the most productive agriculture and the best internal communications in North Sulawesi, makes this result doubly significant. One important implication, also recognized by Edeling in 1875, is that relative food shortages could suppress demographic growth by working through the medium of infant mortality, increasing the susceptibility of the very young to disease, without becoming visible to casual observers in the form of widespread hunger among the adult population.

Carrying capacity revisited

To return, finally, to the question of carrying capacity with which I began. It was pointed out that estimates of carrying capacity in swidden farming systems are customarily arrived at by taking the amount of land cultivated each year, multiplying it by the sustainable fallow period in years, and comparing the result with the total amount of land potentially available for cultivation. For Minahasa the earliest date at which the available sources allow a rough calculation of carrying capacity in this sense is 1862, the first year in which an attempt was made to measure the total area under rice cultivation.[17] The results indicated that 21,926 *bouw* of land were occupied by swidden ricefields, with another 1,932 *bouw* under irrigation.[18] No comparable information is available for other food crops, but since most maize and other 'secondary' crops were planted on the same land as the dry rice the resulting underestimation should not be too severe.[19] Using data on the number of coffee trees in Minahasa at the end of 1861, it is possible to calculate that an additional 7,248 *bouw* were occupied by 'orderly coffee gardens' and 2,002 *bouw* by coffee trees planted in forest swiddens and in scattered plots in and around the villages. Assuming that all of this coffee land was unavailable for food crop farming, the total amount of land under permanent cultivation (all coffee plus irrigated rice) was therefore 11,182 *bouw* or about 79 km². No allowance is made here for other less important tree crops such as coconut or nutmeg, but this is probably more than compensated for by the fact that part of the coffee cultivation was in fact integrated into the swidden fallow cycle and did not imply outright loss of potential rice land.

Regarding the typical length of the fallow period, Dutch sources from the 1860s give estimates ranging from three to ten years.[20] Since these estimates probably refer to the most populous and intensively cultivated districts – earlier sources mention up to 15 years of fallow in more remote areas and even a preference for opening swiddens in primary forest – it is probably safest to take the highest figure of 10 years as the basis for our calculation.[21] It will be evident that this is still a short fallow period by swidden farming standards, shorter indeed than the sustainable minimum of 12 years proposed by J.A. van Beukering in his influential 1947 study on the '*ladang* problem' in Indonesia (Van Beukering 1947:9). Given that the volcanic soils of Minahasa are among the most fertile in the outer islands, however, it does not seem unrealistic. In certain localities permanent dry-field agriculture was even possible without irrigation. In 1940 – by which time the population of Minahasa was more than three times what it had been in 1862 – botanist P.M.L. Tammes claimed that there were some dry fields in the Tomohon area 'which have been cultivated twice each year for more than a hundred years and still produce a reasonable crop' (Tammes 1940:192). On the basis of a ten-year average fallow period, then, the total amount of land involved in periodic cultivation in 1862 was eleven times 21,926 *bouw*, or about 1,712 km². Add to this the area under

17 KV Manado 1862 (ANRI Ambon 1563).

18 1 *bouw* = 0.071 km² ([reply to questionnaire on local weights and measures, 1884], ANRI Manado 14).

19 Much maize, however, was also planted in small separate gardens (KV Manado 1857, in ANRI Manado 52).

20 KV Manado 1861 (ANRI Manado 95), 1862 (ANRI Ambon 1563), 1863 (ANRI Manado 52), 1866 (ANRI Manado 52).

21 Register aantekeningen kommissaris voor Manado 1846 (ANRI Manado 47); Generaal Verslag Manado 1825 (ANRI Manado 48).

permanent cultivation and we arrive at a sum total of 1,791 km^2 of land under long-term use for farming purposes.

The total land area of Minahasa, by comparison, is about 5,200 km^2. So even allowing for the existence of a certain amount of uncultivable land on volcanic slopes and in swampy areas, and for the many swiddens which were no doubt missed by the surveyors, and for the fact that 1862 was a relatively poor year for both rice and maize production, it does not seem likely that Minahasa was close to the limits of its carrying capacity as defined by the land requirements of its existing agricultural system. Yet as I have shown in this contribution, limited food production, if not continuously then at least during years of markedly adverse weather, was still a significant factor – and again, I do not say the only factor – restricting population growth.

Roy Rappaport has pointed out that carrying capacity can be defined in two quite separate ways.[22] In anthropology, it usually means the population which can be supported in a particular area without inducing degradation of the environment. The carrying capacities calculated by Van Beukering, Conklin, Dove, Freeman, and Ellen, corresponding to the maximum population densities which could theoretically be accommodated by their respective agricultural systems without reducing the fallow period below a certain level, all fall either explicitly or implicitly into this category. But if the limits to the food supply can in fact be reached in poor years even in the absence of environmental degradation, then a second definition of carrying capacity becomes more relevant. In animal ecology, the same term is used in the simpler sense of the maximum population which can survive in a given area, regardless of whether or not the processes preventing growth beyond that maximum involve degradation of the environment. Rappaport mentions epidemic disease as one potential limiting factor which may vary with population density without being directly linked to agriculture or fallow intervals (Rappaport 1984:88).

The limiting factor emphasized here, of course, is linked with agriculture but at the same time is not, on the face of it, a function of the existing population density. This suggests that we might arguably go even further than redefining the concept of carrying capacity and question its very relevance, in any form, to the issue of population growth and density in pre- and early-colonial Indonesia. Carrying capacity, after all, is a matter of population density. But we have little direct evidence that the incidence and demographic impact of food shortages were related to the amount of land available per head of population. At most, a high population density probably encouraged famine-related epidemics. The significant factors in the demographic equation seem to have been the amount and above all the reliability of rainfall, in conjunction with the

22 Rappaport 1984:88. See also Ellen 1978:181.

ability of the soil to yield a good crop, not so much per unit area as per unit of labour time applied to it, under the prevailing climatic conditions.

Whether or not carrying capacity can help explain pre-colonial demography, it still remains relevant to environmental history as such. From this point of view my results tend to confirm the old established view that under traditional conditions, without modern medicines and producing mainly for their own consumption, swidden farmers in Indonesia are unlikely to have placed unsustainable pressure upon their local ecosystems. Even in environmentally favourable areas such as Minahasa, subsistence factors as well as disease tended to prevent the population density from reaching such a level that fallow periods became too short to allow restoration of soil fertility. This does not mean, of course, that the natural vegetation was not permanently affected. The ten-year average fallow period inferred for Minahasa, for instance, would certainly not have permitted full regeneration of the rain forest between each cycle of cultivation.

To conclude, however, I would suggest that there is after all some reason to stop short of dismissing carrying capacity as irrelevant to the demography of Indonesian subsistence societies. This has to do with the point made above about soil conditions mediating between climatic factors and crop yield. Even when the area of land cultivated is determined by labour considerations rather than by the total land area available for cultivation, the quality of that land may still be correlated with population density. The various studies of shifting cultivation cited above agree that soil, slope, and drainage conditions have a significant effect upon yields, and that they vary strongly from place to place even at a very local scale.[23] And even small differences in yields might conceivably be crucial in determining which households or settlements can store or harvest enough food to withstand dangerous climatic events, and which cannot. Moreover, in many swidden systems it is customary for a single household to cultivate two or more separate plots, chosen for their suitability for different crops or varieties, precisely as part of a strategy to minimize the risks and consequences of crop failure (Ellen 1978:145; Dove 1985:381). The choice of plots available to each cultivator might well be dangerously reduced by population pressure long before there was any question of an absolute land shortage. If these local factors could somehow be accounted for in the calculation, it might theoretically be possible to arrive at a new and lower carrying capacity figure much closer to the actually observed population density.

Does this observation amount to a new insight, or is it simply a nuanced restatement of the problem which anyone who attempts to estimate a

[23] Conklin 1957:36-8; Metzner 1977:36-61; Metzner 1982:28, 35; Ellen 1978:108-17; Dove 1985:35, 46-8, 75, 299.

maximum carrying capacity must face when determining how much land is to be regarded as 'uncultivable' and excluded from the calculation? Either way, two things are clear. Firstly, a sensitivity to inter-annual yield variations, and hence to climatic anomalies, is crucial to an understanding of the historical relationship between agriculture and population growth and density. Secondly, and unfortunately, applying the concept of carrying capacity in an explanatory way to the demography of traditional swidden farming societies in Indonesia is likely to demand a level of geographical specificity and statistical detail which historical sources can seldom if ever provide.

Abbreviations

ANRI	Arsip Nasional Republik Indonesia, Jakarta
ARA	Algemeen Rijksarchief, Den Haag
Ass. Res.	Assistant Resident
AV	Algemeen Verslag
Gov.	Governor
Gov. Gen.	Governor-General
KITLV	Koninklijk Instituut voor Taal-, Land- en Volkenkunde, Leiden, Manuscript collection
KV	Kultuurverslag
MV	Maandverslag
MvK	Archive of the Ministerie van Koloniën, ARA
n.a.	not available
Res.	Resident

References

Adriani, N. and A.C. Kruyt
1950-51 *De Bare'e sprekende Toradjas van Midden-Celebes (de Oost-Toradjas).* Amsterdam: Noord-Hollandsche Uitgevers Maatschappij. 3 vols.

Allen, B., H. Brookfield and Y. Byron
1989 'Frost and drought through time and space, part II: The written, oral and proxy records and their meaning', *Mountain Research and Development* 9:279-305.

ASEAN
1982 *The ASEAN compendium of climatic statistics.* Jakarta: ASEAN Sub-committee on Climatology, ASEAN Committee on Science and Technology, ASEAN Secretariat.

Bemmelen, W. van
1916 'Droogte-jaren op Java', *Natuurkundig Tijdschrift voor Nederlandsch-Indië* 75:157-79.

Beukering, J.A. van
[1947] *Het ladangvraagstuk; Een bedrijfs- en sociaal economisch probleem.* [Batavia]: Departement van Economische Zaken. [Mededeelingen van het Departement van Economische Zaken in Nederlandsch-Indië 9.]

Braak, C.
1946 'Klimaat', in: C.J.J. van Hall and C. van de Koppel (eds), *De landbouw in den Indischen Archipel*, Vol. 1, pp. 63-82. 's-Gravenhage: Van Hoeve.

Conklin, H.C.
1957 *Hanunóo agriculture; A report on an integral system of shifting cultivation in the Philippines.* Rome: Food and Agriculture Association of the United Nations. [FAO Forestry Development Paper 12.]

Crab, P. van der
1862 *De Moluksche eilanden; Reis van Z.E. de Gouverneur-Generaal Charles Ferdinand Pahud, door den Molukschen archipel.* Batavia: Lange.

Dove, M.R.
1985 *Swidden agriculture in Indonesia; The subsistence strategies of the Kalimantan Kantu'.* Berlin: Mouton.

Ellen, R.F.
1978 *Nuaulu settlement and ecology; An approach to the environmental relations of an eastern Indonesian community.* The Hague: Nijhoff. [KITLV, Verhandelingen 83.]

Freeman, J.D.
1955 *Iban agriculture; A report on the shifting cultivation of hill rice by the Iban of Sarawak.* London: Her Majesty's Stationery Office.

Henley, D.
1994 'Population and environment in precolonial northern Sulawesi'. Paper presented at the 13th Biennial Conference of the Asian Studies Association of Australia, Perth, 13-16 July 1994.
1995 'Trade and industry in precolonial Sulawesi'. Paper presented at the School of Oriental and African Studies, University of London, 5 October 1995.

Hoëvell, G.W.W.C. Baron van
1891 'De assistent-residentie Gorontalo, voor zoover die onder rechtstreeksch bestuur is gebracht', *Tijdschrift van het Koninklijk Nederlandsch Aardrijkskundig Genootschap* 8 (tweede serie):26-43.

Kruyt, A.C.
1903 'Gegevens voor het bevolkingsvraagstuk van een gedeelte van Midden-Celebes', *Tijdschrift van het Koninklijk Nederlandsch Aardrijkskundig Genootschap* 28 (tweede serie):190-205.

Metzner, J.K.
1977 *Man and environment in eastern Timor; A geoecological analysis of the Baucau-Viqueque area as a possible basis for regional planning.* Canberra: Australian National University. [Development Studies Centre Monograph 8.]
1982 *Agriculture and population pressure in Sikka, Isle of Flores; A contribution to the study of the stability of agricultural systems in the wet and*

dry tropics. Canberra: Australian National University. [Development Studies Centre Monograph 28.]

Ormeling, F.J.
1955 *The Timor problem; A geographical interpretation of an underdeveloped island*. Jakarta/Groningen: Wolters. [PhD thesis, University of Indonesia, Jakarta.]

Rappaport, R.A.
1984 *Pigs for the ancestors; Ritual in the ecology of a New Guinea people*. New Haven/London: Yale University Press.

Reid, A.
1987 'Low population growth and its causes in pre-colonial Southeast Asia', in: N.G. Owen (ed.), *Death and disease in Southeast Asia; Explorations in social, medical and demographic history*, pp. 33-47. Singapore: Oxford University Press. [ASAA Southeast Asia Publications Series 14.]
1988 *Southeast Asia in the age of commerce 1450-1680. Vol. 1: The lands below the winds*. New Haven/London: Yale University Press.

Reinwardt, C.G.C.
1858 *Reis naar het oostelijk gedeelte van den Indischen archipel, in het jaar 1821*. Edited by W.H. de Vriese. Amsterdam: Muller.

Schefold, R.
1988 *Lia; Das grosse Ritual auf den Mentawai-Inseln (Indonesien)*. Berlin: Reimer.

[Scherius, R.]
1847 'Eenige bijdragen tot de kennis en den toestand der afdeeling Gorongtalo (eiland Celebes)', *Verhandelingen en Berigten betrekkelijk het Zeewezen en de Zeevaartkunde* 7 (tweede serie):399-421.

Smits, P.J.
1909 'Het klimaat der hoogvlakte van Tondano', *Natuurkundig Tijdschrift voor Nederlandsch-Indië* 68:18-109.

Tammes, P.M.L.
1940 'De biologische achtergrond van het bevolkingsvraagstuk op Noord-Celebes en de Sangihe- en Talaud-Archipel', *Tijdschrift voor Economische Geographie* 31:177-98.

Appendix 1.
Gorontalo: a qualitative record of extreme events, 1820-1869.

Year	Extreme weather	Food shortages	Epidemic disease
1820	–	'very great' famine[1], 'caused by the smallpox' (March-September)[2]; food import[3]	smallpox epidemic[4]
1821	'heavy drought', April-September[5]	by September 'the whole rice crop had completely failed, to the great disadvantage of the inhabitants who were thereby threatened with famine. Fruit and vegetables were unavailable. [...] Already some poor people had [..] died of hunger'[6]; Dutch resident forbids export of foodstuffs[7]	–
1822	n.a.	n.a.	n.a.
1823	wet year (?); floods (October, November)[8]	'not only rice, but all necessities scarce' (June)[9]; 'the rice crop here has failed' (November)[10]; food export forbidden; rice imported from Manado	–

1 Here and elsewhere, only the explicit Dutch term *hongersnood* is translated as 'famine'.
2 Res. Gorontalo to Res. Ternate, 18 March and 24 April 1820 (ANRI Ternate 153).
3 Res. Gorontalo to Res. Ternate, 30 September 1820 (ANRI Ternate 153).
4 Res. Gorontalo to Res. Ternate, 18 March and 10 September 1820 (ANRI Ternate 153).
5 Reinwardt 1858:512; Res. Gorontalo to Res. Ternate, 23 September 1821 (ANRI Ternate 153).
6 Reinwardt 1858:512.
7 Res. Gorontalo to Res. Ternate, 26 September 1821 (ANRI Ternate 153).
8 Res. Gorontal to Res. Ternate, 20 February 1824 (ANRI Gorontalo 1).
9 Res. Gorontalo to Res. Ternate, 22 June 1823 (ANRI Gorontalo 1).
10 Res. Gorontalo to Res. Ternate, 22 November 1823 (ANRI Gorontalo 1).

Year	Extreme weather	Food shortages	Epidemic disease
1824	drought[11]	'prevailing lack of foodstuffs' (August); sago and rice imported by sea[12]; indigenous claims of deaths by hunger denied by Dutch assistant resident[13]	–
1825	–	–	–
1826	'great drought'[14]	'famine owing to the great drought (which went so far that the natives gathered all sorts of roots and leaves with which to feed themselves, and people even died of hunger)'; coconut oil unavailable because nuts serving as famine food [15]	–
1827	–	–	–
1828	–	–	–
1829	–	–	–
1830	–	–	cholera[16]
1831	?	'poor rice harvest'[17]	–
1832	drought[18]	–	'malevolent fevers'[19]

11 Res. Gorontalo to Res. Manado, 26 September 1824 (ANRI Manado 114).
12 Res. Gorontalo to Res. Manado, 26 September 1824 (ANRI Manado 114).
13 Res. Gorontalo to Res. Manado, 6 October 1824 (ANRI Manado 114).
14 Ass. Res. Gorontalo to Res. Manado, 8 December 1826 (ANRI Manado 114).
15 Ass. Res. Gorontalo to Res. Manado, 8 December 1826 (ANRI Manado 114).
16 Ass. Res. Gorontalo to Res. Manado, 6 October 1830 (ANRI Manado 115).
17 Ass. Res. Gorontalo to Res. Manado, 13 September 1831 (ANRI Manado 115).
18 Ass. Res. Gorontalo to Res. Manado, 20 October 1832 (ANRI Manado 115).
19 Ass. Res. Gorontalo to Res. Manado, 20 October 1832 (ANRI Manado 115).

Year	Extreme weather	Food shortages	Epidemic disease
1833	drought[20]	–	dysentery, cholera(?) and other fevers[21]
1834	–	–	–
1835	–	–	–
1836	–	–	–
1837	–	–	–
1838	–	–	–
1839	'long drought'[22]	famine[23]	smallpox (December), *buikloop* (dysentery?) and *koorts* ('fever')[24]
1840	–	–	smallpox (January-February)[25]; at least 5,000 dead[26]
1841	'long drought'[27]	'famine of more than six months'[28]	–
1842	heavy rains and flooding in July[29]	–	–
1843	continuous drought, May-September[30]	rice and maize harvest 'very unfavourable'; 'great scarcity of foodstuffs' in Gorontalo town (September)[31]	–
1844	–	–	–

20 Res. Manado to Gov. Gen., 19 October 1833 (ANRI Manado 77).
21 Res. Manado to Gov. Gen., 19 October 1833 (ANRI Manado 77).
22 Ass. Res. Gorontalo to Res. Manado, 25 November 1841 (ANRI Manado 116).
23 Ass. Res. Gorontalo to Res. Manado, 11 January 1840 and 25 November 1841 (ANRI Manado 116).
24 Ass. Res. Gorontalo to Res. Manado, 14 January 1840 (ANRI Manado 116).
25 Ass. Res. Gorontalo to Res. Manado, 20 February 1841 (ANRI Manado 116).
26 Scherius 1847:401-2, 407.
27 Ass. Res. Gorontalo to Res. Manado, 25 November 1841 (ANRI Manado 116).
28 Verslag Afdeling Gorontalo 1841-43 (ANRI Manado 50); Ass. Res. Gorontalo to Res. Manado, 25 November 1841 (ANRI Manado 116).
29 Ass. Res. Gorontalo to Res. Manado, 10 July 1842 (ANRI Manado 117).
30 Ass. Res. Gorontalo to Res. Manado, 22 September 1843 (ANRI Manado 117).
31 Ass. Res. Gorontalo to Res. Manado, 22 September 1843 (ANRI Manado 117).

Year	Extreme weather	Food shortages	Epidemic disease
1845	'tremendous rains' in January-March; 'unbroken drought of almost seven months', April-November[32]	drought 'destroyed more than two thirds of the rice and maize fields' so that 'the price of foodstuffs climbed to an extraordinary level'[33]; 'pressing food shortage, so that the people are incapable of performing any work' (November)[34]	many deaths among livestock (water buffalo)[35]
1846	renewed drought, January-April[36]	famine (January)[37]; 'total absence' of rice in Limboto, so that seed rice had to be imported (February)[38]; 'lack of proper food over a period of several months, as a result of the intense heat and absence of rain' (April)[39]; seed rice still short in December[40]	'stomach disease' causes a 'tremendous number of deaths' in first part of year[41]
1847	continuous heavy rains, February-July[42]	'repeated crop failures' in northern and eastern districts; foodstuffs short and many people searching for sago in forests on north coast (July)[43]	–

32 Verslag Gorontalo 1845 (KITLV H 111).
33 Verslag Gorontalo 1845 (KITLV H 111).
34 Daghregister Gorontalo, 20 November 1845 (ANRI Manado 172).
35 Daghregister Gorontalo, 7 September 1846 (ANRI Manado 5).
36 Res. Manado to Gov. Moluccas, 15 April 1846 (ANRI Manado 83). See also Daghregister Gorontalo, 21 March 1845 (ANRI Manado 172).
37 Ass. Res. Gorontalo to Res. Manado, 17 January 1846 (ANRI Manado 119).
38 Ass. Res. Gorontalo to Res. Manado, 2 January 1846 (ANRI Manado 119).
39 Res. Manado to Gov. Moluccas, 15 April 1846 (ANRI Manado 83).
40 Daghregister Gorontalo, 9 December 1846 (ANRI Manado 5).
41 Res. Manado to Gov. Moluccas, 15 April 1846 (ANRI Manado 83). See also Daghregister Gorontalo, 21 March 1846 (ANRI Manado 172).
42 Ass. Res. Gorontalo to Res. Manado, 31 July 1847 (ANRI Manado 119).
43 Ass. Res. Gorontalo to Res. Manado, 31 July 1847 (ANRI Manado 119).

Year	Extreme weather	Food shortages	Epidemic disease
1848	–	–	–
1849	–	–	–
1850	heavy rain in first half of year interferes with planting on wet fields[44]; drought in second half[45]	'almost complete harvest failure' in second half of year[46]; 'lack of foodstuffs', sago and rice imported[47]	n.a.
1851	n.a.	February: part of the population 'dispersed in the forests due to the failure of its crops and the resulting famine'[48]; harvest for this year also 'very unfavourable'[49]; sago and rice imported.[50]	n.a.
1852	n.a.	sago and rice imported[51]	n.a.
1853	'continuous, unusually intense drought' in second half of year[52]	rice harvest 'for the most part a failure, and measures have had to be taken to prevent shortages of rice'[53]	–

44 Daghregister Gorontalo, 1st and 2nd quarter 1850 (ANRI Manado 27, 120).
45 Drought inferred from crop failure (KV Gorontalo 1852, in ANRI Gorontalo 3) and from reports of a 'long, intense drought' in Java in the same year (Van Bemmelen 1916:163). Except for 1835, all of the other drought years listed by Van Bemmelen for Java in the period 1833-69 also correspond to droughts in Gorontalo.
46 KV Manado 1850 (ANRI Manado 51); KV Gorontalo 1852 (ANRI Gorontalo 3).
47 Res. Manado to Gov. Moluccas, 21 October 1854 (ANRI Gorontalo 18).
48 Kroniek van Boalemo (ANRI Gorontalo 18).
49 KV Gorontalo 1952 (ANRI Gorontalo 3).
50 Res. Manado to Gov. Moluccas, 21 October 1854 (ANRI Gorontalo 18).
51 Res. Manado to Gov. Moluccas, 21 October 1854 (ANRI Gorontalo 18).
52 KV Manado 1853 (ANRI Manado 51); MV Manado, October 1853 (ANRI Manado 54).
53 MV Manado, October 1853 (ANRI Manado 54).

Year	Extreme weather	Food shortages	Epidemic disease
1854	drought, January-April[54]; heavy rain and flooding, May-July[55]	drought conditions 'very unfavourable for the rice crop, so that the shortage and high price of foodstuffs [...] put the population under great pressure' (March)[56]; rice still in short supply in August[57]; maize shortage in November[58]	'measles and above all dysentery have claimed many victims'[59]; cholera also reported (October)[60]; mortality 'exceptionally high' and 'far in excess of fertility'[61]
1855	exceptional drought and heat in second half of year[62]	rice and maize affected by drought in September, 'so that scarcity of food made itself felt'[63]; by December 'food was short everywhere, and a large part of the population was searching for foodstuffs in the forest'[64]	dysentery epidemic continues until February[65]; 'epidemic fevers of a catarrhal and gastric nature', August-September[66]
1856	much rain in 'dry' season[67]	–	disease outbreak kills more than 500 water buffalo[68]
1857	n.a.	n.a.	n.a.

54 MV Manado, April and May 1854 (ANRI Manado 54).
55 MV Manado, May and September 1854 (ANRI Manado 54).
56 MV Manado, April 1854 (ANRI Manado 54).
57 MV Manado, September 1854 (ANRI Manado 54).
58 MV Manado, December 1854 (ANRI Manado 54).
59 KV Gorontalo 1854 (ANRI Manado 54).
60 MV Manado, September 1854 (ANRI Manado 54).
61 KV Gorontalo 1854 (ANRI Manado 54); MV Manado, December 1854 (ANRI Manado 54).
62 MV Manado, September-December 1855 (ANRI Manado 54).
63 MV Manado, September 1855 (ANRI Manado 54).
64 MV Manado, December 1855 (ANRI Manado 54).
65 MV Manado, February and March 1855 (ANRI Manado 54).
66 MV Manado, September 1855 (ANRI Manado 54).
67 MV Manado, July-October 1856 (ANRI Manado 54).
68 MV Manado, October 1856 (ANRI Manado 54).

Year	Extreme weather	Food shortages	Epidemic disease
1858	n.a.	n.a.	n.a.
1859	–	–	smallpox (December)[69].
1860	–	food import[70]	smallpox, January-May: at least 2,800 dead[71]
1861	wet year; flooding (June–November)[72]	rice and maize scarce and expensive (November)[73]	–
1862	–	rice and maize scarce and expensive (January-April)[74]	–
1863	–	–	heavy fevers (January-April)[75]
1864	harsh dry season[76]	rice scarce and expensive, June-September; rice and sago imported[77]	–
1865	harsh dry season[78]	partial rice crop failure; sago imported[79]	–
1866	–	–	cholera (from October onward): 1,438 deaths reported[80]

69 MV Manado, December 1859 (ANRI Manado 54).
70 MV Manado, November 1860 (ANRI Manado 54).
71 MV Manado, May 1860 (ANRI Manado 54). Van der Crab (1862:379) gives a figure of 3,000 dead.
72 MV Manado, August-December 1861 (ANRI Manado 54).
73 MV Manado, December 1861 (ANRI Manado 54).
74 MV Manado, January-June 1862 (ANRI Manado 54).
75 MV Manado, January-June 1863 (ANRI Manado 54).
76 KV Gorontalo 1864 (ANRI Manado 44).
77 KV Gorontalo 1864 (ANRI Manado 44).
78 KV Manado 1865 (ANRI Manado 52).
79 KV Manado 1865 (ANRI Manado 52).
80 MV Manado, March 1867 (ANRI Manado 54).

Year	Extreme weather	Food shortages	Epidemic disease
1867	unusually wet year; rain and floods destroy part of rice and maize crop[81]	–	cholera, 'fevers' and *buikloop* (dysentery?), January-September; another 542 reported dead)
1868	–	–	–
1869	–		–

81 MV Manado, October 1867 and January 1868 (ANRI Manado 54).

Appendix 2.
Minahasa: a semi-quantitative record of extreme weather, food production, disease and mortality 1849-1870.

Year	Extreme weather	Rice harvest (pikul of padi)[1]	Maize harvest (cobs)[2]	Epidemic disease	Mortality[3]	Population change[4]
1849	–	n.a.	n.a.	–	2,907	+960
1850	drought[5]	'mixed results'[6]	n.a.	'terrible fevers'[7]	2,939	+189
1851	–	'generally favourable'[8]	'generally favourable'[9]	'terrible fevers'[10]	[2,512][11]	[+480][12]
1852	–	'favourable'[13]	'favourable'[14]	'heavy fevers'[15]	3,147	-64

1 KV Manado, 1853-1869 (ANRI Manado 39, 48, 51, 52, 53, 95, 198; Ambon 1543, 1563). 1 *pikul* = 62 kg.
2 KV Manado, 1853-1869 (as above).
3 Algemeen/Administratief Verslag Manado, 1849-1870 (ANRI Manado 51, 52, 53).
4 Algemeen/Administratief Verslag Manado, 1849-1870 (ANRI Manado 51, 52, 53).
5 Inferred from poorer than average harvest in Minahasa (KV Manado 1850, in ANRI Manado 51), crop failure in Gorontalo (KV Gorontalo 1852, in ANRI Gorontalo 3), and reports of a 'long, intense drought' in Java in the same year (Van Bemmelen 1916:163). Except for 1835, all of the other drought years listed by Van Bemmelen for Java in the period 1833-70 also correspond to dry events in Minahasa.
6 KV Manado 1850 (ANRI Manado 51).
7 KV Manado 1850 (ANRI Manado 51).
8 KV Manado 1851 (ANRI Manado 51).
9 KV Manado 1851 (ANRI Manado 51).
10 KV Manado 1851 (ANRI Manado 51).
11 'This figure was not submitted on schedule and its accuracy is therefore doubtful' (KV Manado 1851, in ANRI Manado 51).
12 'This figure was not submitted on schedule and its accuracy is therefore doubtful' (KV Manado 1851, in ANRI Manado 51).
13 KV Manado 1852 (ANRI Manado 51).
14 KV Manado 1852 (ANRI Manado 51).
15 KV Manado 1852 (ANRI Manado 51).

Year	Extreme weather	Rice harvest (pikul of padi)[16]	Maize harvest (cobs)[17]	Epidemic disease	Mortality[18]	Population change[19]
1853	'unusually long and intense drought' in second half of year[20]	324,721	> 43,590,000	'epidemic stomach disease', August-December[21]	2,810	+461
1854	–	324,677	86,882,418	dysentery, measles and malaria, January-November[22]	12,821	-9,473
1855	'long and intense drought' in second half of year.[23]	262,234	> 24,319,865	–	2,456	+928
1856	unusually wet dry season[24]	529,087	94,546,966	–	3,167	–66
1857	–	468,591	131,884,752	–	2,490	+1,210
1858	–	622,206	214,657,043	–	2,499	+1,200
1859	–	635,205	237,202,053	–	2,426	+1,566

16 KV Manado, 1853-1869 (ANRI Manado 39, 48, 51, 52, 53, 95, 198; Ambon 1543, 1563). 1 pikul = 62 kg.
17 KV Manado, 1853-1869 (as above).
18 Algemeen/Administratief Verslag Manado, 1849-1870 (ANRI Manado 51, 52, 53).
19 Algemeen/Administratief Verslag Manado, 1849-1870 (ANRI Manado 51, 52, 53).
20 KV Manado 1853 (ANRI Manado 51).
21 MV Manado, August-December 1853 (ANRI Manado 54).
22 MV Manado 1854 (ANRI Manado 54).
23 KV Manado 1855 (ANRI Manado 51); MV Manado, July-December 1855 (ANRI Manado 54).
24 KV Manado 1856 (ANRI Manado 52).

Year	Extreme weather	Rice harvest (pikul of padi)[25]	Maize harvest (cobs)[26]	Epidemic disease	Mortality[27]	Population change[28]
1860	–	665,013	n.a.	malaria (January-April)[29]	3,276	+575
1861	'unusually high rainfall throughout the entire year'[30]; (interferes with burning of swiddens for 1862)[31]	444,559	161,888,316	–	2,791	+534
1862	continuous rainy weather in first half of year.[32]	413,526	118,690,195	–	3,862	–165
1863	–	620,507	179,988,387	–	2,879	+1,042
1864	–	576,074	187,250,221	–	2,400	+1,587
1865	harsh drought, January-March[33]	531,359	220,279,295	–	2,524	+1,433
1866	–	534,436	236,337,149	–	2,614	+1,326

25 KV Manado, 1853-1869 (ANRI Manado 39, 48, 51, 52, 53, 95, 198; Ambon 1543, 1563). 1 pikul = 62 kg.
26 KV Manado, 1853-1869 (as above).
27 Algemeen/Administratief Verslag Manado, 1849-1870 (ANRI Manado 51, 52, 53).
28 Algemeen/Administratief Verslag Manado, 1849-1870 (ANRI Manado 51, 52, 53).
29 MV Manado, January-April 1860 (ANRI Manado 54).
30 KV Manado 1861 (ANRI Manado 95).
31 KV Manado 1862 (ANRI Ambon 1563).
32 KV Manado 1862 (ANRI Manado 95).
33 KV Manado 1865 (ANRI Manado 52); MV Manado, January-March 1865 (ANRI Manado 54).

Year	Extreme weather	Rice harvest (pikul of padi)[34]	Maize harvest (cobs)[35]	Epidemic disease	Mortality[36]	Population change[37]
1867	'continuous rains through almost the whole year'[38]	673,042	345,493,762	–	2,549	+1,586
1868	–	641,524	333,288,255	–	2,238	+1,875
1869	–	569,818	314,785,410	–	2,909	+1,486
1870	n.a.	n.a.	n.a.	n.a.	1,735	+1,719

34 KV Manado, 1853-1869 (ANRI Manado 39, 48, 51, 52, 53, 95, 198; Ambon 1543, 1563). 1 pikul = 62 kg.
35 KV Manado, 1853-1869 (as above).
36 Algemeen/Administratief Verslag Manado, 1849-1870 (ANRI Manado 51, 52, 53).
37 Algemeen/Administratief Verslag Manado, 1849-1870 (ANRI Manado 51, 52, 53).
38 KV Manado 1867 (ANRI Manado 53).

HAN KNAPEN

Epidemics, droughts, and other uncertainties on Southeast Borneo during the eighteenth and nineteenth centuries

Introduction

One of the most striking aspects of the history of Borneo is the large number of disasters which struck the island at regular intervals (see appendix). Famine, drought, warfare, floods, forest fires, crop failures, and epidemics are common themes in historical sources. While one might expect disasters on a populous and geologically unstable island like Java, for Borneo this is rather surprising. Supposedly covered for centuries by prolific tropical rain-forest, Borneo gives a general impression of stability and security rather than unpredictability.

Disasters on Borneo have often been attributed to man-made changes in the environment: forest fires are caused by irresponsible shifting cultivators, floods by climatic change as a result of forest removal, cholera epidemics by contamination of rivers or unhygienic behaviour, crop failures by lazy farming practices, and so on. Although it may be true that some disasters are indeed induced by humans, this does not mean that calamities in general are something of recent times. I will argue that risk and uncertainty have always been a fact of life on Borneo, and I will examine how the Bornean population adjusted to them. The kinds of uncertainty involved were not constant but changed over time. The opening up of the interior and the increasing communication between coastal and inland populations early in the nineteenth century have to be considered in the light of the continuously changing risk conditions. As I will show, the return of Dutch colonialism after the British interregnum formed an important watershed in this respect.

* Research in the National Archives of Indonesia in 1995 and 1996 was made possible by a grant from the Netherlands Foundation for the Advancement of Tropical Research (WOTRO).

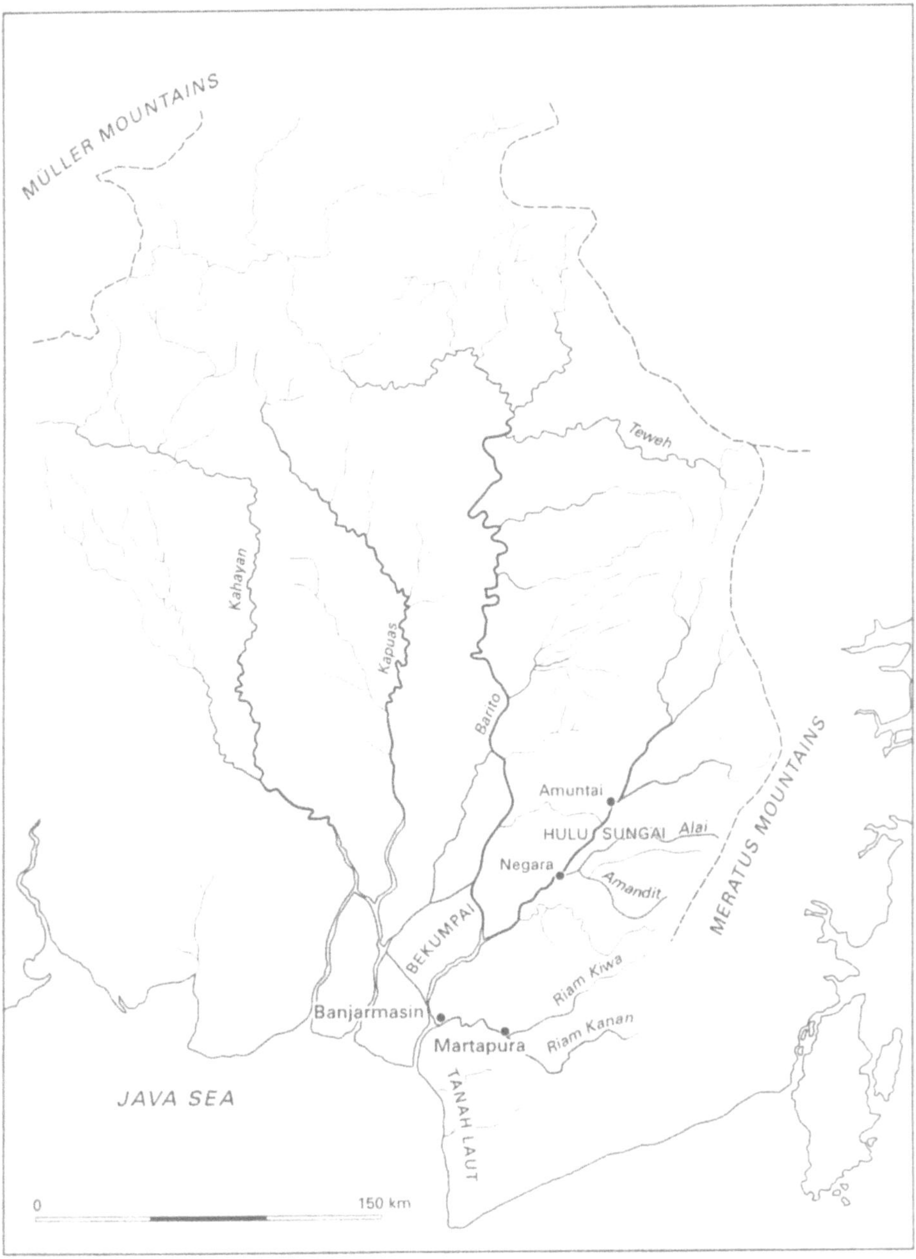

Southeast Borneo

Uncertainties on Borneo before the nineteenth century

'Uncertainties' can be divided into two categories: natural variability and social insecurity. The first category encompasses phenomena such as earthquakes, volcanic activity, climatic variations, epidemics, and other fluctuations in the natural environment outside human control. Among the second group we find, for example, unpredictable behaviour by those in power, political instability, warfare and other acts of violence, and economic uncertainty. Of course, this division is somewhat artificial since the two categories are often interlinked, for example when climatically induced famine causes plunder or even warfare (Boomgaard 1995).

Climate

Natural variation and unpredictability is much more pronounced on Borneo than one might expect. Though volcanic activity and earthquakes were uncommon and insignificant in the historical past, climatic variation, at least in Southeast Borneo, was not.[1] In general it can be said that Southeast Borneo receives most rain during the west monsoon from November to March, while much less falls during the east monsoon (April to October). However, variations from year to year are very pronounced. Unfortunately no figures earlier than the mid-nineteenth century are available, but the earliest data (for Banjarmasin from the 1850s onwards) clearly illustrate the climatic irregularities. In 1858, for example, Banjarmasin had 211 rainy days, but in 1860 only 114 (Figures 1 and 2).[2] During years of excessive rainfall, floods resulted in loss of crops, houses, food stocks, and even lives. A severely dry year, on the other hand, caused other problems, such as pest plagues, livestock deaths, massive forest fires, and low river levels which obstructed trade and interfered with gold washing.

Besides 'wet years' and 'dry years', another category of climatic aberrations can also be discerned. In some years the total number of rainy days was unexceptional, but their distribution over the year did not conform to the typical pattern. Two particular situations merit attention here. In the first the west monsoon started earlier than usual, making it difficult for shifting cultivators to burn open new fields and often resulting in rice shortages (Figure 3). In the second the whole pattern of rainfall became upset, for example when there was little rainfall during the west monsoon and an unexpectedly large amount during the east monsoon (Figure

1 Southeast Borneo is defined here as the drainage basins of the Barito, Kapuas Murung and Kahayan and including Tanah Laut (see map).

2 These data, like those for Figures 1-4, are based on rainfall data from Krecke (1859) and primary data from medical reports in the ANRI. The average rainfall levels in Figures 1-4 are based on the oldest data available (1850-61; figures for 1859 missing), which do not include any really extraordinary years. Unfortunately, only the number of rainy days were counted. The measurement of precipitation in millimetres only started at the end of the 19th century.

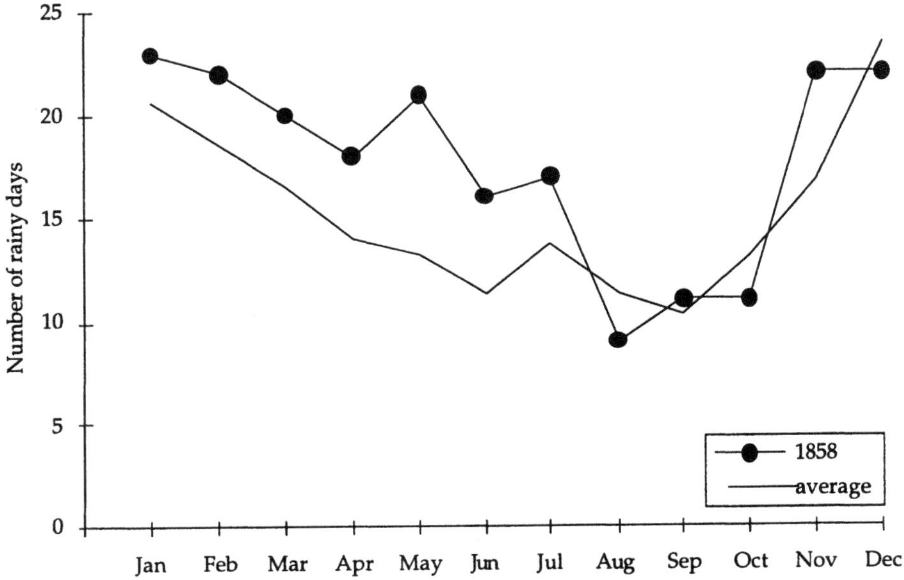

Figure 1. Number of rainy days in Banjarmasin: example of a 'wet year'.

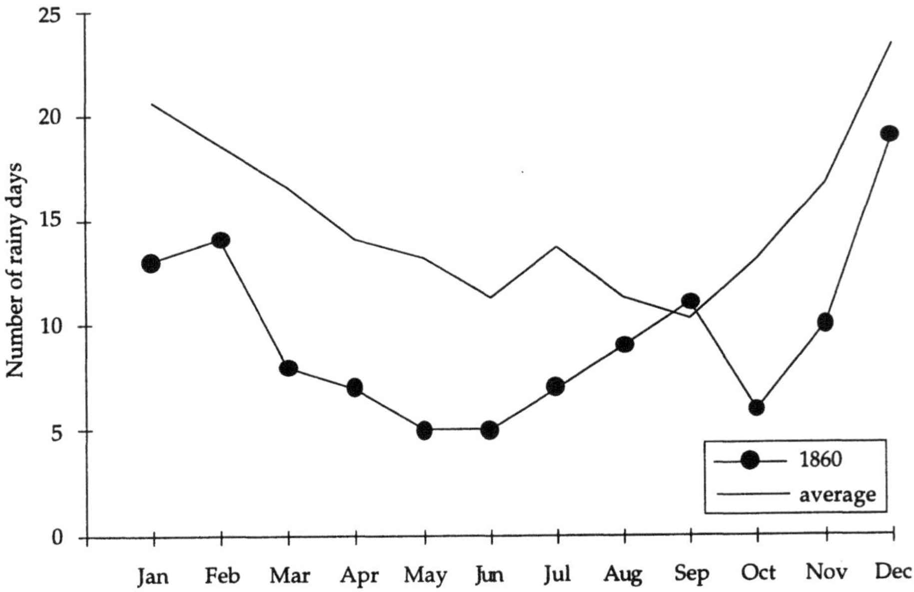

Figure 2. Number of rainy days in Banjarmasin: example of a 'dry year'.

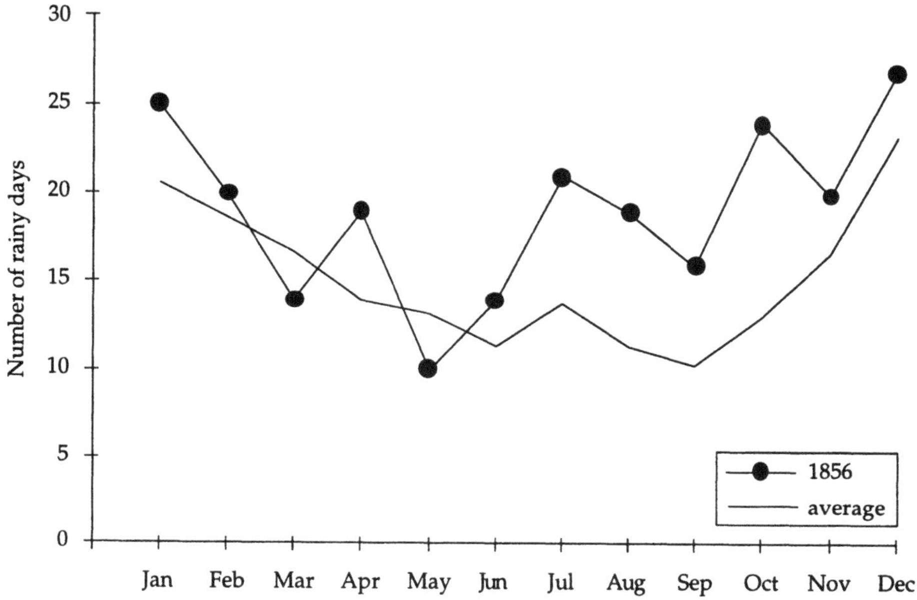

Figure 3. Number of rainy days in Banjarmasin: example of an early west monsoon.

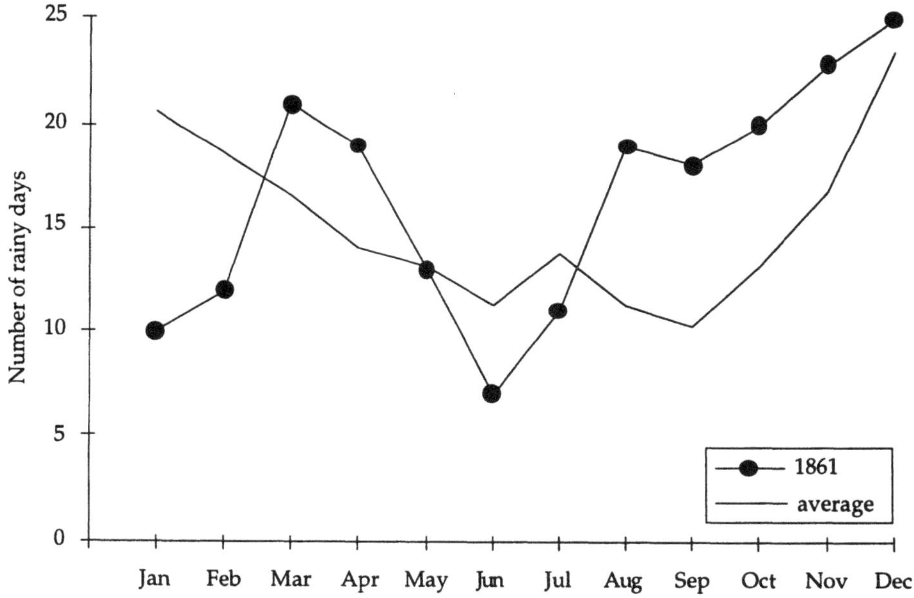

Figure 4. Number of rainy days in Banjarmasin: example of a severe climatic irregularity.

4). This phenomenon, as already described in the *Hikayat Banjar* for the Hindu kingdom Negara-Dipa (probably fifteenth century) in the Hulu Sungai, usually caused much social disruption and severe food shortages.[3] For instance, *ladang* rice would suffer from lack of rain in January or February, while rain during the harvest season (around April) would prevent the standing crop from ripening properly. If in the same year the west monsoon also started early (as in Figure 4), harvests would be uncertain for two consecutive years.

Before the mid-eighteenth century we have little precise information on the occurrence of atypical rainfall patterns since archival sources are scarce prior to the permanent Dutch settlement in Banjarmasin in 1747. The Dutch only commented on the Banjarese climate when it seriously threatened the production or shipment of pepper, as in 1637, or when drought made visitors dependent on the Banjarese for their drinking water, as for instance in 1635.[4] We also know that one of the most serious droughts of the seventeenth century occurred in 1660.[5] For most years before 1747 little meteorological data can be found in the archives, although Berlage's tree-ring data from Java, which also indicate a high level of interannual variability, provide some clues (Berlage 1931). And when in 1762 Batavia complained about declining shipments of pepper from Banjarmasin, Resident of Banjarmasin Lodewijk Wilkens de Lille simply explained that the supply of pepper always fluctuated because of the extremely variable Bornean climate: 'some years it rains heavily every day for 10 months in a row, while in other years it remains dry for as many months without a single drop of rain.'[6]

Except for extreme climatic disruptions, most Southeast Borneo societies were well adapted to climatic uncertainties. They grew, for instance, many rice varieties and practised staggered planting (see Christensen and Mertz 1993). More importantly, Southeast Borneo had three different major agricultural systems: shifting cultivation on *ladang*, rice grown in swamps during the dry season and rice grown in the tidal areas along the coast at the end of the wet season.[7] This usually meant that even under abnormal climatic conditions, at least one or two of these systems would be successful.

[3] The story goes that during the time of Negara-Dipa, adherence to foreign customs and wishes would have the result that (among other things) 'the monsoons do not behave properly, that is, the north-east monsoon behaves like the south-west monsoon or the south-west monsoon behaves like the north-east monsoon' (Ras 1968:265, 443).

[4] ARA VOC 1118, daily register by Pool, 26-9-1635; GM I:658.

[5] ARA VOC 1236, report by Michielsz., 27-5-1661, folio 280.

[6] ARA VOC 3089, letter by De Lille, 15-8-1763, folio 20-2. A second reason mentioned by De Lille was that the sultan and his family extorted the pepper planters too much, which made farmers reluctant to plant the crop.

[7] This, of course, is a very general classification, and variation within each category was great. Schophuys (1936:107), for instance, distinguishes six primary types of *sawah* growing, each during a different part of the year.

Trade between the different agricultural zones ensured that one region's rice surplus compensated for another's shortfall.[8]

Disease

Some epidemic diseases fit best into the category of natural uncertainties, although influenced by human action. The best example, for the period before 1800, is smallpox. Smallpox was periodically introduced into Southeast Borneo from the outside, in particular by traders from Java. When smallpox first arrived on Borneo is unknown, but this probably happened before the coming of the European trading companies to the island (Knapen 1998). Oral histories from Sarawak suggest that smallpox was already prevalent in western Borneo in the seventeenth century (Sandin 1967:15). The first documented smallpox epidemic in Southeast Borneo is dated 1734 (for other recorded epidemics, see appendix).[9]

The impact of smallpox was enormous. Not only was the number of direct victims high (whole villages could be wiped out), but epidemics also caused much economic disruption. Upon the first appearance of the disease, people fled in large numbers into the interior and suspended every economic activity to avoid outside contact.[10] Agriculture and trade came to a standstill, while in political terms Banjarmasin typically fell into complete disorder. The sultan, the princes, and their families were usually the first to flee upstream, whereafter all rivers to the interior were closed for ships in both directions and every communication halted.[11] Since the royal family stayed in the interior until an epidemic was entirely over it could take up to two years before life returned to normal.[12]

Borneans had few ways of protecting themselves against smallpox. Variolation was presumably introduced very late in the Indonesian archipelago, probably by the Chinese in the mid-eighteenth century (Hopkins 1983:115-24; Knapen 1998).[13] Vaccination by way of cowpox vaccine was not available until the beginning of the nineteenth century. The only option which the Borneans had was to isolate themselves by declaring their villages forbidden territory, closing off rivers with rattan cords or tree trunks, excluding those infected from their village, fencing in their houses or, as a last resort, fleeing or migrating upriver. Nomadic groups like the Punan, of course, had much less trouble isolating themselves from the outside world. There are even indications that their practice of 'silent barter'

8 For example: ANRI ZOB 14/2, daily register by Goldman, 15-5-1825 and 18-5-1825.

9 ARA VOC 2314, letter by Panambahan Kusuma Alam, 1734, folio 41.

10 For example: ARA VOC 2799, letter by Van Suchtelen, 19-9-1752.

11 ARA VOC 3151, letter by Palm and Ruhde, 1-3-1765, folio 27.

12 ARA VOC 3580, Letter by Van Walbeeck and Van den Worm, 20-6-1780.

13 This technique required the pus or powdered skin crusts of a person suffering from smallpox to be inhaled through the nose or rubbed into a skin wound. The person thus treated would in most cases develop a mild form of smallpox which resulted in lifelong immunity.

Dayak *ladang* and a tree with beehives on the Barito River, Kalimantan (Müller 1839:plate 48).

(trading exchange without face-to-face contact with the traders) was a consequence of recurrent epidemics (Knapen 1998).

Warfare and head-hunting

Warfare, piracy, head-hunting, human sacrifice, and other forms of violence were all widespread on Southeast Borneo during the eighteenth century and can be categorized as social uncertainties (though in some cases possibly induced by natural phenomena such as drought and resulting crop failures). Sea pirates had kept the coastal area almost free of inhabitants for centuries, forcing people to live in more remote places.[14] Occasionally there were also substantial wars during the seventeenth and eighteenth centuries, involving large numbers of people. Usually these wars had to do with attacks by foreign kingdoms or clashes between pepper growers in the Hulu Sungai and the sultan.[15] But Dayak groups in the vicinity of Banjarmasin were sometimes also in conflict with the Banjarese, mainly because of their desire for head trophies.[16]

Outside the realm of the sultan, warfare and head-hunting in Borneo was generally a small-scale affair causing only a limited number of direct casualties (Knapen 1998). However, there were also indirect casualties as a consequence of the disruption of local economies due to warfare. There are some historical references to dry rice fields or fruit gardens being destroyed because of warfare, groups being forced to migrate to ecologically less favourable areas or planted rice fields being deserted by their cultivators (for instance, Schwaner 1853:119, 193). But the dispersed settlement pattern in Southeast Borneo and the even more dispersed swidden fields (especially before the 1820s, when the Dutch started their policy of concentrating villages and *ladang*), made effective destruction of agricultural or other food sources difficult. Besides, the Bornean rain-forest could supply those driven from their homes and fields with sufficient edible forest products, such as sago or roots and tubers, to prevent the outbreak of famine as a consequence of warfare.[17] In this respect Borneo

[14] For example: ANRI ZOB 96, letter by Bloemsz., 18-5-1808.

[15] For example in 1712 (ARA VOC 1826, letter by Van den Bosch, 15-7-1712, folio 54).

[16] In January 1691, for example, it was reported that 'the Beajuses were at war with the Moors' (Gemelli Careni 1732:261). Another incident took place in 1751, when 2,000-3,000 Ngaju Dayak (according to the Banjarese) attacked Banjarmasin, killing 8 or 9 Banjarese and 'taking the heads of the bodies away, which is what they are principally after'. The figure of 2,000-3,000 is probably strongly exaggerated and served to underline the request of the sultan for a well-armed vessel to subdue the Dayak (ARA VOC 2799, letter by Doeve, 23-6-1752).

[17] There is considerable discussion whether people can live independently in undisturbed tropical rain-forests without agriculture, or whether the Dayak will actually revert to alternative starch sources which they usually despise (Bailey and Headland 1991; Rousseau 1990:132). On historical record are certainly numerous examples of people surviving in the forest for short periods, whether on food found in scattered old gardens, planted or wild stands of sago, or naturally occurring roots like *gadung* (*Dioscorea hispida*) (ARA MR 1879 no. 197 by Meijer; ARA VOC 3706, missive by Van den Worm to Batavia, 25-3-1786; ANRI ZOB 5/10, appendix to general report 1825, section 'Landbouw'; ANRI ZOB 10/10, monthly reports 1862,

differs from the more populous regions of Southeast Asia, such as Java, where the intensive agricultural systems could easily be destroyed.

Besides dispersed settlement patterns, another consequence of the continuous threat of head-hunting was that in the seventeenth and eighteenth centuries the involvement of most inland Dayak groups in the collection of forest products was still limited. In 1789, for example, Resident Christoffel Hoffman commented that the upriver areas were rich in beeswax but that, because of the raiders, little of it was brought to market. Most wax came from the Hulu Sungai and other areas close to the coast, not from the Dayak areas.[18] The same was true for rattan. Archival sources suggest that before the nineteenth century most cut rattan came from areas not far into the interior: the direct vicinity of Banjarmasin, the Hulu Sungai and Riam Kiwa, and Riam Kanan.[19] These areas were under the control of the sultan of Banjarmasin, who completely monopolized trade, determined rattan prices, collected tolls, and ordered his subjects (though with varying success) to cut rattan when he wanted.[20] The interior Dayak groups living along the Barito, Kapuas, and Kahayan Rivers supplied little of Banjarmasin's rattan and collected mainly for local use. In 1790 sergeant F.J. Hartman, one of the first Dutch visitors to the Upper Barito, noted that the Dayak of Murung found it prohibitively dangerous to collect forest products because of recurrent raids upon the area by Kayan (Pari) from eastern Borneo (in Leupe 1864:377). In this light it becomes clear why the Dutch often had difficulties procuring sufficient rattan, gold or other products from Southeast Borneo even though these were abundant in the interior.[21]

Political domination

A final long-standing form of uncertainty is political uncertainty, in particular the erratic and unpredictable behaviour of indigenous and colonial governments. Systems of taxation could change overnight, different colonial powers succeeded each other, and sultans and their families

---

June; ANRI ZOB 10/12 monthly reports 1864, February. See also De Roy 1706:22; Perelaer 1870:180-1; Lumholtz 1920, II:320).

18 ARA VOC 3859, letter by Hoffman, Matthijszen and Du Pont, 23-6-1789, folio 21.

19 In the 1630s rattan reached Banjarmasin through Martapura, at that time an area where pepper growing was expanding rapidly (GM I:630; Van Dijk 1862:35). In 1735 rattan came from the Riam Kiwa and Riam Kanan area (ARA VOC 2345, report by Brouwer and De Marre, 10-6-1735, folio 47). In 1794 it was explicitly stated that rattan came from the Hulu Sungai, and was bartered against chinaware (ARA OIC 102, letter by Jorissen, 14-10-1794, section 78). See also Knapen 1996.

20 Van Dijk 1862:53; ARA VOC 2837, letter by Te Lintelo, 18-7-1754, folio 148; ARA VOC 3761, letter by Hoffman, 14-6-1787, folio 33. Until the early 1870s a similar sultan's monopoly in rattan existed in Kutai (ARA MR 1876 no. 131, travelogue by Gersen, 17-2-1876).

21 ARA VOC 1248, letter by Hurdt, 26-10-1664, folio 2297; ARA VOC 2133, letter by Van Broijel and De Broun, 5-10-1729, folio 291; ARA VOC 2345, letter by Brouwer and De Marre, 10-6-1735, folio 47; ARA VOC 3552, letter by Palm and Ruhde, 27-10-1776, folio 12; ARA MvK 3081, report by Boeckholtz, 1-7-1802, folio 53.

could monopolize trade and change prices and tolls as they wished. Given the strategic location of his realm, controlling the mouths of the major rivers, the sultan was in a good position to control the prices of the products shipped downriver. If necessary he could try to enforce his wishes by blockading the rivers.[22] The population, however, was not entirely powerless in the face of this kind of extortion. It could respond, for example, by protest or other forms of resistance.[23] More often, however, Dayak groups would simply move away from coastal powers to live an isolated life in the interior.[24] 'In order to have sufficient revenues and given the relatively modest volume of trade', Rousseau (1989:49) remarked regarding the situation in East Borneo, 'the sultans were sometimes tempted to overtax the Dayak, who would then leave for the interior'. Staying away from outside control was the easiest solution to the threat of political domination. As in East Borneo, 'the sultan controlled only the Dayak who lived a few days' travel from the capital' (Rousseau 1989:47).

The consequences of uncertainties before the nineteenth century

What was the effect of these uncertainties on the lives of the people on Borneo? In general it can be assumed that individuals and groups make choices in response to their felt needs and in consideration of the perceived risks. A great deal of irrationality and emotion can be involved in the judgement of risks. Minor dangers may cause disproportionate and irrational concern, while very obvious and dangerous hazards are sometimes simply ignored. In addition, since nobody has exact prior knowledge of all hazards, people have to make decisions without being aware of all the risks involved (Douglas and Wildavsky 1982; Kleindorfer et al. 1993).

Unfortunately we know little about the subjective perception of risk by eighteenth-century Borneans, but references in oral and written histories to the terror of epidemics, climatic aberrations, floods, wars, political domination, and encounters with foreigners suggest a strong imprint on the life and minds of the people. It is safe to assume that under pre-nineteenth-century conditions, when the hazards were obviously significant and life-threatening, people generally acted with caution. Whether or not risk is a

[22] In 1729, for example, the sultan tried to control the pepper growers 'by closing the rivers, and exacting the pepper from the upriver populace at very low prices' (ARA VOC 2120, letter by Broijel, Landsheer and De Broun, 23-12-1729, folio 6393 recto). Besides pepper, the sultan also monopolized trade in many forest products and in gold, for which he paid very low prices (ARA VOC 3023, letter by De Lille and Beck, 23-7-1761, folio 38).

[23] For instance, the pepper growers of the Hulu Sungai would ship dirty pepper, mixed with sand and small stones, to Banjarmasin as a form of protest (ARA VOC 2133, report by Van Broijel and De Broun, 5-10-1729, folio 290). On other occasions they would start a rebellion against the representatives of the sultan (ARA VOC 1801, daily register of Van den Bosch and Indius, 29-10-1711; GM VII:173; VOC 3089, letter by De Lille, 15-8-1763).

[24] For example: GM IV:343.

cultural construct and hazard may be judged differently by various groups or individuals, few Borneans would readily play Russian roulette by, for example, going downstream with a raft full of forest products and risking their goods or lives in a likely ambush. As with gambling, there is a relation between risk and potential return. Though this relation may have many irrational facets, in general most people will not take extreme risks when the anticipated returns (whether material or social) are small. Only when the possible rewards are exceptionally high might they consider risking their lives. In the eighteenth century, venturing far from one's village to collect forest products involved huge hazards and was not something one would do for minor returns such as, for example, a little imported cloth or foodstuffs. Considering the numerous sources of uncertainty outlined above, it becomes understandable why many Dayak groups at that time lived an isolated life. Since no foreign products were really needed by the Dayak (salt could be made by boiling water from salt springs, iron tools were forged from local iron ore, cigarettes made from locally grown tobacco), they tended to avoid external contacts since the risks at stake were simply too high.

This does not mean, however, that no outside contacts existed, that people had an inherent aversion to trade, or that people only wanted to fulfil their basic needs after which they would revert to indolence. Paradoxically, exotic goods were much esteemed by the Dayak and have long played an important role in their culture. Foreign cloth, beads, gongs, and ceramics were perceived as indispensable for both daily life and ritual. By almost all possible means the Dayak would try to obtain the capital needed to buy these items. Rhinoceros horn or bezoar stones from the remote interior have been traded to China at least for a millennium, while Dayak probably took forest products, gold or rice in their boats to Banjarmasin since the establishment of the town in the sixteenth century. The value attached to the imported prestige and ritual goods was in fact directly related to their scarcity and to the risks involved in procuring them through trade or collecting export goods to exchange for them.

Besides the possession of imported trade goods, activities involving travel, such as going on distant trading voyages, making strategic alliances with neighbouring groups, acquiring foreign political titles from a sultan or colonial power, or taking part in head-hunting expeditions were also much respected. The social status which men acquired by venturing into the outside world to collect forest products, heads or foreign goods can also be understood as a function of the large number of hazards involved. The Dayak had to take great risks for each outside contact, sometimes gambling away their money, freedom, health or even their lives. Under the extremely uncertain conditions in Borneo, all were bound to think twice before venturing into the unknown and many would simply opt for isolation.

A broad geographical distinction can be made between those Dayak groups preferring isolation and those trying their luck in external contacts. Generally speaking, only the Dayak groups living close to centres of Malay settlement were involved in more or less regular communications with the outside. They exposed themselves to numerous risks, but of course with the hope for potential success, fortune, and prestige. Groups in the remote interior, on the other hand, chose to refrain from communication with the outside world, which meant that they had to make their own salt, continue to use stone axes and so on. Of course, there was no clear-cut boundary between isolationists and gamblers. A continuum existed between the coast and centre of the island. The possibility of migration up- or downriver allowed villages to decide how much risk they wanted to take at any given point in time.[25] This involved a kind of ongoing cost-benefit analysis: how much risk do we want to take for how much potential gain, whether in terms of food, valuables, medicines, military protection, or social prestige? I do not suggest that the two groups were inherently different in their cultural attitudes. The same people could choose for either commerce or isolation at different moments in time. But the history of Southeast Borneo shows that under conditions of great uncertainty many inhabitants tended to avoid substantial hazards and had to wait for less uncertain times to satisfy their economic and social aspirations.

I conclude that during the seventeenth and eighteenth centuries, the balance in a large part of Southeast Borneo was in favour of isolationism. But even at this early stage there were important exceptions. A large number of Ma'anyan Dayak, for example, started growing pepper in the Hulu Sungai and around Martapura in the seventeenth century. These people did dare to take the risks of exposure to smallpox, exploitation by the sultan, or dependence on the vagaries of the world market for cash crops – although other members of the same group decided to flee inland (Hudson 1976). They could earn a fortune when their plantations flourished, and in addition they were offered protection by the sultan against outside aggressors. But when famine followed the drought of 1660, people had to resort to leaves and tubers for food because they had

[25] There are numerous examples of people permanently moving away as a result of extortion, epidemics, harvest failures, and the like. In 1789, for instance, Hartman noted that the Southeast Bornean Dayak 'are not fond of houses, household goods, and fields. If they are forced to do something which they do not particularly like, then they will leave the one dwelling-place and look for another, deep in the forest or on another river, in the manner of all upriver natives' (ARA Archive Hoge Regering Batavia, 44, letter by Hartman, 23-6-1789, folio 3 recto). Even in the mid-nineteenth century the Ngaju of the Kahayan and Kapuas would move outside Dutch control when they regarded the head taxes as too high (ANRI ZOB 6/9, General report 1848, folio 6 verso). See also ARA Archive Hoge Regering Batavia, 44, travel account by Kruysweg of his journey to the Dusun in 1789; ARA OIC 102, letter by Jorissen and Dietz, 22-11-1794; ARA AZR 274, letter by Jorissen, 2-10-1796; ANRI ZOB 96, letter by Bloemsz.,18-5-1808, folio 2; ANRI ZOB 5/15, general report 1835-7, folio 160-1.

neglected rice cultivation in favour of growing pepper for cash.[26] The farmers knew that growing pepper involved great risks because of climatic variability. One flood could destroy their entire investment, as could fire during a period of serious drought. The sultan, moreover, turned out to be incapable of offering adequate protection and many villages and pepper gardens were destroyed in attacks by people from Pasir on the other side of the Meratus Mountains to the east.[27] Finally, continuous political and economic exploitation by the sultan also burdened the planters, forcing them into frequent rebellion and ultimately into withdrawal from pepper growing.[28]

Changing uncertainties during the nineteenth century

Warfare

One important source of uncertainty, warfare, disappeared from the scene at the beginning of the nineteenth century. In 1817, immediately upon their return after the British interregnum, the Dutch launched a major effort to eliminate head-hunting and the practice of human sacrifice. This was not based on humanitarian considerations alone: it served economic goals as well. The collection and trade of forest products, the exploitation of mineral resources, the establishment of plantations and other economic activities which could raise revenues were all impossible under conditions of continuous plunder and warfare.

The fight against river and sea piracy was the first concern of the Dutch. Not all pirates came from distant areas like the Sulu Archipelago. The coastal Bekumpai of Southeast Borneo, for instance, were also capable seamen and made the river mouths and banks unsafe for incoming and outgoing vessels.[29] After the Dutch restoration, cruisers patrolled the Java Sea and from the 1830s onwards pirates were rarely seen off Southeast Borneo.[30] In 1817 the lower reaches of the Barito and Kapuas Rivers were practically under the control of the Bekumpai, who had filled the political vacuum created when the power of the sultan of Banjarmasin was taken over by the Dutch.[31] The latter, of course, viewed this development

26 Michielsz. wrote that rice had become very expensive and this had caused 'great hunger', prompting the population to shift to 'leaves and roots of trees', but that they were 'otherwise reasonable well-provided with foodstuffs' (ARA VOC 1236, report by Michielsz., 27-5-1661, folio 280).

27 For example: ARA VOC 3810, letter by Hoffman, Matthijszen and Du Pont, 8-6-1788.

28 For example: ARA VOC 1801, daily register by Van den Bosch and Indius, 29-10-1711; GM VII:173; VOC 3089, letter by De Lille, 15-8-1763.

29 ANRI ZOB 19/4, letters by Goldman, 4-5-1820, 20-3-1821; ANRI ZOB 21/4, letter by Goldman, 26-6-1821.

30 ANRI ZOB 5/15, general report 1835-7, folio 26; ANRI ZOB 5/16, general report 1838, folio 13; ANRI ZOB 6/2, general report 1840, folio 23 and 81; ANRI ZOB 6/8, general report 1847, folio 12.

31 ANRI ZOB 15/2, letters by Halewijn, 10-11-1822.

with dismay and ultimately went to war against the Bekumpai (1824-1825) in order to end their plundering and reduce their economic strength. As early as 1826 one Dutch traveller described the presence of small trading boats on the Middle Barito as 'convincing proof of the security which one finds in these areas where in the past no small vessels dared to venture'.[32]

After the Bekumpai War the Dutch effort to eliminate head-hunting also gained momentum and head-hunts became rare events within a decade. A simple and effective way by which the Dutch achieved their goal was to offer presents such as foreign cloth, corals, copper rings, and tobacco to Dayak leaders who prevented their subjects from head-hunting.[33] Those who reported cases of head-hunting or, even better, handed over the offender, could also count on a generous reward from the colonial government. When there was a shortage of rice the Dutch were in an even stronger position, since imported Javanese rice would only be offered to loyal Dayak villages.[34]

The Dutch reacted vigorously if, despite such preventive measures, head-hunting nevertheless took place somewhere in Southeast Borneo. A river might be blockaded for months to force those involved to surrender the culprits. In November 1834, for example, Resident C.G. Goldman closed the Kahayan for all vessels and demanded that every Dayak leader report to Banjarmasin, where the perpetrators of a series of head-hunting raids were to be handed over. Three months later these conditions were fulfilled and the river was reopened.[35] Those found guilty of murder were either banned to Java for life, or sentenced to death.

There remained one serious problem for the Dutch, namely the invasion of alien Dayak groups into Dutch-controlled territory. The Kayan of the Mahakam basin, for instance, regularly made head-hunting sorties along the Upper Barito, Kapuas, and Kahayan. Only by maintaining a continuous military presence could the intruders be kept out. Especially the Teweh River, one of the major routes linking the homeland of the Kayan people with Southeast Borneo, had to be kept clear of head-hunters by numerous military expeditions.[36]

Except in the most remote parts of the Dutch territory, head-hunting became extremely rare during the 1830s and the Dayak now offered *kerbau*, pigs or other animal sacrifices instead of human beings. 'The strict measures in force, namely the death penalty', as one 1837 report put it,

32 ANRI ZOB 13, letters by Halewijn, 17-11-1826.

33 ANRI ZOB 19/1, letters by Hartmann, 31-7-1828. According to the Dutch, the Dayak leaders were responsible for head-hunting raids carried out by their subjects. See also ANRI ZOB 15/7, letter by Halewijn, 1-5-1829.

34 ANRI ZOB 19/7, letters by Hartmann, no. 1, 2-2-1829.

35 ANRI ZOB 20/4, letters by Resident Goldman, no. 13, 8-2-1835.

36 ANRI ZOB 19/8, letters by Hartmann, 30-7-1830; ANRI ZOB 6/9, general report 1848, folio 2 verso; ANRI ZOB 7/3, general report 1854, 15-2-1855, folio 2d and 4c.

'have completely ended these atrocities'.[37] Probably the colonial officers in Banjarmasin had good reason to present a rather rosy picture to their principals in Batavia, but nevertheless I think that along the Kahayan, Kapuas, and Barito, with the exception of the remote upriver areas (for the Barito this means above the Teweh River), head-hunting was curbed much earlier than is generally accepted.[38]

It can be said that after the Banjarmasin War (1859-1863) safety and security of goods and persons had increased dramatically all over Southeast Borneo, except for some isolated areas where resistance against the Dutch colonial power continued. The great majority of the population had little to fear any more from head-hunters and pirates. A Pax Neerlandica was already effective in Southeast Borneo as early as 1830.[39]

Economic uncertainties

Not all uncertainties, however, decreased during the nineteenth century. This is most obvious when looking at economic developments. Simultaneously with the Dutch pacification efforts, the Bekumpai started moving upriver in large numbers to acquire forest products, gold, and other products to be sold on the Banjarmasin market. Originally presumably Ngaju Dayak, living as farmers scattered in the coastal zone of the Barito and Kapuas basins, from the 1820s onwards these Bekumpai became intermediary traders while at the same time more and more of them converted to Islam.[40] The traditional agricultural system of the Bekumpai in the tidal coastlands had been relatively secure, but the yields of trade were not so certain. Income from trade was unpredictable for two reasons: because of fluctuating prices on the world market, and because the supply of some forest products depended on climatic conditions.

The wax 'harvest', for example, depended heavily on the time of flowering of rain-forest plants, which in turn depended on climatic circum-

37 ANRI ZOB 5/15, general report 1835-37, folio 9. Also ANRI ZOB 6/1, general report 1839, folio 64.

38 There were some ups and downs in the effectiveness of the Dutch campaign against head-hunting. For example during the Java War, when funds were short, head-hunting increased again. Head-hunting was also rampant during the Banjarmasin War. Schwaner (1853:193) states that in the 1840s head-hunting was mainly limited to the Siang, Murung, and Ot Danum areas, while there were sometimes hostilities between the Middle Kapuas and Dusun Dayak.

39 For East Borneo this change did not come until about 1900. 'Although no precise data are available', comments Rousseau (1989:45), 'there appears to have been a significant increase in trade around the turn of the twentieth century, because of an increased demand for some products, and because of pacification and the disappearance of head-hunting, which facilitated trade.'

40 Some Bekumpai were possibly traders as early as the late seventeenth century, operating as middlemen of the sultan and trading with the Hulu Sungai (Schwaner 1853). The Bekumpai probably started trading with the Upper Barito area on a limited scale at the end of the eighteenth century, when the Dayak territories were taken over by the Dutch in 1787. Hartman, at least, mentions that in 1790 the Bekumpai had two collecting points (*verzamelplaatsen*) for trade goods on the Teweh and Montallat River (in Leupe 1864:369-71).

stances. When storms and heavy rains during the east monsoon damaged the blossoms for two consecutive years in 1848 and 1849, the wax trade almost came to a standstill.[41] The next year, by contrast, an extremely dry east monsoon favoured the collection of wax and supply to the market of Banjarmasin was abundant.[42] Trade in rattan depended less on climatic variability, but here world market prices could be halved or doubled within a few months (see Figure 5). Wholesale traders in Banjarmasin tried to compensate for this by stockpiling rattan in warehouses when the price was low. But this too was not without risks, since moisture or rats frequently rendered the stored rattan useless.[43]

Disease

After economic uncertainties, diseases were a second source of increasing unpredictability for the inland population during the nineteenth century. As trade expanded, the risk of contracting diseases grew correspondingly. The Dayak would be infected, via coastal traders, with epidemic diseases like smallpox. Visiting colonial or military officers were another new source of infection (Knapen 1998). Whereas in the eighteenth century an epidemic of smallpox was usually followed by approximately 12 smallpox-free years, during the first half of the nineteenth century this interval shortened drastically to about six years after 1820 (see appendix). I assume that this was initially the result of the opening up of the interior, after which greater numbers of Southeast Borneans came into more frequent mutual contact, resulting in a rapid growth of the susceptible population.[44] Absolute population growth, which accelerated in Southeast Borneo somewhere during the mid-nineteenth century, is an additional explanation for the shortening cycle. After 1860 it appears that smallpox no longer disappeared completely from Southeast Borneo between epidemics, and had thus become endemic to the region.[45]

41 ANRI ZOB 6/9, general report 1848, folio 19 recto; ANRI ZOB 6/10, general report 1849, section on wax. For the same phenomenon in 1788: ARA VOC 3906, letter by Matthijszen and Du Pont, 16-5-1790, folio 44; in 1826: ANRI ZOB 5/8, report on Bekumpai and Dusun 1826, 13-2-1827.

42 ANRI ZOB 6/11, general report 1850, sections on agriculture and wax.

43 ANRI ZOB 19/6, letter by Hartmann, 27-12-1828; ANRI ZOB 21/14, letter by Hartmann, 6-1-1831; ANRI ZOB 10a/2, monthly reports for May and June 1867.

44 The existence of smallpox in a community depends on the number of susceptible people available for the disease to infect. For smallpox to become endemic, a population of at least 200,000 people regularly in touch with one another is thought to be necessary (Fenner et al. 1988:118). In areas of low population density like Borneo, smallpox could not be sustained *in situ* and had to be introduced from outside, resulting in regularly spaced epidemics. Populations which were not immune as a result of previous outbreaks suffered severe mortality, until the number of susceptible individuals became too low and the disease eventually died out (Hopkins 1983:8).

45 After 1861 there were dozens (at least) of reported cases every year. There is no doubt that this was a new development and not a misinterpretation which can be explained by the improving accuracy of the archival sources. Already in earlier times a single case of smallpox would invariably petrify both Borneans and colonial officials (Knapen 1998), and the latter

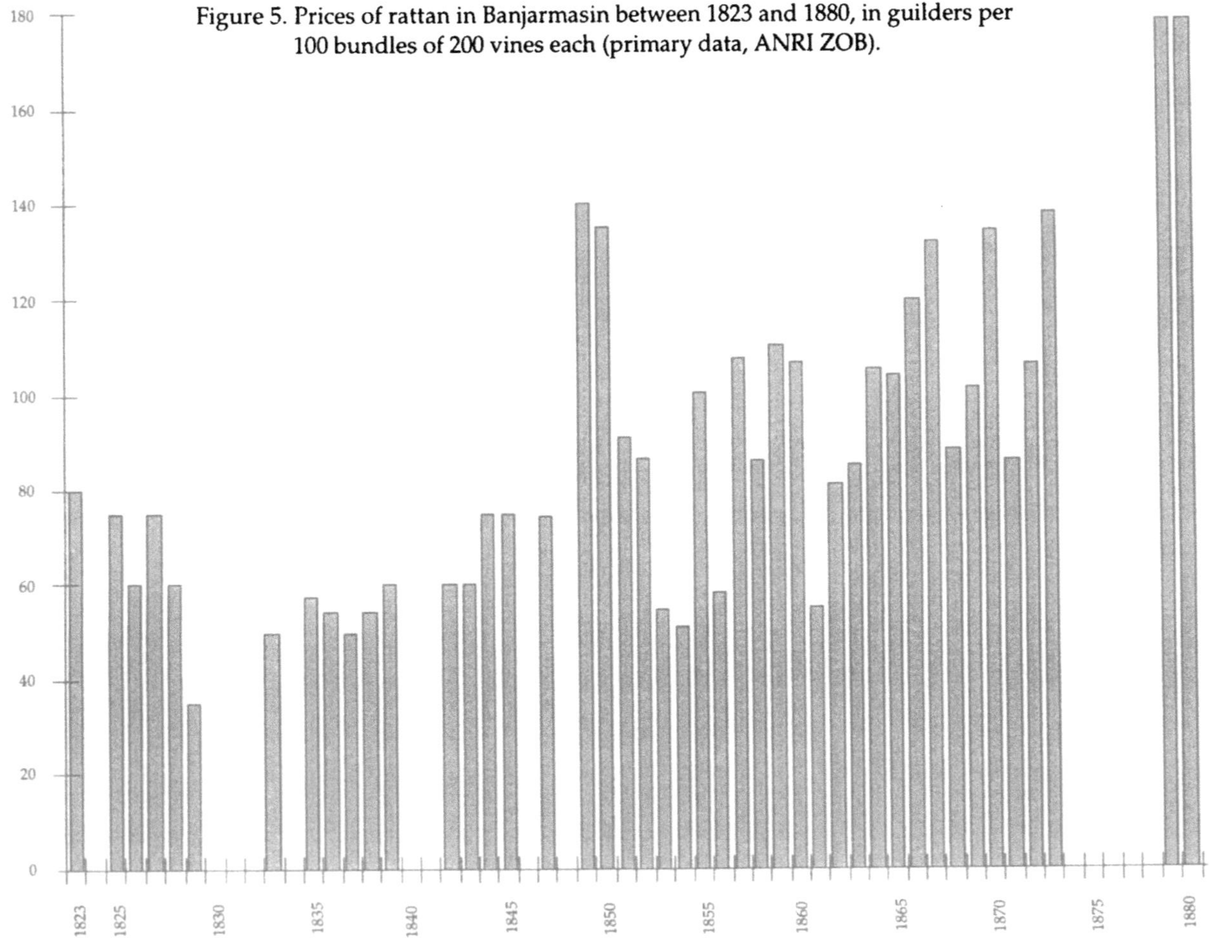

Figure 5. Prices of rattan in Banjarmasin between 1823 and 1880, in guilders per 100 bundles of 200 vines each (primary data, ANRI ZOB).

Vaccination of the population had a high priority for the Dutch colonial government. To early-nineteenth-century officials it was obvious that population growth was the key to making Borneo into a profitable island instead of a revenue-consuming territory. As long as it remained practically uninhabited, labour costs would be high and extraction of its valuable resources unfeasible. It was in an attempt to overcome this problem of scarce labour that the Dutch began their cow-pox vaccination campaign in the late 1820s. Especially for the coastal population, among whom most of the vaccinations were carried out, this meant an important increase in safety. The more remote Dayak groups had to wait another fifty years or so before the vaccinators operated on a regular basis in their villages.

A number of new contagious diseases appeared on Southeast Borneo for the first time during the nineteenth century, most significantly cholera and measles. Cholera arrived in Banjarmasin in July 1821, and its immediate effects on the population were similar to those of smallpox. The sultan and his family fled to the mountains and remained completely out of contact with their subjects, most Banjarese and Chinese locked themselves up in their houses, and economic life came to an almost complete standstill. Despite river blockades intended to prevent the spread of the disease, within a few weeks it started moving inland and causing the same terror and flight in the interior.[46] Cholera became one more factor disrupting the trade in forest products. As Chew (1990:102) has noted for nineteenth-century Sarawak, the trader's reliance on Dayak for collecting these products 'was never more apparent than in the fall-off in trade when cholera struck, as it did periodically'. Flight to upriver areas continued to occur during epidemics, as in previous times. But commercialization spread rapidly upriver, encompassing more and more of the Southeast Borneo population.

Animal as well as human diseases found their way to Borneo through trade, causing loss of both domestic livestock and wild animals.[47] Lethal epidemic diseases among horses, for example, raged in 1830-1831, 1839,

---

reported almost every case in their letters to Batavia. If Fenner's estimate is correct (see note 44), the almost continuous presence of smallpox in Southeast Borneo would imply that the population of the region was at least 200,000 by the 1860s.

46 ANRI ZOB 19/4, letters by Goldman, 1-8-1821, 16-8-1821, 19-8-1821, 17-10-1821. See also Knapen 1998.

47 I have no information on pre-nineteenth-century epidemics amongst animals (epizootics) in Southeast Borneo, but this may very well reflect the fact that the Dutch did not think them worth recording, or even that they were not recognized at all. For epidemics among horses, however, I had expected some references in eighteenth-century archival sources since the sultan, who owned hundreds of animals, was partly dependent on the Dutch for new horses and would surely have informed the resident if he needed to replace stock lost to disease. This suggests that equine epidemics are relatively recent phenomena in the area. It would also be interesting to know something about historical changes in plant diseases, especially those affecting food crops, but unfortunately information on this subject is even more scarce.

1842, 1869 and 1873.[48] The 1842 outbreak, involving what was described as an 'incomprehensible disease', almost caused the extinction of all horses in Southeast Borneo. The first reports of diseases among *kerbau*, cattle, pigs, and other animals are from 1871, and probably relate to the increase in livestock breeding (in itself a result of increasing safety?) and the growing import of disease-carrying animals from Java. Rinderpest killed a large proportion of the cattle population of Southeast Borneo between 1871 and 1872 before spreading to the wild pig population, which it affected in areas as far afield as the Upper Kahayan and Kapuas.[49] In 1878 rinderpest struck once more, again hitting the cattle population first and later killing pigs in large numbers.[50]

As in the case of human diseases, it seems that the incidence of animal diseases increased during the second half of the nineteenth century. Although livestock breeding appeared at first to offer increased economic security, it brought its own risks and uncertainties. One could get rich by selling *kerbau* this year, while waiting a few months might be disastrous. And when newly introduced diseases spread to wild pigs, resulting in meagre returns for Dayak hunters, their impact was felt even in the most remote places and amongst those refraining from outside contact.

Demographic change and resource pressure

It was noted above that somewhere around the middle of the nineteenth century, population growth in Southeast Borneo accelerated. This led among other things to increased pressure on resources, especially land. In the Hulu Sungai a shortage of agricultural land had already arisen by the 1860s resulting in conflicts over land, movement of *ladang* farmers into more mountainous areas, and migration to other regions.[51] In 1866 Resident K.W. Tiedtke reported that in Alai District the population 'has now turned to laying out *ladang* fields on the slopes of the mountains'.[52] The same process was going on in the district of Amuntai.[53] In 1865 farmers in the districts of

[48] ANRI ZOB 19/8, letter by Hartmann, 1-8-1830; ANRI ZOB 6/1, general report 1839, folio 104; ANRI ZOB 6/4, general report 1842, folio 70; ANRI ZOB 8/13, cultivation report 1869, folio 39-40; ANRI ZOB 10a/8, monthly report September 1873.

[49] ARA MR 1871 nos. 241, 408; ANRI ZOB 10a/7, monthly reports March and May 1872; 10a/8, monthly report May 1873.

[50] ANRI ZOB 10a/13, monthly reports January, July, September and October 1878; ARA MR 653 16-11-1878.

[51] Since the Hulu Sungai only came under direct Dutch rule after the Banjarmasin War in 1860, it is difficult to establish exactly when conflicts over land began in this area. Possibly the laws promulgated by Sultan Adam around 1835, containing many articles on titles to land and the settlement of land conflicts, were the result of increasing scarcity of wastelands. Joekes (1917:365) noted in his commentary on these laws that 'disputes leading to murder and manslaughter were not uncommon'. Clearly, individual titles to land were developing in the Hulu Sungai in the 1930s.

[52] ANRI ZOB 8/8, cultivation report 1866, folio 3.

[53] ANRI ZOB 8/8, cultivation report 1866, folio 17; see also ANRI ZOB 8/9, cultivation report 1865, folio 10.

Alai and Amandit also started making irrigated *sawah* fields to compensate for the scarcity of land suitable for swidden cultivation.[54]

In the 1870s the Landraad (court) of Banjarmasin had its hands full, 'especially around the time when *padi* planting has to start', settling disputes 'concerning the property rights to the lands suitable for rice growing'.[55] Access to land was no longer guaranteed and landlessness appeared. From the 1880s onwards the problem was partly alleviated by migration out of the Hulu Sungai.[56] At the fringes of the Hulu Sungai disputes over other resources arose. For instance, wood for shipbuilding became scarce (Bangert 1860:159), ironwood trees were subject to competing claims (Mallinckrodt 1933:87), and rattan collection by outsiders almost led to overexploitation.[57]

Agricultural uncertainties

Agricultural patterns, finally, changed during the nineteenth century. The adoption of irrigated *sawah* cultivation in the Hulu Sungai around 1860 may have had positive consequences since it added to the stability of the agricultural system, but in general the agricultural base of Southeast Borneo was getting more vulnerable. People started moving into areas where agriculture involved more risks, such as land closer to the seashore or swamp rice areas where there was less certainty of an adequate dry period (for instance, below Negara).

More importantly, many Bornean farmers started collecting forest products and growing small quantities of cash crops on their *ladang* after the rice crop. Under normal conditions this only broadened their economic base and so reduced their vulnerability to uncertainties. In principle such activities can readily be combined with rice agriculture and offer additional security in the event of a rice crop failure. But because prices of gutta percha and rattan rose to extreme heights during the 1870s, many people took the risky bet of concentrating on forest-product collection to the extent that food production was affected. Ricefields were neglected and part of the population started depending on imported rice. None of the three agricultural systems which had compensated for each other's shortfalls in the past could now alleviate the recurrent rice deficits in other areas. Southeast Borneo, formerly a rice exporter, now imported rice from Java and Singapore. The yearly shipments of large quantities of rice into

54 ANRI ZOB 8/5, general report 1865, folio 28.

55 ANRI ZOB 9/12, general report 1878, folio 28. Also: ANRI ZOB 139/4, report on rights to wastelands in the district of Amandit, folio 12; ANRI ZOB 8/8, cultivation report 1866, folio 24.

56 For example: ANRI ZOB 9a/8, general report 1887, folio 11; Heersink 1987:32. Weinstock (1983:145) even claims that overpopulation and land shortage stimulated the Bekumpai to move upriver in search of land about 150 to 200 years ago.

57 ANRI ZOB 139/4, report on rights to wastelands in the districts Amandit and Negara, folio 23.

Southeast Borneo after 1870 (see appendix) were the direct result of the forest-product boom.

Developing local commercial aspirations led to many experiments with new cash crops. One may say that in the nineteenth century most of them failed because of unexpected events. Sugar-cane is a good example. Local varieties, well adapted to the conditions of Southeast Borneo, had already been grown for a long time on a limited scale, in swidden fields for private consumption. In 1833 the colonial government stimulated the first commercial plantations in Tanah Laut by giving credit to enthusiastic local entrepreneurs and attracting Dayak cultivators.[58] A factory was build in 1834 to produce large quantities of sugar, but it never managed to make a profit. A severe drought in 1838 and a plague of rats in 1840 caused the destruction of almost all sugar-cane in the fields. In addition the government withdrew its credit, while new regulations forbade the use of debt slaves in the factory. As a result, the factory had to be closed in 1842 and commercial sugar-cane planting came to an end.[59] In 1888 new trials were carried out, with new sugar-cane varieties from Java. This time a plant disease destroyed the crops, and the idea of growing sugar-cane was abandoned again a few years later.[60] Only in the twentieth century would technical improvements turn sugar-cane into a viable commercial crop, using improved sugar-cane varieties and pesticides.

Similar unfortunate examples can be given for almost every other commercial crop introduced to the Hulu Sungai, Tanah Laut, and Martapura areas during the nineteenth century. Cotton was promoted by the Dutch as a possible export crop as early as 1827.[61] In practice, however, climatic variability was a serious problem. In 1848 and 1849, for example, the *kapas* suffered from excessive rain, while the next year the plants flowered too early because of a lack of rain, both cases resulting in a very meagre harvest.[62] The Dutch placed more hope upon modern cotton varieties, and experiments with imported seeds were carried out from 1849 onwards, but acceptance was slow. In the late 1850s cotton growing was becoming a little more extensive and even spread to Dayak areas, but caterpillars in 1858, an excessively rainy west monsoon in 1859, and finally the Banjarmasin War destroyed all cotton crops. When the replanted

58 ANRI ZOB 19/13, letters by Goldman, 10-12-1832, 10-7-1833; ANRI ZOB 14/3, daily register by Goldman, 25 and 27-1-1833; ANRI ZOB 14/5, daily register by Goldman, 5-10-1834; ANRI ZOB 5/15, general report 1835-7, folio 91. Sugar-cane was regarded as an ideal crop, since it grew well on Borneo and there was plenty of firewood available for processing it (see also Tobias 1828:19).

59 ANRI ZOB 6/2, general report 1840, folio 98; ANRI ZOB 6/4, general report 1842, folio 67.

60 ANRI ZOB 9a/8, general report 1887, folio 15-6; ANRI ZOB 9a/9, general report 1888, folio 16; ANRI ZOB 9a/10, general report 1889, folio 13-4.

61 ANRI ZOB 15/6, letter by Halewijn, 6-7-1827.

62 ANRI ZOB 6/9, general report 1848, section on cotton; ANRI ZOB 6/10, general report 1849, section on cotton; ANRI ZOB 6/11, general report 1850, section on cotton.

1867 made farmers reluctant to plant the crop again.[63]

Instead, they shifted to indigo. European entrepreneurs too invested in indigo plantations and factories for processing this dye, but floods had destroyed most of the crop in September 1870 and the venture ended in bankruptcy.[64] Finally tobacco was also tried, with a comparable array of setbacks.[65]

As early as the 1840s some colonial officers realized that stimulating the production of commercial crops was not the way to make Borneo profitable, since such endeavours were simply too risky.[66] Instead their attention shifted to a more secure resource: minerals, and in particular coal, for which demand was increasing rapidly as steamships became a common sight in the archipelago.

Ngaju Dayak of the Middle Kahayan: A historical reconstruction

We can illustrate the general trends of the nineteenth century using the example of a hypothetical Ngaju Dayak group living along the middle reaches of the Kahayan River. Until the end of the eighteenth century this group led an isolated existence practically out of reach of the sultan and the Dutch, only occasionally engaging in trade in order to obtain prestige goods. When security of life and property improved early in the nineteenth century, more and more Bekumpai traders started coming to their area in search of forest products. Since venturing further from their villages became much less dangerous, the Dayak used the opportunity to collect rattan, *getah perca*, and gold to earn more cash. Attractive modern goods could be bought from the traders, while the Dutch, for their part, demanded their taxes.

Indebtedness now increased among the Dayak, stimulating them to search harder for increasingly scarce forest products. Fluctuations in the

63 ANRI ZOB 7/1, general report 1852, section on cotton; ANRI ZOB 10/6, monthly reports September-November 1858; ANRI ZOB 10/7, monthly reports January and February 1859; ANRI ZOB 8/9, cultivation report 1865, folio 31-2; ANRI ZOB 8/11, cultivation report 1867, folio 64.

64 ANRI ZOB 8/11, cultivation report 1867, folio 18-21; ANRI ZOB 10a/5, monthly reports September 1870; ARA MR 1870 no. 420; ANRI ZOB 9/5, general report 1874, folio 46-7.

65 ANRI ZOB 8/11, cultivation report 1867; ANRI ZOB 8/13, cultivation report 1869, folio 34; ANRI ZOB 9a/9, general report 1888, folio 16-7; ANRI ZOB 9a/10, general report 1889, folio 9 and 14; ANRI ZOB 9a/11, general report 1890, folio 14.

66 In 1845, for example, A.L. Weddik recommended that attempts to grow coffee in Southeast Borneo should be terminated because 'the soil there is not very suitable for coffee cultivation, and moreover the climate is very variable because of the equatorial location, so that the coffee beans seldom ripen thoroughly and the quality is always very low'. Weddik recommended that attention be shifted to coal, iron, gold, platinum, and diamonds instead of cash crops. 'It is not from agriculture that Borneo will yield significant profits – at least not that part of the island with which I have become familiar. The wealth of mi.neral deposits in the ground offers better prospects, and is very considerable, even incredible.' (ANRI WAB 32/1, notes on the products of South and East Borneo, letter of 16-4-1845.)

search harder for increasingly scarce forest products. Fluctuations in the prices of these products made some Ngaju desperate during bad years, especially when low prices coincided with other disasters such as smallpox epidemics. Dayak groups loyal to the colonial government, however, received help in the form of vaccination. Falling rattan prices were compensated for by collecting or planting more rattan, or by gathering or growing other products for the world market. Rice cultivation, meanwhile, tended to decrease in favour of earning cash. Regular shortages of rice were the result, which had to be compensated by rice imports and the extension of credit by the colonial government.

An example of the precarious situation into which the Dayak ultimately manoeuvred themselves can be seen in the 1870s. After the rattan boom and collapse came the *getah perca* boom. But this was followed by a severe climatic anomaly, with rain during the east monsoon and drought during the west monsoon. The rice barns were empty and famine was imminent. In order to obtain money to buy rice, only two options were left to the Dayak: either to collect more *getah perca* (of which the producing tree was already becoming extinct) or to sell one's labour to the Dutch, who had been eagerly looking for 'hands' for at least two centuries. Now, after decades of illusions, the Dutch finally had the labour to dig a canal linking the Kahayan River with Banjarmasin and thereby to push the trade in forest products up to unprecedented levels. Even the most remote parts of Borneo were now becoming part of the global economy, exposing the local population both to new opportunities and to new risks.

Concluding remarks

In past centuries Borneans continuously had to adapt themselves to the prevailing hazards on the island. I hope to have demonstrated that a study of uncertainties may help to explain various historical practices and processes on Borneo, including migration from the coast to the interior, the custom of silent barter, the use of stone axes in some areas until the twentieth century, and the preference of some Borneans for nomadism or for the isolated existence of what the Indonesian government now calls an 'alienated tribe' or *suku terasing*. Such a 'historical risk analysis' should pay attention to regional differences in uncertainties, for instance between those faced by coastal and interior populations. On a smaller scale, differences between local risk environments may help to explain why an area like the Hulu Sungai was economically so successful in the past while others, such as Tanah Laut or the Middle Barito, were not.

There is no proof that unpredictability was something new to the nineteenth century. Serious droughts, smallpox epidemics, floods, and excessive demands from sultans or colonial powers were already common during the seventeenth and eighteenth centuries and even earlier. Many Borneans

period, restricting themselves to a subsistence economy adapted to the uncertainties of the region. They isolated themselves from outside markets, foreign diseases, and alien powers, not out of conservatism or 'backwardness', but because the risks at stake were simply too high. Those who did not, like some coastal Malay and Ma'anyan Dayak who grew pepper during the eighteenth century, generally suffered from much more uncertainty than before as a result of unstable market prices, taxation, warfare, and diseases. Not without reason, the *Hikayat Banjar* warned against accepting foreign habits and authority since this 'would bring misery over the country: there would come many diseases, the north-east monsoon would be like the south-west monsoon and the south-west monsoon would be like the north-east monsoon, the prices of foodstuffs would go up and the government would fall into disorder' (Ras 1968:443; Dove, this volume).

One event in particular significantly reduced the uncertainties on Borneo during the early nineteenth century: the elimination of warfare and head-hunting. Later the Pax Neerlandica was complemented by the possibility of broadening the economic base by collecting forest products or planting cash crops, and by the successful battle against smallpox. Against this background of increasing security, enterprising and aspiring individuals would take the first steps into new commercial enterprises. On the other hand, many of these early undertakings failed because natural variability was too large and some uncertainties not yet eliminated. In such cases people would once again withdraw from outside involvement, much to the disappointment of the Dutch. In general, however, it can be said that the nineteenth century saw a trend towards expanding external contacts as some of the hazards involved in such contacts were reduced.

It should be stressed here that it was probably not the cultural attitudes of the Borneans that changed during the nineteenth century. As in the past, the people of Southeast Borneo appreciated luxury and wealth and based status on individual success in commerce, warfare, and contacts with outside groups. In the past, when risks were high, outside communication was nevertheless rare since individual evaluation of the hazards usually resulted in negative decisions: the chances of success were too small in relation to the enormous risks one had to take. During the nineteenth century, predictability and likewise the possibility of success increased, luring more people to take advantage of the new opportunities.

It seems that the Southeast Bornean population was far from being ignorant of the outside world or inert to innovations. Nor was its behaviour inherently risk-averse. Rather, that behaviour can better be characterized as rational, calculating, and opportunistic. When one uncertainty was taken away, people tended to take new risks instead. Their strategies sometimes resulted in misery and sometimes in fortune, depending on the economic, climatic, medical or political situation of the moment. Interestingly, the same process of risk substitution can still be observed today. All along the Kahayan River, tens of thousands of Borneans recently joined an enormous gold rush. Each invested millions of rupiah in the necessary equipment. And all over Southeast Borneo one can still see farmers converting their *ladang* into rubber gardens, anticipating high rubber prices in the near future. The risks of destructive droughts or epidemics are now greatly reduced, but in their place have come the uncertainty of finding a rich gold vein and the unpredictability of rubber prices.

Abbreviations

ANRI	Arsip Nasional Republik Indonesia, Jakarta
ARA	Algemeen Rijksarchief, Den Haag
AZR	Archive of the Raad der Aziatische Bezittingen en Etablissementen
GM	Generale Missiven
MvK	Archive of the Ministerie van Koloniën, ARA
MR	Mail Reports, MvK ARA
OIC	Archive of the Oost Indisch Comité, ARA
VOC	Archive of the Dutch East Indies Company (VOC), ARA
WAB	Archive of the Residentie Westerafdeeling van Borneo, ANRI
ZOB	Archive of the Zuider- en Oosterafdeeling van Borneo, ANRI

References

Bailey, R.C. and T.N. Headland
1991 'The tropical rain forest; Is it a productive environment for human foragers?', *Human Ecology* 19-2:261-85.

Bangert, C.
1860 'Verslag der reis in de binnenwaarts gelegene streken van Doessoen Ilir', *Tijdschrift voor Indische Taal-, Land- en Volkenkunde (TBG)* 9:134-218.

Berlage, H.P.
1931 'Over het verband tusschen de dikte der jaarringen van djatibomen (*Tectona grandis* L.f.) en den regenval op Java', *Tectona* 24:939-53.

Boomgaard, P.
1995 'Fluctuations in climate, food availability, war, and mortality in Indonesia in the 17th century'. Paper for the EUROSEAS Conference in Leiden, June 1995.

Chew, D.
1990 *Chinese pioneers on the Sarawak frontier 1841-1941*. Singapore: Oxford University Press.

Christensen, H. and O. Mertz
1993 'The risk-avoidance strategy of traditional shifting cultivation in Borneo', *Sarawak Museum Journal* 44-65:1-18.

Douglas, M. and A. Wildavsky
1982 *Risk and culture; An essay on the selection of technical and environmental dangers*. Berkeley/Los Angeles: University of California Press.

Dijk, L.C.D. van
1862 *Neêrland's vroegste betrekkingen met Borneo, den Solo-Archipel, Cambodja, Siam en Cochin-Chin.* Amsterdam: Scheltema.

Fenner, F., et al.
1988 *Smallpox and its eradication.* Geneva: World Health Organization.

Gemelli Careni, J.F.
1732 'A voyage around the world', in: A. Churchill (ed), *A collection of voyages and travels [...]*, vol. 4, pp. 1-572. London: Walthoe.

Generale Missiven
1960-88 *Generale missiven van Gouverneurs-Generaal en Raden aan Heren XVII der Verenigde Oostindische Compagnie.* Edited by W.Ph. Coolhaas [vol. 1-8] and J. van Goor [vol.9]. 's-Gravenhage: Nijhoff. [RGP, Grote Serie, 9 vols.]

Heersink, C.G.
1987 Islam & islamisering in Bandjermasin van eind 18e tot begin 20ste eeuw. [MA thesis, Vrije Universiteit Amsterdam.]

Hopkins, D.R.
1983 *Princes and peasants; Smallpox in history.* Chicago/London: Chicago University Press.

Hudson, A.B.
1976 'Padju Epat; The Ma'anjan Dayak in historical perspective', *Indonesia* 4:8-42.

Joekes, A.M.
1917 'Oendang-oendang soeltan Adam (1835)', in: *Adatrechtbundels* 13, pp. 343-72. 's-Gravenhage: Nijhoff.

Kleindorfer, P.R., H.C. Kunreuther and P.J.H. Schoemaker
1993 *Decision sciences; An integrative perspective.* Cambridge: Cambridge University Press.

Knapen, H.
1996 'Outside influences on the trade and collection of rattan in Southeast Borneo, 1600-1900'. Paper presented at the 4th Biennial Conference of the Borneo Research Council, Brunei Darussalam, 10-15 June 1996.
1998 'Lethal diseases in the history of Borneo; Mortality and the interplay between disease environment and human geography', in: V.T. King (ed.), *Environmental challenges in South-East Asia.* Richmond: Curzon Press. [Forthcoming.]

Krecke, F.W.C.
1859 'Waarnemingen gedaan te Banjermassing (Borneo)', *Meteorologische Waarnemingen in Nederland en Zijne Bezittingen* (1859):1-64 (appendix).

Leupe, P.A.
1864 'Beschrijving van eenen togt naar de bovenlanden van Banjermassing enz., in het jaar 1790', *Kronijk van het Historisch Genootschap* 20-5 (vierde serie):331-404.

Lumholtz, C.
1920 *Through central Borneo; An account of two years' travel in the land of the head-hunters between the years 1913 and 1917*. New York: Scribner's. 2 vols.

Mallinckrodt, J.
1933 'Grensregeling van beschikkingskringen bij de Boekit-Dajak (1875)', in: *Adatrechtbundels* 26, pp. 86-9. 's-Gravenhage: Nijhoff.

Müller, S.
1839 *Land- en volkenkunde*. Leiden: La Lau. [Verhandelingen over de natuurlijke geschiedenis der Nederlandsche overzeesche bezittingen 1.]

Perelaer, M.T.H.
1870 *Ethnographische beschrijving der Dajaks*. Zalt-Bommel: Noman.

Pijnappel, J.
1860 'Beschrijving van het westelijke gedeelte van de Zuid- en Oosterafdeeling van Borneo (de afdeeling Sampit en de Zuidkust)', *Bijdragen tot de Taal-, Land- en Volkenkunde* 7:243-346.

Ras, J.J.
1968 *Hikajat Bandjar; A study in Malay historiography*. The Hague: Nijhoff. [KITLV, Bibliotheca Indonesica 1.]

Rousseau, J.
1989 'Central Borneo and its relations with coastal Malay sultanates', in: P. Skalnik (ed.), *Outwitting the state*, pp. 41-50. New Brunswick: Transaction Publishers.
1990 *Central Borneo; Ethnic identity and social life in a stratified society*. Oxford: Clarendon Press.

Roy, J.J. de
1706 *Hachelijke reystogt van Jacob Jansz de Roy na Borneo en Atchin, in syn vlugt van Batavia, derwaards ondernoomen in het jaar 1691 en vervolgens [...]*. Leiden: Vander Aa.

Sandin, B.
1967 *The Sea Dayaks of Borneo before White Rajah rule*. London: Macmillan.

Schophuys, H.J.
1936 *Het stroomgebied van de Barito; Landbouwkundige kenschets en landbouwvoorlichting*. Wageningen: Veenman. [PhD Thesis, Landbouwuniversiteit Wageningen.]

Schwaner, C.A.L.M.
1853-54 *Borneo; Beschrijving van het stroomgebied van den Barito en reizen langs eenige voorname rivieren van het zuid-oostelijk gedeelte van dat eiland in de jaren 1843-1847*. Amsterdam: Van Kampen. 2 vols.

Tobias, J.H.
1828 'Macassar; Beschrijving van Banjer massing', *De Nederlandsche Hermes* 3-13:1-38.

Weinstock, J.A.
1983 *Kaharingan and the Luangan Dayaks: Religion and identity in Central-East Borneo*. [PhD thesis, Cornell University, Ithaca.]

Appendix

Southeast Borneo: overview of disasters and rice import/export, 1747-1891

	1 rain	2 drought	3 forest fire	4 cholera	5 small-pox	6 measles	7 rats/ mice	8 rice export	9 rice import	10 famine	11 war
1747	-	-	.	-	-	-	.	-	+	-	-
1748	++	++	.	-	-	-	.	-	+	-	-
1749	++	-	.	-	-	-	.	-	+	-	-
1750	-	++	.	-	++	-	.	-	+	-	-
1751	-	-	.	-	++	-	.	-	+	-	-
1752	++	-	.	-	-	-	.	-	+	-	+
1753	++	-	.	-	-	-	.	-	+	-	+
1754	-	-	.	-	-	-	.	-	+	-	-
1755	-	-	.	-	-	-	.	-	+	-	-
1756	++	-	.	-	-	-	.	-	+	-	-
1757	-	-	.	-	-	-	.	-	+	++	-
1758	-	-	.	-	-	-	.	-	+	++	+
1759	-	-	.	-	-	-	.	-	+	-	-
1760	++	-	.	-	-	-	.	-	+	-	-
1761	-	++	.	-	-	-	.	-	+	-	+
1762	-	-	.	-	-	-	.	-	+	+	+
1763	-	++	.	-	-	-	.	-	+	-	+
1764	++	-	.	-	++	-	.	-	+	-	-
1765	++	-	.	-	++	-	.	-	+	-	-
1766	-	-	.	-	-	-	.	-	+	-	-
1767	-	-	.	-	-	-	.	-	+	-	-
1768	-	-	.	-	-	-	.	-	+	-	-
1769	-	-	.	-	-	-	.	-	+	-	-
1770	-	++	.	-	-	-	.	-	+	-	-
1771	-	-	.	-	-	-	.	-	+	-	-
1772	-	-	.	-	-	-	.	-	+	-	-
1773	-	-	.	-	-	-	.	-	+	-	-
1774	-	-	.	-	-	-	.	-	+	-	-
1775	-	-	.	-	-	-	.	-	+	-	-
1776	-	-	.	-	-	-	.	-	+	-	-
1777	-	-	.	-	-	-	.	-	+	-	-
1778	-	-	.	-	++	-	.	-	+	-	-
1779	-	++	.	-	++	-	.	-	+	-	-
1780	++	-	.	-	-	-	.	-	+	-	+
1781	++	-	.	-	-	-	.	-	+	-	-
1782	-	-	.	-	-	-	.	-	+	-	-
1783	-	-	.	-	-	-	.	-	+	-	-
1784	-	-	.	-	-	-	.	-	-	-	-
1785	-	-	.	-	-	-	.	-	-	-	++
1786	-	-	.	-	-	-	.	-	-	++	++
1787	++	-	.	-	?	-	.	-	-	-	-
1788	-	-	.	-	-	-	.	-	-	-	-
1789	++	-	.	-	-	-	.	-	+	-	-
1790	-	-	.	-	-	-	.	-	-	-	-
1791	-	++	.	-	-	-	.	-	-	-	-
1792	-	-	.	-	-	-	.	-	-	-	-

	1 rain	2 drought	3 forest fire	4 cholera	5 small-pox	6 measles	7 rats/ mice	8 rice export	9 rice import	10 famine	11 war
1793	++	-	.	-	-	-	.	-	-	-	-
1794	-	-	.	-	-	-	.	-	-	-	-
1795	-	-	.	-	-	-	.	-	-	-	-
1796	-	-	.	-	-	-	.	-	-	-	-
1797	++	-	.	-	-	-	.	-	-	-	-
1798	-	-	.	-	-	-	.	-	-	-	-
1799	-	-	.	-	-	-	.	-	-	-	-
1800	-	-	.	-	++	-	.	-	-	-	-
1801	++	-	.	-	++	-	.	-	+	++	-
1802	-	-	.	-	-	-	.	-	-	-	-
1803	-	++	.	-	-	-	.	-	-	-	-
1804	-	-	.	-	-	-	.	-	-	-	-
1805	++	-	.	-	-	-	.	-	-	++	-
1806	-	-	.	-	-	-	.	-	-	-	-
1807	-	-	.	-	-	-	.	-	-	-	-
1808	-	-	.	-	-	-	.	-	-	-	-
[1809-1819 (British interregnum): no data available]											
1820	.	.	.	-	-	-	-	-	-	-	-
1821	.	.	.	++	-	-	-	-	-	-	-
1822	.	.	.	-	-	-	-	-	-	-	-
1823	.	.	.	-	-	-	-	-	-	-	-
1824	.	.	.	-	-	-	-	-	-	-	-
1825	.	.	.	-	++	-	-	+	-	-	++
1826	.	++	.	-	++	-	++	-	-	++	-
1827	++	++	.	-	-	-	-	-	-	-	-
1828	++	.	.	-	++	-	-	-	-	-	-
1829	.	.	.	-	++	-	-	-	-	-	-
1830	.	.	.	-	+	-	-	-	-	-	-
1831	.	.	.	-	-	-	-	-	+	-	-
1832	.	.	.	-	-	-	-	-	-	-	-
1833	.	.	.	-	-	-	-	-	-	-	-
1834	.	.	.	-	-	-	-	++	-	-	-
1835	.	.	.	++	++	-	-	++	-	-	-
1836	.	.	.	-	++	-	-	++	-	-	-
1837	-	-	.	-	+	-	++	++	-	-	-
1838	-	++	.	-	-	-	-	-	-	-	-
1839	-	-	.	-	-	-	-	++	-	-	-
1840	-	-	.	-	-	-	++	+	-	-	-
1841	-	-	.	-	-	-	-	-	+	-	-
1842	-	-	.	-	-	-	-	-	+	-	-
1843	++	-	.	-	++	-	-	-	+	-	-
1844	-	-	.	-	++	-	-	++	-	-	-
1845	-	-	.	-	-	-	++	+	-	-	-
1846	-	++	++	-	-	-	-	-	-	-	-
1847	-	-	.	-	-	-	-	-	-	-	-
1848	++	-	.	-	+	-	-	-	-	-	-
1849	++	-	.	-	++	-	-	-	-	++	-
1850	-	++	.	-	++	-	++	-	-	-	-
1851	-	-	.	++	-	-	-	-	-	-	-
1852	-	-	.	+	-	-	-	-	-	-	-

	1 rain	2 drought	3 forest fire	4 cholera	5 small-pox	6 measles	7 rats/ mice	8 rice export	9 rice import	10 famine	11 war
1853	-	++	++	-	-	-	-	-	++	-	-
1854	-	-	-	-	-	-	-	-	+	-	-
1855	-	++	-	++	-	-	-	-	+	-	-
1856	++	-	-	+	-	+	-	-	+	-	-
1857	-	-	-	-	-	++	-	-	+	-	-
1858	++	-	-	-	-	-	-	-	-	-	-
1859	++	-	-	+	++	-	-	-	-	-	-
1860	-	++	-	-	++	-	-	-	++	++	++
1861	++	-	-	-	++	-	-	-	-	++	++
1862	-	-	-	-	+	-	-	-	-	+	+
1863	++	-	-	-	+	-	-	-	++	++	+
1864	++	-	-	++	+	-	-	++	-	-	-
1865	++	-	-	+	+	-	++	-	++	-	-
1866	-	++	-	+	+	++	-	-	++	-	-
1867	++	-	-	-	+	-	++	++	-	-	-
1868	-	++	++	++	+	-	-	-	++	-	-
1869	++	++	-	-	+	-	-	-	+	-	-
1870	++	-	-	-	+	-	++	-	++	-	-
1871	++	-	-	-	+	-	-	-	++	-	-
1872	-	-	-	-	+	-	-	+	++	-	-
1873	-	++	-	-	+	-	-	-	++	-	-
1874	++	-	-	++	+	-	-	-	++	-	-
1875	-	-	-	+	+	-	-	-	++	++	-
1876	-	++	-	+	+	-	-	-	++	-	-
1877	-	++	++	-	+	-	-	-	++	-	-
1878	++	++	++	-	+	-	++	-	++	++	-
1879	++	-	-	-	+	-	-	-	++	-	-
1880	-	-	-	-	+	-	-	-	++	-	-
1881	++	-	-	++	++	-	-	-	++	-	-
1882	-	-	-	++	+	-	-	-	++	-	-
1883	-	++	-	++	++	-	-	-	++	-	-
1884	-	-	-	-	++	-	-	-	++	-	-
1885	-	++	-	+	+	+	++	-	++	-	-
1886	++	-	-	-	+	++	-	-	++	-	-
1887	++	-	-	-	+	-	-	-	++	-	-
1888	-	++	-	+	++	-	-	-	++	-	-
1889	++	++	-	-	+	-	-	-	++	-	-
1890	++	-	-	-	++	-	++	-	++	-	-
1891	-	-	-	-	+	-	-	-	++	-	-

++ event recorded; severe impact or epidemic
+ event recorded; low impact or incidental cases
- event explicitly absent, or highly unlikely (since its occurrence would otherwise have been recorded)
. no data available
1 years with problematic rainfall; for example, in which rain causes crop losses or floods, or impedes trade
2 years with extreme drought during the east monsoon; or an extended east monsoon
3 large-scale uncontrolled forest fires

4-6 occurrence of cholera, smallpox and measles
7 plagues of rats or mice in agricultural fields
8 rice exports from the port of Banjarmasin
9 rice imports into the port of Banjarmasin
10 occurrence of famine; that is, people suffering from real hunger and reverting to crisis foods (impending but alleviated food shortages not included here)
11 hostilities among substantial numbers of people

Sources: primary data from ANRI and ARA; Pijnappel 1860.

JAN WILLEM NIBBERING

Upland cultivation and soil conservation in limestone regions on Java's south coast
Three historical case-studies

Introduction

Much has been written on the historical development of upland farming in Java in a generalizing way.[1] Historical case-studies on particular regions or environments on the island are still scarce, however. In this contribution I will try to bring out patterns of change in farming systems in one particular environment, one of the 'limestone areas' as it is often loosely called, but it often also includes other sedimentary formations less calcareous in nature. We will concentrate on the question when, under what conditions and in what manner, land-management measures, in particular soil-conservation measures, were adopted in these areas as a response to erosion and declining soil fertility. The material for this paper is largely provided by studies held in three such regions, all situated on Java's south coast: the South Serayu Mountains in the province of Central Java, the Gunung Sewu in the Yogyakarta Sultanate, and the South Malang hill range, in the province of East Java. The purpose of this paper is to verify, and, where possible, to refine general statements on historical developments in Java's upland environments by focusing on both similarities and differences that may exist between the regions. Before tackling the three cases, I briefly sketch the general historical development of upland cultivation and soil conservation in Java.

* I would like to thank Solichin Abdul Wahab, Jan de Graaff, David Henley, Teunis van Rheenen, and Freerk Wiersum for providing me with a large part of the material necessary for the preparation of this paper.

1 The terms 'upland farming' and 'upland cultivation' are used after Ruthenberg (1980), Palte (1989) and many others, to indicate rainfed (or 'dry') permanent cultivation of annual crops in the tropics without the use of impounding rain water or irrigation. The term has been derived from the fact that this type of farming predominates in the uplands, usually hilly or mountainous areas with small catchments. As such, the uplands contrast with the relatively flat irrigated plains and plateaux in the periphery of which the uplands are usually situated. As the altitude of these irrigated plains and plateaux may differ, so can the surrounding upland areas: upland areas can therefore also be found at relatively low altitudes and should not be confused with 'highlands' which in the context of Java are the upper slope volcanic areas of over 1,000 m above sea level.

Landscape in the Gunung Sewu. The vegetation suggests a man-made fire regime. Drawing by F.W. Junghuhn, who visited the area in 1837 (KITLV print collection 2256).

The historical development of soil conservation in Java's uplands: the general picture

Of old, Java's population was concentrated in the lowlands and on the intermontane plains (see also Reid in this volume) where irrigated cultivation of rice was the prevailing mode of agricultural production. However, upland clearance and occupation must have been going on well before the nineteenth century, most likely facilitated by the spread of maize (see Boomgaard 1995:3, and Brookfield, this volume); the occupation of the uplands accelerated during the last century and came to an end during the first decades of the present century (Palte 1989:40-9; Boomgaard and Van Zanden 1990:32-3).[2] It involved the clearing of new land in the largely forested and still sparsely inhabited hilly and mountainous upland regions. Where land could be turned into *sawah,* farmers took up wet rice cultivation, but the possibilities were limited and were soon exhausted, forcing the settlers to resort to rainfed or upland cultivation. As long as land was still abundant, fertility management in upland cultivation still relied on fallowing the land.[3] With continuing population growth, however, the pressure on the land increased and intensification of land use became necessary when the expansion of arable area was outstripped by that of population growth. This situation had come to prevail in all of Java after 1920 (Booth 1988:100). In a number of regions this process was precipitated by concessions by the colonial government to private enterprises, and also by the delimitation of forest reserves (Vink 1926:279-80). This delimitation was undertaken because, from early 1900 onwards, the colonial government had also become increasingly concerned about the apparently adverse effects of deforestation on the hydrological cycle and thereby on the supply of irrigation water in the lowlands. Considerable pressure was exerted by the sugar industry on the government to stop this process of deforestation (Altona 1914; Algemeen Syndicaat van Suikerfabrikanten in Nederlandsch Indië 1931). Eventually, the arable area could no longer be expanded. As a result, increasing population pressure led to the permanent cultivation of the land, whereby fallow systems were increasingly replaced by the cultivation of *tegal.*[4] Intensification did not halt here, it went ahead with single cropping systems changing into multiple cropping systems. One of the consequences of the expansion of agricultural land was also the disappearance of village communal grazing grounds (Donner 1987:71, 77).

With the intensification of upland farming, the perils of soil-fertility

2 According to Palte the most important wave took place between 1860 and 1925. This is probably exaggerated, partly because much of the dry-cultivated land area had been underreported before. The registration of dry land improved considerably between 1905 and 1920 (Boomgaard and Van Zanden 1990:32-3, 90).

3 Cordes 1881:137-9; Verwijk 1888:191; Veth 1903, I:182, 434-5, 468, 508.

4 *Tegal* is defined as dry-cultivated land with predominantly annual crops.

decline and soil erosion made themselves felt. Already during the second half of the last century, when the large-scale occupation of upland areas was in full swing, warnings were voiced about the 'stealthily' (sic) mounting danger of erosion to which upland areas became exposed. This awareness grew in particular with the decline of soil fertility and, consequently, the fall in production in government coffee plantations (Holle 1866; Ossewaarde and Wellensiek 1946:99). Whereas the European-owned mountain estates quickly learned their lesson and incorporated soil-conservation measures into their management, these concerns had far less effect on upland cultivation by the indigenous population, and the implementation of soil-conservation measures by smallholders generally lagged behind the soil-conservation efforts made by the mountain estates (Ossewaarde and Wellensiek 1946:99). Some early discussions by colonial officials on the promotion of terracing indicate that, as long as land was still abundant, farmers were unwilling to put much effort into soil-conservation measures (Kievits 1893; Holle 1894). Schuitemaker commenting on the state of indigenous soil conservation several decades later made similar observations in that he found a clear negative correlation between land-use intensity and 'care for the topsoil' (Schuitemaker 1949:161, 170), but at the same time he pointed out that in some areas farmers had started simple measures of soil conservation, on their own initiative. These measures varied from place to place and included oblique drains, grass lines, earth bunds, terraces, hedges of *Eupatorium* on terrace risers, and stone terrace walls (Schuitemaker 1949:153-4, 157, 160-1, 165-6, 171-2, 174).

It was only in 1933 that the agricultural extension service started to pay serious attention to the propagation and improvement of soil-conservation techniques among the indigenous population (Joosten 1941:1065).[5] Some regional variation in the adoption of soil-conservation measures was noted. Joosten (1941) and De Haan (1952, quoted in Wiersum 1980), for instance, reported that soil-conservation measures in West Java were adopted much more readily in areas with volcanic soils than in areas with marl soils. Schuitemaker also observed this difference in other parts of Java, and showed it to be partly related to the intensity of cultivation which in turn was related to average farm size, with young volcanic soils generally having smaller farms than areas with tertiary sedimentary formations. In these non-volcanic areas farmers were often still found to practise grass fallow on a large part of their dry-cultivated land (Schuitemaker 1949:157, 162, 165-6). In fact, he judged the erosion situation in such areas to be not very dramatic: the soil was relatively well protected under the grass for a number of years and for the agricultural extension service the low production pressure on the land would allow

[5] This is not to say that, locally, terracing had not been promoted in earlier days, such as in the Ciamis area by Holle, and as it seemed with lasting success (Schuitemaker 1949:170).

enough time to induce farmers in a more gradual fashion to adopt systematic soil-conservation practices. At any rate, the immediate construction of fully fledged terraces would be too much of an effort for the local population in those areas. Only where cropping became permanent, more radical measures were deemed necessary and the population had usually already taken up certain conservation measures some of which I mentioned above (Schuitemaker 1949:156-7, 160-1). Schuitemaker did not, however, touch upon the question to what extent the lower intrinsic fertility levels in sedimentary areas accounted for larger farm sizes requiring more extensive forms of agriculture than in the volcanic areas. Another reason for the lower pace of adoption of soil-conservation measures in non-volcanic areas may have been that the fertile volcanic soils probably also offered more opportunities for cash crops, which gave farmers a higher output value and allowed them to purchase off-farm inputs, both good reasons for improving the crop-growing environment by way of soil conservation. Farmers cultivating poorer soils did not have these opportunities.

During the Japanese occupation and the early years of independence little was done to stimulate soil-conservation measures. But towards the end of the 1950s soil conservation and regreening programmes were taken up again by the government and stepped up considerably in the 1970s. Nowadays, after years of relative neglect of upland farming in favour of wet rice irrigation in the lowlands, the problems of upland farming are getting more attention. Better infrastructure, the availability of fertilizers and increasing demand for agricultural products from the uplands have offered new opportunities for upland farming. However, soil conservation and the upkeep of soil fertility in the upland areas of Java still pose problems yet to be solved.

The case-studies

All three case-studies dealt with in this article concern regions in the tertiary hill and mountain ranges of marl and limestone rock that border on much of Java's south coast (Map 1). Nevertheless, their physical environments show some clear differences which have influenced the historical development of upland cultivation in each area, one of them being rainfall which decreases along a west-east axis. However, what the three areas share historically, is that well into the nineteenth century they all used to be peripheral, isolated, and sparsely inhabited areas with a predominantly forest or woodland vegetation. Subsequently, they experienced a rapid influx of settlers from more crowded areas in Java (and in one case also from the island of Madura), which caused a quite drastic conversion of a wooded area into the landscape of dry-cultivated fields that we witness today. Nowadays these are all agriculturally marginal areas with low

Table 1. Several data for average farms in the three case-study areas: present situation

Region	South Serayu Mountains	Gunung Sewu		South Malang hill range	
Hamlet/village	Merden Kidul	Kepek, Legundi, Mendak		Putukrejo	Kedungsalam
Farm resources					
Farm size (ha)	0.38	1.27		0.72	1.02
- cultivated	0.35	0.85		0.72	0.66
- fallow/not cultivated	0.03	0.42		-	0.35
Livestock					
Cattle	-	1.1		1.0	1.2
Goats/sheep	2.6	1.2		0.3	1.2
Fertilizer (kg)		*valleys*	*hillsides*		
Dung	2300	2060	750	?	?
Urea	121	206	82	268	64
TSP	89	55	15	263	16
Crop yields per ha (kg)					
Hill rice	-	840	340	1200	1220
Maize	670	350	?	2050	1160
Cassava	3000	4890	2780	3270[a]	
Tree densities (per ha)[b]					
Home garden/ mixed garden[c]	?	696		945	
tegal: valley	48	200		100	
tegal: hillside	-	433		206	
fallowed hillside	-	110		110	

a. Particularly cassava yields were considered to be low due to little rainfall during the 1990-91 season in this area.
b. Trees with a breast diameter of 5 cm or more (Gunung Sewu and South Malang hill range).
c. Home gardens in Gunung Sewu; mixed gardens in South Malang hill range.
Sources: Merden Kidul (Season 1984-85?): Van Dijk et al., 1987; Palte 1989; Kepek, Legundi, Mendak (season 1988-89): Nibbering, 1991b; Putukrejo, Kedungsalam (season 1990-91): Efdé 1996; Muslich Mustadjab 1991; Van Rheenen 1995.

population-growth rates due to out-migration, but there are differences too. In the following pages an account will be given for each case of the developments that have led to the present situation. Table 1 gives some general information on average farm size, livestock, fertilizer use, and crop yields for the three case-study areas at present.

The South Serayu Mountains

Data for this case have been derived from an analysis by Palte (1989) which was in turn based on fieldwork by Van der Poel and Schinkel in 1984-1985 (Van Dijk et al. 1987). The study centered on Merden Kidul, a settlement in the South Serayu Mountains, in *kabupaten* (district) Banjarnegara, *kecamatan* (subdistrict) Purwanegara, located at about 25 kilometres southwest of the town of Banjarnegara.

The South Serayu Mountains, a tertiary and partly pre-tertiary formation, form a chain of hills and secondary mountains, which extend from the Serayu River to the southern coastal plain. Annual rainfall ranges between 3,000 and 4,000 mm. Most of this falls in seven wet months (over 300 mm per month) from October to April. During the relatively drier season from May to September, rainfall still reaches between 100 and 200 mm per month. The altitude of Merden Kidul ranges between 150 and 300 metres. The mean annual temperature is therefore high. The land is strongly dissected by a dense system of valleys and gullies with steep slopes. Where the vegetative cover is affected, severe erosion may occur. This danger is worsened because of the predominant soft bedrock material of marls and shales. By nature the red lateritic soils that make up most of the soils are of low quality. Where they remain only thin or are already completely washed down the fertility of the land is very poor.

The residency map of Banyumas (Topografische Inrichting 1860) shows that in the mid-nineteenth century the South Serayu Mountains were still largely covered with natural forests. Yet amidst these forests a few small settlements surrounded by clearings had already been founded, of which Merden Kidul is one.

A few decades later the situation in the area had changed drastically. The forest had vanished and had given way to extensive expanses of *ara-ara* fallow with wild grasses and shrubs as shown on a map of 1904 (Topografisch Kantoor 1904). To what extent this change is only due to activities of the local population remains unclear.[6] According to oral statements by

[6] Palte argues that much wood may have been cut to serve as fuel for the railways and sugar factories in the Serayu valley, but he provides no evidence for this. Another possibility explaining this change may have been large-scale forest clearance for the compulsory cultivation of coffee which is certain to have occurred further to the south and to the east in the South Serayu Mountains (Palte 1989:179).

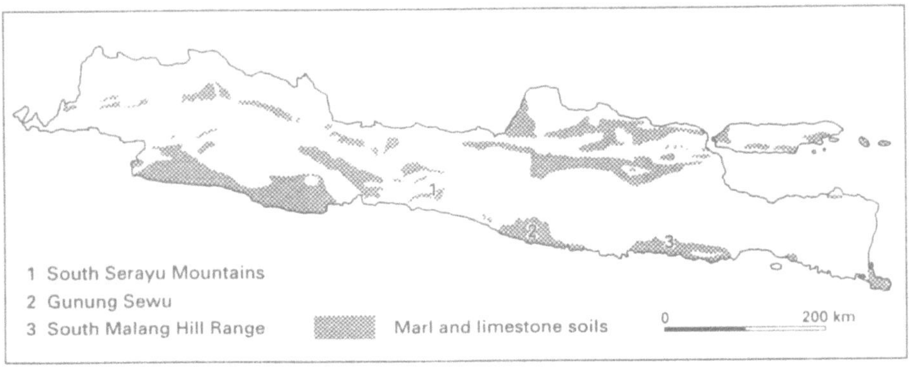

Map 1. The distribution of marl and limestone soils on Java and location of the three study areas (based on the soil map by Mohr 1938).

elderly people in the village, the local population practised a swidden type of cultivation with hill rice (*pari gaga*) as the main crop, on the *ara-ara* expanses. In the neighbourhood of the hamlets with their home gardens, land use was somewhat more varied with a few coconut groves. Because of the rather rough topography and the limited possibilities for irrigation, only a few small *sawah* had been laid out on some valley bottoms.

By the 1930s the situation had not altered much, except for a more extensive reclamation of the wild grasslands in pace with the increase in population. Older farmers will recall the extremely heavy work involved in clearing a field of the dense *alang-alang* (*Imperata cylindrica*) growth. Hill rice was still the main staple crop. The temporary fields on the steep slopes were not terraced, which caused severe run-off during the period they were tilled. Because population growth led to a decrease of the fallow period and manuring was not yet customary, the fertility of the soils declined rapidly.

When in the course of the 1950s fallow was no longer possible because of population growth and the permanent cropping on *tegal* became the custom, maize and cassava had completely replaced rice as basic food crops. At that time also the construction of simple terraces commenced. At the outset, horizontal drainage trenches were dug out along the contour lines, to break the speed and power of rainwater flowing down, but eventually the farmers started to build real terraces (see Figure 1). Because these were still sloping downwards, the terraces were not very effective in terms of soil protection.

Nowadays, the population density in the area has reached 560 per km^2. Virtually all land is taken into cultivation and land consists of largely open dry-cultivated fields. Productivity of the land is low (see Table 1), and small landholdings do not produce enough to live from. Maize and cassava are almost entirely used for home consumption. Little of the agricultural produce is sold. Despite the fact that most holdings are small, the construction of terraces is not yet completed. There are still many *tegal* fields which have not even reached yet stage c in Figure 1. The dominant cropping pattern is one crop of maize intercropped with cassava. Although rainfall would allow a second maize crop, it is generally refrained from. Farmyard manure and nowadays also mineral fertilizer are applied to the crop. Farmyard manure consists largely of dung produced by goats which are all stall-fed. The farmers do not keep cattle (or water buffaloes), because of insufficient forage supplies and because they have no *sawah* to plough. The few former *sawah* have disappeared since they were buried under deposits of washed-down soil.

Of old, utility wood used to grow spontaneously on the fields, but only in a low density. The intentional planting of perennials, especially of fruit trees, stimulated by regreening programmes did take place, but remained

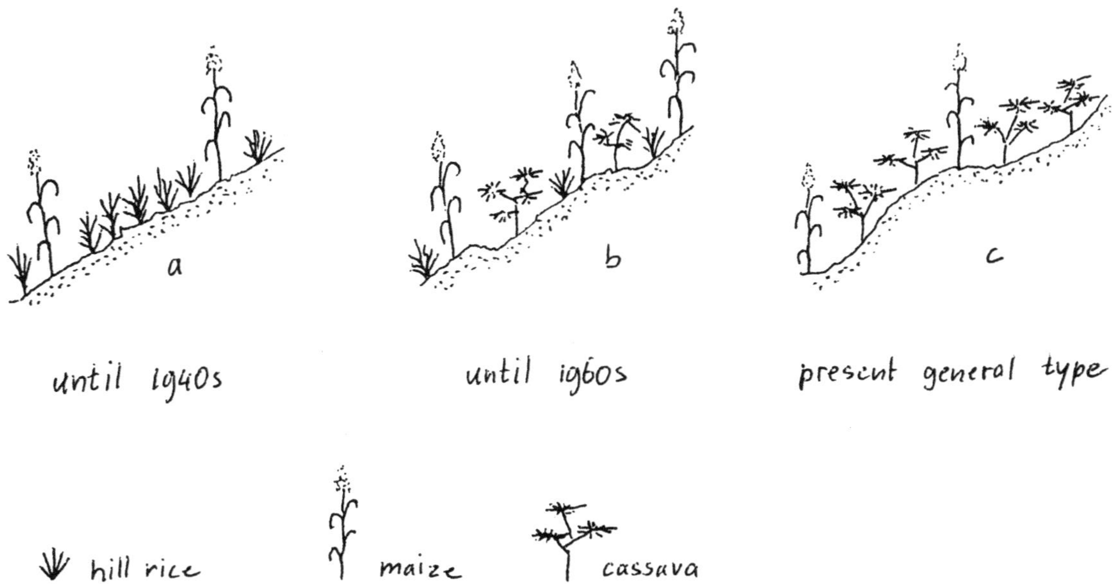

Figure 1. Development of terracing in Merden Kidul, South Serayu Mountains (adapted from Van Dijk et al. 1987).

confined to the area surrounding the houses. Nowadays wood trees are also planted on purpose in the fields, in a somewhat higher density than before.[7] A few dispersed trees stand among the annual food crops. Only around the scattered houses and in some valley bottoms do we find a more dense cover of perennials. These trees serve primarily the direct needs of the household. Since the fields are not well terraced, and the vegetative cover is generally thin, erosion continues to be severe, both through run-off, as witnessed by the gullies in the fields, through landslides, and the covering of the valley bottoms with soil material.[8]

According to Palte, the reasons why terracing is done only provisionally are lack of labour and the farmers' lack of motivation to use their land more intensively, for instance, by double-cropping maize. The farmers are too busy with (seasonal) off-farm activities, often performed elsewhere, which they consider more remunerative, and which they sorely need to survive. This is also the reason why they do not allow too many trees on their land, because the immediate consequence would be a reduction in their food supply. Clearly, these farmers are trapped in various vicious circles (Van Dijk et al. 1987:52-9; Palte 1989:129-32, 139-45, 151-64, 183-90, 197).

The Gunung Sewu

Data for this case were derived from fieldwork conducted by the author in 1988-1989 in this area (Nibbering 1991a, 1991b, 1993). The material used was more substantial than for the other case-studies, because the very subject of the original study was the historical development of upland cultivation in this area. As a result, the picture that emerges from the following account is more dynamic and less 'linear' than the ones reconstructed for the other two cases.

The Gunung Sewu (literally the Thousand Hills) constitute the southern half of *kabupaten* Gunungkidul in the Special Area of Yogyakarta. The data for this case were collected in the hamlets Kepek, Mendak, and Legundi in the *kecamatan* Paliyan and Panggang in the western half of the Gunung Sewu, at a distance of about 80 km southeast of the city of Yogyakarta.

The Gunung Sewu is made up of a complex of 100 to 300 metres in height consisting of hard reef limestone. It has a characteristic karst topography with an irregular pattern of hundreds of conical hillocks that rise some 50 metres above their base. Drainage occurs through a system of sink holes which are located in valleys and caves and lead down to subterranean

7 This finding was based on a comparison of aerial photographs from 1973 with the situation during the fieldwork.

8 As much as 12 ha of the 160 ha of this small settlement has severely been damaged by landslides.

streams. Water supplies are obtained almost entirely from small lakes which have formed in the valleys and are sealed by accumulated clay that prevents leakage into the underlying karst. The brownish-red soils rest directly on the hard limestone. They have a high clay content. They are permeable and moderately water retaining. The depth varies from very shallow on the steep slopes to several metres in the narrow U-shaped valleys. The valley bottoms provide only 10 to 20% of the cultivable area. Average annual rainfall varies between 2,000 and 2,200 mm, but reliable falls are negligible from June to September. The combination of shallow soils and porous limestone results in water shortages fairly soon after rainfall ceases.

The Gunung Sewu used to constitute the most peripheral part of the Gunung Kidul. The extensive teak (*Tectona grandis*) forests that still existed in Gunung Kidul Regency at the beginning of the nineteenth century and which were exploited by the Sultanate were, apart from some scattered patches in its western part, located predominantly north and west of the Gunung Sewu. Although forced labour may have been recruited from the Gunung Sewu for corvée duties in the teak forests (Houben 1987:162, 376), the area itself remained untouched by compulsory crop cultivation.

The natural vegetation in the Gunung Sewu area was most probably mixed evergreen rain forest. In 1837 Junghuhn (1850, I:330) observed a type of vegetation typical of tropical savannahs, with *alang-alang*, *glagah* (*Saccharum spontaneum*) and scattered *Albizzia* trees, which must have developed under man-made fire regimes. This indicates that the area was already subject to human impact which may have occasioned early forms of accelerated soil erosion well before the land was actually brought under cultivation.[9] The three hamlets of this case-study already figure on maps dating from this period (De Stuers 1847; Versteeg 1857).

From the mid-nineteenth century up to 1940, the settlement pattern in the Gunung Sewu became more dense and the cultivated area expanded as a consequence of population growth. Subsequently, land use intensified when land became more scarce. Crop cultivation expanded from the valley bottoms onto the hillsides, reducing the woodland vegetation on the hillsides. Then fallowing in valley bottoms disappeared, and with it, the *alang-alang* fallow. Cultivation of hillsides intensified.[10] Early this century the principal crops were rice, maize, cassava, and tobacco grown in multiple cropping systems (Daneš 1910:256). Hillrice, however, remained

9 Daneš at the beginning of this century observed that the land surface in the Gunung Sewu was intensively dissected by micro-karst forms (*karren*) which appeared particularly nicely on hillsides where the grass had been newly burned and the loam had been washed away somewhat (Daneš 1910:256).

10 Junghuhn 1850, I:342-5; Veth 1903, III:498; Daneš 1910:256; Instituut voor Volksvoeding 1941:6.

the main staple until the 1920s. In some valley bottoms rice could also be grown on rainfed *sawah* (*tadahan*) made out of shallow, stagnant pools (see Map 2).

From 1900 onwards farmers started terracing their fields, first on the valley bottoms and on the saddles which connected the valleys. Starting in the 1930s terracing also spread onto the hillsides. Initially, terraces were made from tree trunks. Later on, as the number of trees decreased, farmers turned to the more laborious method of piling up stones. When loose rubble too became scarce, they started to cut into the jagged limestone outcrops to obtain fresh material. By the 1940s terracing had already become widespread (Dames 1955:131). Terracing of hillsides put an end to the recurrent flow of nutrients from the hillsides released by regular burning of its vegetation. When also fallowing declined and less fertile soils had to be brought under cultivation, the need for manuring arose. Cattle became increasingly necessary as a source of manure, and also as a means to till the land and especially to break the roots of *alang-alang* and other vegetation on fallowed lands. As more and more outcrops emerged as a result of erosion and hillsides had to be terraced, the practice of ploughing hillsides was gradually replaced by hoeing. The areas of grassy and shrubby wasteland that came into being after deforestation were used for grazing cattle (Van Valkenburg and White 1923:132, 139). Apart from their use in farm operations, cattle became a supplementary source of income. In the first decade of the present century the area was a large net exporter of young breeding cattle to neighbouring regions. Even a kind of transhumance developed whereby cattle from the more densely populated northern parts of Gunungkidul District were sent to the Gunung Sewu to graze during the dry season (Instituut voor Volksvoeding 1941:8).

However, with population continuing to grow, the area under fallow declined markedly after 1940 and fallow periods were shortened. At the same time very steep slopes were brought under cultivation. Most of the remaining trees in the fields disappeared in this period too because of the expansion in the area under cultivation and as a result of increased charcoal production in periods of economic stress. Whereas the cultivation of cassava had been largely restricted to the hillsides in previous years and its surplus was traded with people in the north, planting densities of cassava increased conspicuously in the valleys. Rainfed *sawah* disappeared, primarily to allow for cassava cultivation, which does not tolerate flooding. Pre-harvest food shortages had now become the rule for a large part of the population (Bailey 1961; Timmer 1961) and acute malnutrition occurred. The cattle industry became less profitable due to an increasing shortage of fodder and staggering inflation rates during the 1950s and 1960s (Selosoemardjan 1962:242-3). Dry-season transhumance from the north stopped and cattle began to be stall-fed because grazing

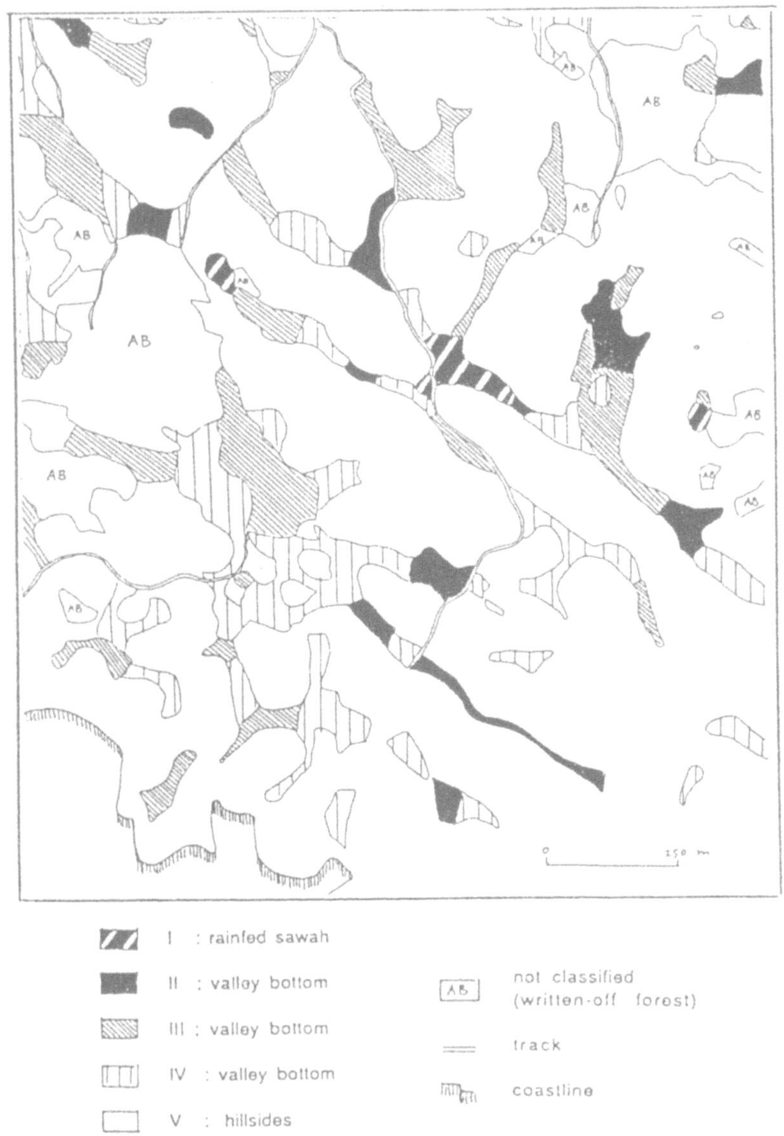

Map 2. Fragment of land tax classification map, Gunung Sewu (adapted from Landrentekaart, 1937).

lands were diminishing. Soil problems had come to be regarded as particularly serious. It was observed in the early 1950s that previous deforestation had caused both acceleration in erosion and unfavourable water conditions in the valleys (Dames 1955:131).

Agricultural production was reduced by recurrent droughts. In 1963 an extraordinarily severe rat plague occurred, totally devastating the rice crop and reducing the cassava crop by one half as well as existing food stocks. In order to survive, impoverished farmers were forced to sell their cattle, land, and even houses. Many people moved out of the area to find food or work. A large number of households joined the government transmigration programme during the years that followed.

Various developments that have taken place in the area since the early 1970s, however, such as better infrastructure, the availability of off-farm inputs (mineral fertilizers, improved crop varieties), an increasing urban demand for agricultural produce, the lift of a ban on growing teak by smallholders, lower population growth, remittances from migrants, and all sorts of development projects, including regreening campaigns, have created important economic opportunities for the rural people of Gunung Sewu, thus enabling them to improve the productivity of their land and provide them with markets for their produce. Farmers began applying inorganic fertilizers to their crops in the early 1970s and the quantities used have gradually risen ever since. Crop yields have risen as a result. Farmers have also developed an interest in planting trees to secure a supply of firewood, fodder, timber, fruits, and cash, as it has now become easy to sell firewood on urban markets. With teak forest no longer close at hand – the state forest had not been able to withstand the disturbances caused by the surrounding population – and with little regeneration of native trees on farmland, farmers have adopted the habit of planting trees, particularly on hillsides. Tree densities are for instance much higher than those found in the South Serayu Mountains (see Table 1). The number of cattle has risen again. Constraints on fodder supply have been mitigated by planting elephant grass and hauling of maize stalks and leaves from the lowlands by truck during the dry season. Land under permanent shrubs or trees tends to increase as a result of farmers engaging in more non-agricultural activities or circular migration and as a result of increased tree cropping.

Nowadays, the valley bottoms are cultivated permanently. They are manured and cropped with hill rice, maize, cassava, and groundnuts. The terraces in the valley bottoms developed into large permanent constructions resembling amphitheatres towards the saddles (Figure 2A). Gunung Sewu farmers think of their valley bottoms as their *sawah* and treat them accordingly. But the hillsides, too, are well terraced. Although decreasing, regular fallowing is still widespread in the more rugged parts of the Gunung Sewu. However, even here, every time farmers clear fallow land,

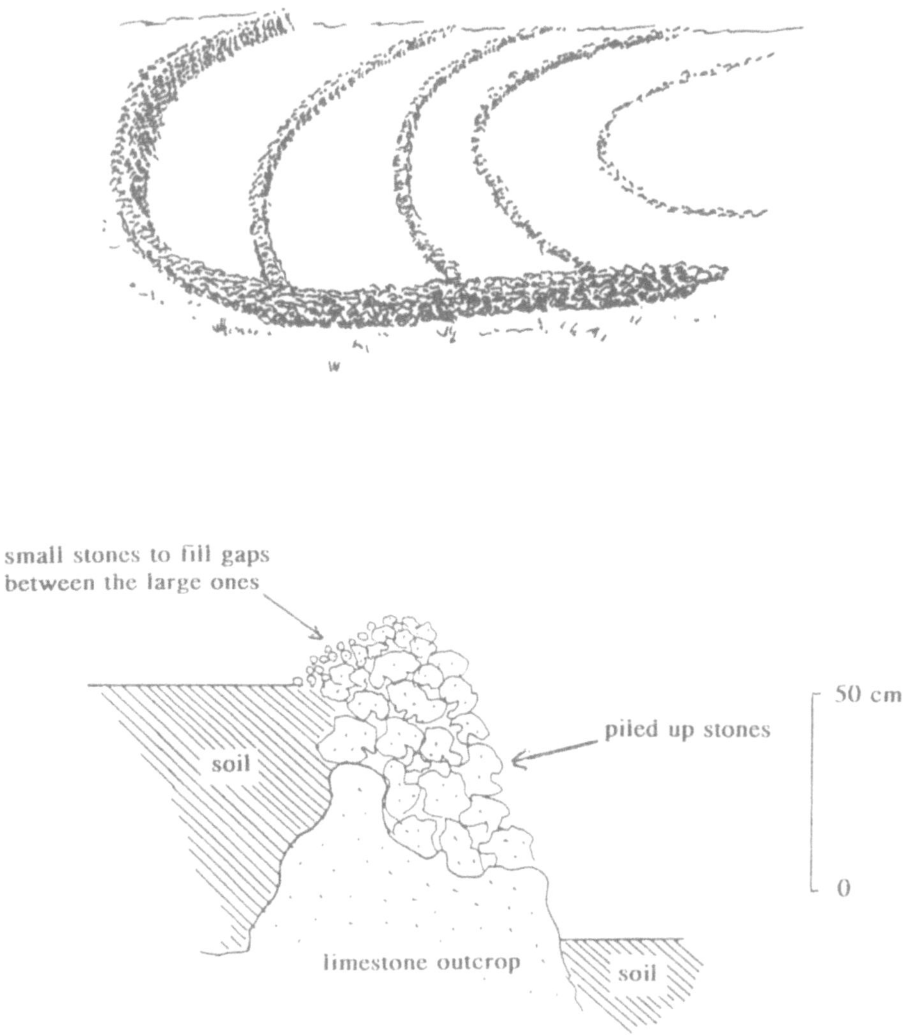

Figure 2. A. Terraces in valleys; B. Cross-section of a hillside terrace, Gunung Sewu (drawings by the author).

they restore the terraces meticulously with fresh material. If they did not, the organic material accumulated under the fallow vegetation of *Chromolaena odorata* would quickly wash away.[11] The terraces on the hillsides are sophisticated in that the interstices between the big rocks are carefully closed off with smaller material (Figure 2B). Farmers have a strong incentive for bringing their terraces in order after a fallow period, they usually start with hill rice, a crop with the highest value and the greatest demands on the soil. For this, they have to clear large patches of vegetation during the dry season, or – and this practice is on the increase – they take up the cultivation of commercial horticultural crops, such as chillies, beans, cowpeas, for which they clear much smaller patches of land throughout the rainy season. In both cases, maize and cassava will be grown in the years that follow until the land is fallowed again. Present terracing practices in the Gunung Sewu may not have halted erosion completely, as narrow bands of lighter colour on the limestone outcrops just above the soil indicate, but farmers go to great lengths to keep soil loss to a minimum (Nibbering 1991a:111-21, 1991b:76-126, 140-53, 224-34, 1993:21-4).

The South Malang hill range

Material for this case was derived from data assembled for the Lesti Watershed Malang Selatan (LESMAS) study conducted by the Konto River Project (1990), from data collected by the INRES-project[12] (INRES 1989; Solichin Abdul Wahab 1991a, 1991b, and other studies), and my own observations and interviews in the area.

By the South Malang hill range is meant the hilly area in the south of *kabupaten* Malang, in particular its western part. In the framework of the INRES-project, studies were carried out in the villages Putukrejo, *kecamatan* Kalipare, and Kedungsalam, *kecamatan* Donomulyo, which are

[11] The fallow vegetation has changed over the years. From the 1960s onwards, a shrub species, *Chromolaena odorata* (syn. *Eupatorium odoratum*), has spread into the area. It crowded out *Lantana* and *Acacia villosa* as well as *alang-alang* and other grasses. Overgrazing may have contributed to the spread of *Chromolaena* at the expense of *lantana* as has been observed in the Philippines by Sajise (1984:147), while *Chromolaena* succeeded in shading out the *alang-alang* growing on the flatter sections of the hillsides when regular burning of hillsides halted (169 1985:165). The shrub now covers virtually all unused or fallowed land, where it also manages to survive under a light canopy of planted trees. As soon as the land ceases to be cultivated, its wind-blown seed quickly germinates. The shrubs grow up in dense thickets up to three metres tall before other plants can establish. Farmers are pleased with the advent of *Chromolaena* because of its enormous capacity for litter production, which, according to them, makes the soil thicker and richer again far better than does other known fallow vegetation. This shrub is also said to be easier to clear than *alang-alang*.

[12] An INterdisciplinary RESearch training project carried out by Brawijaya University, Malang, in collaboration with Wageningen Agricultural University and Leiden University from 1989 to 1992, funded through the Netherlands University Fund For International Cooperation (NUFFIC). The project conducted farming-system research in the case-study area.

situated southwest of the city of Malang at about 50 to 65 km respectively.

The South Malang hill range is made up of a Miocene sedimentary limestone facies, partly covered by volcanic material. In the north the area consists of undulating terrain with relatively flat and deep soils. The village of Putukrejo is situated at an altitude of about 500 m above sea level. In the south the terrain becomes much more dissected and is characterized by the presence of large areas of very shallow, highly erodible soils on the hillsides, and thick clayey soils with vertic properties in the depressions and on the valley bottoms. The southern part of the village of Kedungsalam is situated here, its altitude ranging between 0 and 400 m. The northern part of Kedungsalam constitutes a transitional zone to the flatter areas in the northern part of the South Malang hill range. In this transitional zone, the topography allows for some irrigation on limited narrow tracts of flat stream and river-valley bottoms. The entire western part of the hill range is characterized by relatively dry conditions with six consecutive months of less than 100 mm of rainfall. The average annual rainfall is about 1700-1900 mm. Water quickly infiltrates into the soil and comes to the surface at a number of places with impermeable layers. The prolonged dry season and the low water-holding capacity of the soil quickly result in water shortages, seriously limiting crop growth.[13]

Well into the nineteenth century, the area was still largely forested and practically uninhabited (Elson 1984:2). The earliest settlement is said to have been Sumbermanjing founded early in the nineteenth century, 15 km away from the study area. At the sites of the study villages, the first clearings and land occupation occurred between 1880 and 1890. The then existing natural vegetation was described as a dense forest vegetation with a mixture of tropical rain forest and monsoon forest species. The village of Putukrejo was founded by settlers from Central Java, soon to be followed by migrants from the island of Madura. The Madurese had initially been recruited as labourers on the nearby Alas Tledek estate, where coffee (*C. canephora*), rubber, and also some cocoa were grown on the flattest part of the present Putukrejo village area. In a later stage people from East Java also joined the settlers. The area of Kedungsalam largely attracted settlers from Central and East Java. Kedungsalam received the status of *desa* (village) by the colonial government in 1888. Since the founding father of the village became its first *lurah* (village head), the growth of the settlement and the concomitant occupation of the land must have been quite rapid.[14] Around 1900 most of the land had been cleared. The first settlers

13 INRES 1989, III-1:7-8; Konto River Project 1990:9, 21, 30; Efdé 1996:25.

14 Not just for the two village areas, but for the entire region the land occupation rate during this period was enormous. From 1855 to 1905 the arable area in the Residency of Pasuruan increased by 340 per cent. This was mainly a result of the expansion of arable land in its Malang Regency. The increase in arable area in the Residency of Pasuruan was thereby the

were engaged mainly in growing hill rice and also in tapping *arèn* (sugar palm or *Arenga pinnata;* Solichin Abdul Wahab 1991a:1-2, 1991b:1-2).

In its heydays the Alas Tledek estate in the Putukrejo area covered 378 ha. It was one of the many plantations for which the forest in the South Malang hill range would be cleared and the land given in leasehold by the colonial administration.[15] During the Japanese occupation the plantations of many of these estates were cut down in order to be replaced by castor (*Ricinus communis*) and food crops to sustain the Japanese war effort. This also happened to the Alas Tledek estate. All of its produce had to be handed over to the Japanese army. After independence, the land was claimed by the government of the new republic and distributed among the local people, who turned it completely into food-crop land. The area is now part of the village of Putukrejo (Solichin Abdul Wahab 1991a:2-5).

In the early 1940s farmers were confronted with a decline in the productivity of their land. In Putukrejo the practice of fallowing land had already ceased. In Kedungsalam it was still in use to some extent on hillsides. Ploughing became harder at some places because of the increasing stoniness of the soil; the colour of the soil was becoming lighter; and yields were decreasing, particularly on the hillsides. The cultivation of hill rice virtually disappeared and was taken over by intercropping of maize with cassava. The area was gravely affected by land degradation and droughts and malnutrition was rampant during the 1950s and 1960s like in other areas where cassava had become the staple.[16]

On the Alas Tledek estate various soil-conservation measures such as terracing and digging drainage ditches had been carried out by its management. Farmers in Putukrejo started copying them to some extent. In Kedungsalam, however, a much greater effort had to be deployed by the population on the steep hillsides. During the 1950s entire rock walls were built, very much resembling the constructions in the Gunung Sewu. Although we lack written records describing how these terraces developed, oral information from elderly farmers and present reforestation activities in the nearby forest land with *taungya* systems[17] resembling early terrace building, suggests that these terraces had developed, rather gradually, from stones and rubble laid aside along the contour during tillage opera-

greatest of all residencies in Central and East Java. It even exceeded its population growth which rose only by 211 per cent in the same period. Not surprisingly, the residency also had the lowest agricultural density (580 persons per km^2) in 1905 (Meijer Ranneft 1915:75, 77; quoted by Palte 1989:50). However, the real increase must have been lower due to underestimations before 1880 (Boomgaard and Van Zanden 1990:32-3).

15 About 43,000 ha in total in the Lesti watershed area and the South Malang hill range together (Konto River Project 1990:12).

16 Bailey 1961:217; Napitupulu 1968:63; Solichin Abdul Wahab 1996:202.

17 In these systems, locally called *tumpangsari,* the population is allowed to crop the land for a number of years among the newly planted trees in return for their labour.

tions. Farmers also received outside support,[18] seed material for fruit trees and gliricidia cuttings for the improvement of their terraces in Kedungsalam by planting them on the terrace edges (INRES 1989, III-1:13; Solichin Abdul Wahab 1996:54). Gliricidia trees also had to be planted on fallowed hillsides which had become entirely covered with *alang-alang*. Farmers more and more had to rely on farmyard manure to fertilize their fields. The arrival of gliricidia[19] therefore also brought an important source of fodder to the area used for stall-feeding livestock. But it was only in the 1970s when mineral fertilizers became readily available that the soil fertility problem in the South Malang hill range was somewhat alleviated (Edmundson and Edmundson 1983).

Since the 1970s other developments have favourably affected upland cultivation and soil conservation in the area. It benefitted from better accessibility, when the stone road constructed in 1932 which linked the two villages with the northern lowlands was asphalted in the 1970s. In Putukrejo, although erosion had occurred, it had not removed all the topsoil in the flatter parts of the area. Annual crops continue to be grown here, performing relatively well (INRES 1989, III-1:8, 13; see Table 1). Farmers grow maize and cassava as well as some hill rice. A large part of the cassava crop is sold. In 1978 sugar-cane was introduced into Putukrejo by the Kebon Agung sugar company in Malang. Many farmers have now taken to growing sugar-cane and about 40% of total farmland is planted with this crop, mostly in the former Alas Tledek estate area. Sugar-cane cultivation has also created employment opportunities for small farmers and landless families as well as a large supply of cattle feed (Ifar 1996:64-5, 70).

In Kedungsalam, however, particularly in its southern part, the ridges and the slopes had lost much of their in part volcanically derived soil. Despite the fact that most of the land is terraced now, terracing obviously came late. Unlike on recently cleared hillsides in the nearby forest land, one can observe considerable bedrock exposure on the terraced hillsides on farmland. From the 1970s onwards annual crop cultivation on many hillsides in Kedungsalam has been abandoned, and the terraces are no longer maintained. The low yields on these hillsides, in combination with the emergence of the limestone-burning business and probably also the possibility for farmers to grow food crops on virgin forest land as participants in forest planting schemes, have made that much of these hillsides is now used for growing wood – gliricidia, teak and *Acacia auriculiformis* – for fuelling the limestone kilns. Other hillsides are simply abandoned and

18 College students were mobilized to inform the population in the South Malang Hill Range about gliricidia planting and to urge them to do so (personal communication Professor Soenardi Prawirohatmodjo of the Forestry Faculty of Gadjah Mada University, Yogyakarta, who participated himself in this effort as a student in 1957).

19 The growing of gliricidia received a second impulse in the area when *lamtårå* (*Leucaena leucocephala*), another leguminous multi-purpose tree, was wiped out by a psillid in the mid-1980s.

have become wastelands (*bongkor*) under a natural shrub-and-grass vegetation of *Chromolaena* and *Imperata* and are only visited by farmers who collect firewood and fodder for their cattle. Annual crop cultivation in Kedungsalam largely continues on valley bottoms and in the flatter areas with little input mainly to meet subsistence needs. Dry-land cultivation of the valley bottoms is, despite their fertility, not a very attractive activity because of the unfavourable physical properties of the soil which make them hard to till. This is one of the reasons why farmers have no problems with allowing many trees and perennials on their *tegal*. Particularly in the north of Kedungsalam dry-cultivated fields have been turned into complete *kebun* (perennial or mixed gardens) with coconut, banana, clove, vanilla, *mlinjo* (*Gnetum gnemon*), and the like from which they derive important cash incomes. The vigorous revival of regreening programmes since the 1970s, the improved accessibility of the area and the larger demand for these products made this possible.[20] In the north of Kedungsalam the area under *sawah* has increased as a result of the construction of a diversion dam, while in Putukrejo farmers expand the area under *sawah* whenever they can.[21] Clearly, the consequent changes in land use have greatly reduced soil erosion in the area.

Discussion and conclusions

The three case-studies show that, very broadly, upland occupation and cultivation in all three areas have followed a pathway which corresponds to the general picture described earlier in this article. More specifically, however, there are some interesting elaborations to be made and there are also clear differences between the regions that deserve attention.

It appears that both the South Serayu Mountains and the Gunung Sewu have known a phase of 'extensive' exploitation, which was superseded by more intensive forms later on. Considering this extensive exploitation of still readily available resources in sparsely populated areas one may speak of a temporary 'frontier' situation (see also Colombijn, this volume). In the South Serayu Mountains, government coffee cultivation, which was generally an extensive form of coffee cultivation, had been introduced although perhaps not everywhere. In the Gunung Sewu land and labour were used to some extent for teak production, but much more for extensive cattle grazing which continued well into the twentieth century. The South

20 The latest development is that farmers are contracted by fruit merchants in Malang and other cities to grow citrus (Van Rheenen 1995:66).

21 Some farmers here having small patches of valley bottom land practise controlled sedimentation to improve their land and turn it into *sawah*. By constructing a dike around the area and diverting silt-laden water into the basin formed by the dikes, the water spreads out into the basin and drops its load of silt (INRES 1989, III-1:14). This is an old practice, but its scope is limited in the area.

Malang hill range, on the other hand, does not seem to have known a similar phase. It was one of the last regions on Java to be opened up for upland cultivation, and the people who settled here were confronted with virgin forests. There are two interesting aspects to this general observation. First, in those areas where it occurred, the frequent burning of grassland associated with extensive cattle grazing and shifting cultivation may already have caused significant soil loss well before the land was taken into permanent cultivation.[22] Terra, for instance, attributes the bad erosion conditions in the Gunung Kendeng (in East Java) and the Gunung Kidul in Central Java to swidden cultivation with extensive cattle raising originally practised in these areas (Terra 1958:170-1, 173).[23] Also, government coffee cultivation may have caused severe erosion in those areas very early on, depending on how coffee plantations were established (Palte 1989:51). Secondly, the use of upland areas for extensive grazing in the past appears to be a little-examined subject. There are indications that it was not unimportant in Java (Schuitemaker 1949:162, 164). Once the forest in the South Serayu Mountains was gone, extensive cattle grazing may have occurred here as well. As mentioned earlier, common grazing grounds all disappeared in Java due to the expansion of the agricultural area. In the same vein, little is known about the process of transition Javanese farmers had to go through, shifting from cattle grazing on wastelands to the stall-fed or zero-grazing systems that we know today.

In all three regions the best land was taken into cultivation first, that is, valleys or flat areas were cleared before hillsides, and valleys or flat areas were taken into permanent cultivation before hillsides were. However, not only topography, but also vegetation influenced farmers' decisions on where to cultivate. In the Gunung Sewu, for instance, in times of stress, when farmers had no animals to plough the fallowed, *alang-alang* covered valley bottoms, they preferred to cultivate the *Lantana-*

22 Soil loss under grass and shrub fires is not only caused by exposure of the soil, but it is also due to reduction of its organic matter content by 20 to 30 percent, which renders the soil more susceptible to erosion (Eppink 1978:21).

23 Terra mentions this in his article on farm systems in Southeast Asia when dealing with the 'Indonesian farm system with cattle'. In his general explanation of this system he describes how 'cattle [carabaos] wander about unattended on the harvested ladangs and in the dry season the dry bushy grass is burned down to provide fresh shoots for the cattle, as a result of which the grass acreage is being constantly increased [...] thus there is a steady increase in erosion which has obtained a hold on the burnt grassy plains' (Terra 1958:170). Terra goes on to indicate a number of regions in Indonesia where this system still exists (mainly on the outer islands) alongside with areas where it has disappeared, but not without leaving behind certain traces. For Java he writes the following: 'In the Sunda lands most relics of the old farm system are to be found in the south, viz. taluns, grassy plains, carabao keeping, houses built on piles, rectangular village greens with a grass cover, fenced in villages, and customs recalling the bride price. Little of this is now to be seen in the dissimilar areas of Central Java, but the form of the villages in the Gunung Kendeng, and unlike the remainder of Kedjawen, the very bad erosion conditions in the Gunung Kendeng and the Gunung Kidul may well be connected with the old system of farming' (Terra 1958:173).

covered hillsides of their land instead, the latter being much easier to till. In the Southern Serayu Mountains, however, farmers may never have had this choice, for all hillsides are reported to have come under *alang-alang*. But when considered over time, deviations in cultivation patterns brought about by vegetation were only temporary.

In none of the three cases has terracing – or its incipient forms – required any instigation from outside. This is most clear in the case of the Gunung Sewu where terracing started before the agricultural extension service became active in the area. In the South Malang hill range the plantations probably exerted some influence on the local population in this respect, but this is less likely in its southernmost zone. Terracing in the South Serayu Mountains also seems to have been a self-generated process. Only in later stages did these areas receive outside support for terracing, and, as it seems, the South Malang hill range most of all.

Generally, it is true that farmers started to take soil-conservation measures when upland cultivation became permanent. The observation made by colonial officers that farmers were reluctant to do something about soil erosion as long as they still had extra land available was correct in that sense. But it should not be taken too literally. Some measures, although still very provisional, such as placing tree trunks along the contour, were perhaps already practised when farmers were still shifting their fields. Conversely, even the construction of fully-fledged terraces did not stop farmers from fallowing their land as it can still be witnessed in the Gunung Sewu today.

No doubt, terracing has developed gradually in all three case-study areas. Various factors have played a role in this. Both the need for soil conservation and farmers' perception of this need increased over time, as soils became shallower and poorer each time they were cultivated. Consequently, the willingness on the part of the farmers to do something about it grew as well. At the same time, as erosion proceeded, more and more surface stones and rock outcrops emerged on the land, providing the very means to implement soil-conservation measures, at least in the pure limestone areas. Surface stones had to be removed anyway in order to create more rooting space for the crops. As the population increased more people had to live from the same land area but more labour also became available to do the work. The farmers, however, did not do all the work themselves. They learned how to let nature do part of the job, by raising the stone edges of the terraces, so that material eroded upslope and on the inner side of the terrace would settle on its outer side behind the edges, eventually creating bench terraces of some kind with a stable slope. In the Gunung Sewu and the South Malang hill range farmers have generally not engaged in moving earth from one place to the other, which would have been necessary if complete bench terraces had been constructed on previously unterraced slopes. Essentially, farmers have picked up, cut and moved the stones; Nature has moved the soil. The resulting terraces not only reduced runoff and soil erosion, but at the same time improved infiltration of rain water into the soil over the entire terrace surface, an important advantage in areas with shallow soils and semi-humid climates, when water stress occurs frequently.

Evidence from other limestone areas in Indonesia also suggests that there is a gradual tendency towards terrace construction and terrace improvement associated with intensifying land use induced by population pressure. In some limestone areas in Central Irian Jaya, for instance, small cross-slope stone walls were constructed to check soil-creep, which, however, had nowhere evolved yet towards fully-fledged terraces (Brookfield 1962:248). In swiddens on East Sumbawa, loose stones were arranged in long narrow piles across the slope. These piles not only served to get the stones out of the way, but also to capture some of the soil washed down by the rains. A consequence of this practice is that some swiddens had a partially terraced look (Brewer 1979:95). This description corresponds closely to what terracing in the Gunung Sewu and the South Malang hill range must have looked like in an earlier stage and what the recently cleared state forest lands in the South Malang hill range look like today. In semi-permanently cultivated cassava gardens in the same East Sumbawa area, people often went further and built small rock walls to reduce erosion. Over time, these rock walls had gradually created terraces (Brewer 1979:162). Finally, on Bali's southern limestone peninsula the most densely populated area of our spectrum, permanently cultivated upland fields are all terraced with stone walls (Uhlig 1980:36). It would appear that when cultivation intensifies in limestone regions, terracing is an obvious device on the shallow hillside soils, where building material in the form of loose stones or outcrops are abundant. Ossewaarde and Wellensiek (1946:111-2) already noted that, generally in the Dutch Indies, terrace risers were fortified with stones in regions where labour was cheap and stones abundant.

On the marl soils of the South Serayu Mountains the situation has been different. Here, the parent material produced no rocks hard enough to clear the field of them and to use them as terrace risers. So, from the outset there was no direct need to set aside part of the land for stones removed from the field. Farmers have continued to cultivate all of their land. They did start digging trenches along the contour and these trenches gradually became the basis of very sloping terraces, for which farmers moved soil from the inner part of the terrace to the outer part, and unlike in the two other areas a much smaller role was given to nature. Apparently, farmers have never been willing or urged to plant hedges of trees or shrubs with grass undergrowth to start a process of natural terrace formation as has been tried with *Leucaena* in other areas in Java in the 1930s (Schuitemaker 1949:157, 159, 162), because they were reluctant all along to sacrifice part of their crop land to other uses. It must be added that rainfall is much higher than in the two other areas further east, so that farmers are perhaps reluctant to level their terraces too much to avoid waterlogging and landslides.

Although we deal with upland cultivation in this paper, it seems relevant to devote some attention to the fate of rainfed *sawah* that have disappeared in the South Serayu Mountains and the Gunung Sewu alike in the course of this century. *Sawah* is a highly valued form of land use, particularly in the uplands. One can expect that people will go to great lengths to maintain them. Their disappearance therefore must have had serious environmental or other causes. In the South Serayu Mountains sedimentation has reportedly led to the disappearance of the rainfed *sawah*. No details are given by Palte, but it is likely that deposits of large quantities of coarse and unfertile material over good valley-bottom land have been liable for this. In the Gunung Sewu there is reason to believe that the factors underlying the disappearance of rainfed *sawah* were more complex. Sedimentation of upslope gravelly material may have been one of them, but other factors seem to have been more important such as a reduction of run-off due to terracing on the higher parts of the valley bottoms and the hillsides, as well as the need to include cassava into the cropping system, which militated against wet rice cultivation and favoured hill rice cultivation intercropped with cassava. Paradoxically, in the northern and flatter part of the South Malang hill range sedimentation has been a source of *sawah* building until today, but this process involves silt suspended in tapped water courses, and not the coarse materials washed down from hillsides.

Before ending this comparative discussion it is worthwhile to point out that in both the Gunung Sewu and the South Malang hill range present conditions and practices have much to tell about the past. In both cases it was possible to estimate the effects of past agricultural land use on the vegetation and on the soil by comparing the situation on farmland with that found on patches of land still under relatively undisturbed forest.[24] Also, present clearing activities in the forest land of the South Malang hill range give us a view of the past, as they show how the forest must have been cleared a century ago. For this, the farmers resorted to old felling and burning techniques and on the newly cleared land they practice *gaga* rice-based cropping patterns that disappeared long ago from their own farmland. Furthermore, the Gunung Sewu presents some archaic forms of land use and land management. The combination of fairly intensive permanent valley-bottom cultivation with a fallow-cultivation sequence

[24] In the Gunung Sewu, sacred hills have retained their original vegetation and have never been cultivated. In the South Malang hill range, it is the nearby forest area which is still partly under a natural forest vegetation but which is now rapidly being converted into forest plantations by the Forest Service.

on meticulously terraced hillsides seems anachronistic and paradoxical. It also contradicts what most authors writing on the uplands of Java state about fallow systems in Java. According to them, fallowing belongs to the past (Palte 1989:57; Smiet 1990:291). It does in most areas, but not in all. It is beyond the scope of this paper to discuss the causes of this phenomenon in the Gunung Sewu. Here it suffices to say that it can be explained in terms of the rational allocation of labour and fertilizers in this special biophysical and socioeconomic environment. It does show, however, that the practice of fallowing, or terracing for that matter, is not necessarily associated with one particular stage of Boserup's scheme of agricultural intensification driven by population growth (Boserup 1965). It can rationally also occur side-by-side with more intensive forms of agricultural production practised by the same farmers.

When upland cultivation in Java intensified, a general decline in productivity of the land followed. Diemont et al. (1991:221) indicated that in much of the colonial and also more recent literature soil erosion is considered to be the main cause of the decline of productivity in upland cultivation rather than the exhaustion of nutrients. When in the 1980s it was noted that yields in the uplands had been increasing again, this rising productivity was connected primarily with the increasing use of mineral fertilizers in upland areas (Roche 1988:5-8; Van der Eng 1988). This led to some controversy on what had been more important in the productivity decline in the past, soil erosion or soil exhaustion, and, similarly, what has been more conducive to rising yields in recent decades: soil conservation or soil fertilization (Diemont et al. 1991; De Graaff and Wiersum 1992). Clearly the two are closely connected, fertilization pays off best on the best-protected soils, as may appear from the three case-studies: soil-conservation practices in some areas have become more advanced, due to a combination of pressures and opportunities, and the application rates as well as the impact of external inputs have therefore been greater here than in other areas.

References

Algemeen Syndicaat van Suikerfabrikanten in Nederlandsch-Indië
1931 *Verslag van de reboisatiecommissie ingesteld door de provinciale raad van Oost-Java*. Soerabaja: Algemeen Syndicaat van Suikerfabrikanten in Nederlandsch-Indië.

Altona, Th.
1914 'Rapport nopens het voorloopig hydrologisch onderzoek van het Brantasgebied', *Tectona* 7:245-67, 317-47, 417-33.

Baily, K.V.
1961 'Rural nutritional studies in Indonesia', *Tropical and Geographical Medicine* 13:216-33, 234-54, 289-302, 303-15.

Boomgaard, P.
1995 'Maize and tobacco in upland Indonesia', 1600-1940'. Paper prepared for the workshop on Agrarian Transformation in Upland Indonesia, Halifax.

Boomgaard, P. and J.L. van Zanden
1990 *Food crops and arable lands, Java 1815-1942.* Amsterdam: Royal Tropical Institute. [Changing Economy in Indonesia 10.]

Booth, A.
1988 *Agricultural development in Indonesia.* Sydney: Allen and Unwin. [ASAA Southeast Asia Publications Series 16.]

Boserup, E.
1965 *The conditions of agricultural growth; The economics of agrarian change under population pressure.* Chicago: Aldine.

Brewer, J.D.
1979 *Agricultural knowledge and cultural practice in two Indonesian villages.* [PhD thesis, University of California.]

Brookfield, H.
1962 'Local study and comparative method; An example from central New Guinea', *Annals of the Association of American Geography* 52:242-54.

Cordes, J.W.H.
1881 *De djati-bosschen op Java; Hunne natuur, verspreiding, geschiedenis en exploitatie.* Batavia: Ogilvie.

Dames, T.W.G
1955 *The soils of East Central Java (with a soil map 1:250,000).* Bogor: n.n. [Contributions of the General Agricultural Research Station 141.]

Daneš , J.V.
1910 'Die Karstphänomäne im Goenoeng Sewoe auf Java', *Tijdschrift van het Koninklijk Nederlandsch Aardrijkskundig Genootschap* 27 (tweede serie):247-60.

Diemont, W.H. A.C.. Smiet and Nurdin
1991 'Re-thinking erosion on Java', *Netherlands Journal of Agricultural Science* 39:213-24.

Donner, W.
1987 *Land use and environment in Indonesia.* London: Hurst.

Dove, M.R.
1985 'The agroecological mythology of the Javanese and the political economy of Indonesia', *Indonesia* 39:1-36.

Dijk, H. van, P. van der Poel and R.F. Schinkel
1986 *Social forestry and farming systems research; Two case-studies on Java.* Yogyakarta: Forestry and Nature Conservation Research Project, Universitas Gadjah Mada. [FONC project communication 1987-83.]

Edmundson, W.C. and S.A. Edmundson
1983 'A decade of village development in East Java', *Bulletin of Indonesian Economic Studies* 19-2:46-59.

Efdé, S.L.
1996 *Quantified and integrated crop and livestock production analysis at the farm level; Exploring options for land use of mixed farms on heavy limestone soils south of Malang, East Java, Indonesia.* [PhD thesis, Landbouwuniversiteit Wageningen.]

Elson, R.E.
1984 *Javanese peasants and the colonial sugar industry; Impact and change in an East Java residency.* New York: Oxford University Press.

Eng, P. van der
1988 *Growth and productivity change in Indonesian agriculture 1880-1965.* Groningen: Faculty of Economics, University of Groningen.

Eppink, L.
1978 *Bodemtechniek; Deel E: Erosie, erosiebestrijding en bodembescherming.* Wageningen: Vakgroep Cultuurtechniek Landbouwhogeschool.

Graaff, J. de, and K.F. Wiersum
1992 'Rethinking erosion on Java; A reaction', *Netherlands Journal of Agricultural Science* 40:373-9.

Haan, J.H. de
1952 'The economic and social aspects of erosion control', *O.S.R. News* 3:69-72.

Holle, K.F.
1866 'Een groot gevaar dat sluipend nadert', *Tijdschrift voor Nijverheid en Landbouw in Nederlandsch-Indië* 12-2:122-34.
1894 'Eenige gegevens omtrent den inlandschen landbouw op tegalans en het nut van terrassen', *Tijdschrift voor Nijverheid en Landbouw in Nederlandsch-Indië* 48-1:1-35.

Houben, V.J.C.
1987 *Kraton en Kumpeni; Surakarta en Yogyakarta 1830-1870.* [PhD thesis, Leiden University.]

Ifar, S.
1996 *Relevance of ruminants in upland mixed-farming systems in East Java, Indonesia.* [PhD thesis, Landbouwuniversiteit Wageningen.]

INRES
1989 *Sondeo report.* Malang: Interdisciplinary Agricultural Research Training (INRES) at Brawijaya University. [Internal project document.]

Instituut voor Volksvoeding
1941 *Goenoeng Kidoel-rapport; Onderzoek naar de voeding en voedingstoestand der bevolking in het Regentschap Goenoeng Kidoel (Djokjakarta) in 1938-1941.* Batavia: Instituut voor Volksvoeding. [Mededeeling 5.]

Joosten, J.H.L.
1941 'Ontwikkeling en problemen van de bemoeienis van den Landbouwvoorlichtingsdienst met de bodembescherming in West Java', *Landbouw* 17-12:1063-80.

Junghuhn, F.W.
1850 *Java, deszelfs gedaante, bekleeding en inwendige structuur.* Amsterdam: Van Kampen.

Kievits, J.H.
1893 'Het nut van terrassen', *Tijdschrift voor Nijverheid en Landbouw in Nederlandsch Indië* 45:280-303, 46:136-49, 263-70.

Konto River Project
1990 *Lesti watershed and South Malang area basic inventory*. Malang: Konto River Project/DHV Consultants.

Meijer Ranneft, J.W.
1915 'Volksverplaatsing op Java', *Tijdschrift voor het Binnenlandsch Bestuur* 49:59-87, 165-84.

Mohr, E.J.C.
1938 Bodemkundige schets van Java, 1:2,000,000, in: *Atlas van Tropisch Nederland*. Batavia: Topografische Dienst van Nederlandsch-Indië.

Muslich Mustadjab
1991 'Annual crop cultivation', in: Muslich Mustadjab, Salyo Sutrisno, Umar Burhan, T. van Rheenen and J.W. Nibbering, *Resource allocation by farming households in the South Malang Limestone Area*. Malang: Integrated Research and Training Project (INRES), Universitas Brawijaya. [Internal project document.]

Napitupulu, B.
1968 'Hunger in Indonesia', *Bulletin of Indonesian Economic Studies* 9:60-70.

Nibbering, J.W.
1991a 'Crisis and resilience in upland central Java; A case-study in Gunungkidul District', in: J. Hardjono (ed.), *Indonesia; Resources, ecology, and environment*, pp. 104-132. Singapore: Oxford University Press.
1991b *Hoeing in the hills; Stress and resilience in an upland farming system in Java*. [PhD thesis, Australian National University, Canberra.]
1993 'Rats and droughts; Stability in a changing upland farming system in Java', *The Asian Geographer* 12-1/2:19-32.

Ossewaarde, J.G. and S.J. Wellensiek
1946 'Capita selecta uit de algemeene plantenteelt', in: C.J.J van Hall and C. van de Koppel (eds), *De landbouw in den Indischen archipel*, Vol. 1, pp. 83-200. 's-Gravenhage: Van Hoeve.

Palte, J.G.L.
1989 *Upland farming on Java, Indonesia; A socio-economic study of upland agriculture and subsistence under population pressure*. Amsterdam: Koninklijk Nederlands Aardrijkskundig Genootschap, Utrecht: Geografisch Instituut Rijksuniversiteit Utrecht. [Nederlandse Geografische Studies 97.]

Rheenen, T. van
1995 *Farm household level optimal resource allocation; An explorative study in the limestone area of East Java*. [PhD thesis, Landbouwuniversiteit Wageningen]

Roche, F.C.
1988 'Java's critical uplands; Is sustainable development possible?', *Food Research Institute Studies* 21-1:1-43.

Ruthenberg, H. (ed.)
1980 *Farming systems in the tropics*. Oxford: Clarendon Press.

Sajise, P.E.
1984 'Plant succession and agroecosystem management', in: A.T. Rambo and P.E. Sajise (eds), *An introduction to human ecology research on agricultural systems in South East Asia*. Los Baños: University of the Philippines.

Schuitemaker, B.
1949 'Maatregelen tot het behoud van de bodem in het inheemse landbouwbedrijf op Java', *Landbouw* 21-4/5:153-77.

Selosoemardjan
1962 *Social changes in Jogjakarta*. Ithaca: Cornell University Press.

Smiet, A.C.
1990 'Forest ecology on Java; Conversion and usage in a historical perspective'. *Journal of Tropical Forest Science* 2-4:286-302.

Solichin Abdul Wahab
1991a *The history of desa Putukrejo, as told by pak Sadikin, former village secretary*. Malang: Integrated Research and Training Project (INRES), Universitas Brawijaya. [Internal project paper.]
1991b *The history of desa Kedungsalam, as told by pak Mardiwijoyo and pak Sarmin*. Malang: Integrated Research and Training Project (INRES), Universitas Brawijaya. [Internal project paper.]
1996 *How farmers cope; Case-studies of decision-making in six farm households in the south of Malang, East Java*. [PhD thesis, Landbouwuniversiteit Wageningen.]

Stuers, F.V.A. de
1847 *Gedenkschrift van den oorlog op Java, van 1825 tot 1830*. Uit het Fransch vertaald door H.M. Lange. Amsterdam: Müller.

Terra, G.J.A.
1958 'Farm systems in South East Asia', *Netherlands Journal of Agricultural Science* 9:157-82.

Timmer, M.
1961 *Child mortality and population pressure in the D.I. Jogjakarta, Java, Indonesia; A social-medical study*. [PhD thesis, Vrije Universiteit, Amsterdam].

Topografisch Kantoor
1904 *Detailbladen 1:25.000 der topografische kaart der Residentie Banjoemas*. Blad XV-e. Batavia: Topografisch Kantoor.

Topografische Inrichting
1860 *Topografische kaart der Residentie Banjoemas 1:100.000*. Den Haag: Topografische Inrichting.

Uhlig, H.
1980 'Man and tropical Karst in Southeast Asia', *GeoJournal* 4-1:31-44.

Valkenburg, S. van, and J.Th. White
1923 'Enkele aanteekeningen omtrent het Zuidergebergte (G. Kidoel)', *Jaarverslag van den Topografischen Dienst in Nederlandsch-Indië* 19:127-44.

Versteeg, W.F.
1857 *Kaart van de Residentie Djocjakarta*. Batavia: Van Haren, Namen en Kolff. [Scale 1:235,000.]

Verwijk, J.
1888 'De gaga-bouw voor de afkondiging van de ontginningsordonnantie', *Tijdschrift voor het Binnenlandsch Bestuur* 1:191-4.

Veth, P.J.
1896-1907 *Java, geografisch, ethnologisch, historisch*. Haarlem: Bohn. 4 vols.

Vink, G.J.
1926 'Bedrijfseconomie van den bevolkingslandbouw', in: C.J.J van Hall and C. van de Koppel (eds), *De landbouw in den Indischen archipel*, Vol. 1, pp. 239-85. 's-Gravenhage: Van Hoeve.

Wiersum, K.F.
1980 'Erosie, plattelandsontwikkeling en bos op Java', *Landbouwkundig Tijdschrift* 92-9:338-45.

PETER BOOMGAARD

Hunting and trapping in the Indonesian archipelago, 1500-1950

Introduction

The history of resource use in Indonesia has always been focused on agriculture. To a much lesser extent attention was paid to livestock raising and forest use. Hunting, however, has been neglected entirely by historians. This may be partly explained by a scarcity of historical sources, but also by the fact that the importance of hunting, as a major claim to the use of natural resources, has been underestimated.

In this article I am ultimately interested in the environmental impact of hunting and trapping. This varies according to time and place, and an assessment of the impact of hunting should be preceded by a discussion of the factors responsible for this variation. Population growth and the influence of the state, both indigenous and colonial, are factors that come readily to mind. So do changes in demand for the various proceeds of the hunt, and changes in hunting technology. Cultural factors, including religious ones, also influenced hunting behaviour.

This article is not much more than a somewhat embellished research agenda, exploring a number of issues that are related to hunting and its impact on the environment. It is based almost entirely on library research, and I certainly cannot claim to have seen all the relevant books and articles. This applies particularly to the voluminous literature on the Outer Islands. It may also be expected that a thorough search in Indonesian, Dutch, and British archives will yield more data. Nevertheless, there is a possibility that the results will be disappointing, given the fact that the big trading companies (EIC, VOC) and the colonial governments were not all that interested in data on hunting.

Population growth

It seems a truism to state that the intensity of hunting influences the density of game, which implies that higher human population densities led to a drop in the density of game animals. Observations regarding the Indonesian archipelago over the last 75 years or so seem to confirm this

Elephant hunters in Sumatra; photograph by M. Mazaraki, ca 1905 (KITLV photo collection 35784).

statement. Nevertheless, it looks as if this may not be true for earlier periods of Indonesian history. Most – but not all – of what follows in order to explain why early population growth may not have resulted in lower densities of game, is based on speculation, but that is the stuff hypotheses are made of.

Tropical rain forests do not seem to be habitats preferred by humans. Around 1600, when Western sources began to shed more light on these regions, the heavily forested areas around the equator were virtually uninhabited. Even nowadays, population densities in these areas are low. Part of the explanation is that the soils of the areas concerned are relatively infertile owing to leaching, that there is not much undergrowth owing to a lack of sunlight at ground level, and that the available nutrients are stored in the upper layers of the forest canopy. As long dry or cold periods are usually absent, storage of carbohydrates in roots, bulbs, and tubers and, to a somewhat lesser degree, in nuts and seeds, are less necessary than in drier or colder climes. So not much food fit for humans was to be found on and under the forest floor. As the hunter-adventurer Whitney put it, 'nothing living really "abounds" in the Malaysian jungle except leeches' (Whitney 1905:155).

What applies to humans also goes for the larger herbivorous mammals. Owing to this scarcity of ground-dwelling herbivorous mammals, carnivorous and omnivorous ground-dwelling mammals were also scarce. There were various mammals, to be sure, but they tended to be arboreal. There were periodic lean periods when animals went hungry. Many animals existed at very low population densities and probably suffered local extinction during exceptionally lean periods.[1]

Another factor may have been that the humid tropics are particularly conducive to parasite survival. Many well-known, and during certain historical periods worldwide human and animal diseases, had their starting-point in tropical climates. Even nowadays, many worm infections (schistosomiasis, helminthiasis, tapeworm) are mostly tropical afflictions. Arboviruses (arthropod-borne viruses, carried by mosquitoes, ticks or other arthropods) almost invariably have as their natural hosts creatures of the tropical jungles or swamps (monkeys, rodents, birds). As Burnet and White have it: 'When man enters one of these ecosystems which have been established over millennia of evolution he does so at his own risk.'[2]

Humans did enter the humid tropical forests. I am not referring to the earliest humans (*Homo erectus*), who may have come to the archipelago as early as 1.7 million years ago, but to the Australoid *Homo sapiens*, in evidence from perhaps 40,000 years ago, and the Southern (or Indo-Malaysian) Mongoloid *Homo sapiens*, who started to arrive in the archipelago some

1 Hutterer 1983:179; Rambo 1985:38-40; Harris and Ross 1987:54; Headland 1987; Cohen 1989:75-6; Simmons 1989:58; Whitmore 1990:64-5; Headland and Bailey 1991.

2 Burnet and White 1972:69, 248-9; Jones 1981:6, 31; Cohen 1989:8.

5,000 to 4,000 years ago. It is now generally accepted that the Southern Mongoloids were already agriculturists when they replaced (or merged with) the hunter-gatherer Australoids in large parts of the archipelago. Nowadays, it is only in eastern Indonesia (Moluccas, eastern Lesser Sundas, Irian Barat (New Guinea)) that the Australoids survive as a recognizable type (Bellwood 1985).

Although, for the reasons given above, these early humans did not multiply quickly, multiply they did. This led to the creation of arable lands, grasslands, savannahs, and secondary forests. They thus created ecotones (transitional zones between major biotic communities) which are particularly rich in plant life and also attract a highly diverse number of animals often also interested in the planted crops. These forest fringes were the preferred habitat of, among other animals, tiger, deer, wild boar, and various monkeys and birds.[3]

One could argue, therefore, that during the earlier phases of population growth, the activities of humans, far from leading to lower animal densities, stimulated the proliferation of various species of game. This is in keeping with the so-called 'garden-hunting hypothesis', that suggests that early and present-day hunter-gatherers hunted a number of terrestrial mammals attracted by crops planted in small clearings and therefore reaching much higher concentrations than they did under natural conditions (Linares 1976; Persoon 1994:104).

The question is, of course, whether we will be able to test this hypothesis satisfactorily with historical data. In this volume, Kathirithamby-Wells presents some data for nineteenth- and twentieth-century Malaysia. Here, I will mention the few examples I have found that might point in the direction indicated above. Alfred Russell Wallace, the famous British naturalist who travelled all over the archipelago between 1854 and 1862, visited sparsely populated and densely forested Seram (Moluccas) in 1859-1860. Looking for animals in a particular forest area in Central Seram he became totally disgusted: 'I [...] have never been in a forest so utterly desert of animal life as this appeared to be'. On the other hand, some authors suggested that in certain areas of Sumatra between 1900 and 1920, numbers of deer and wild boar, and therefore of tigers, were increasing due to increased cultivation. One author argued that numbers of wild hogs and *banteng* (wild cattle (*Bos javanicus*)) were increasing in Southwest Java during the same period, when plantations were being laid out in this heavily forested region. Turning to Borneo/Kalimantan, it is remarkable that many animals, though not rare, occurred in low densities. The American hunter Hornaday, who visited Sarawak around 1880, observed that the orang-utan was only to be found in certain regions,

3 Müller 1839:29; Junghuhn 1853-54, I:363; Cordes 1874:333; Otto 1903:69; Volz 1912:376; Westenenk 1962:67-8; Seidensticker and Suyono 1980:55-9; Rambo 1985:38.

'although all those portions which are covered by lofty virgin forests seem to present the same features'. Around 1930, similar remarks were made about the clouded leopard, the rhinoceros, the *banteng*, and, again, the orang-utan in Dutch Borneo. This was attributed to intensive hunting by the indigenous population, but given the low population density this seems to be questionable. Finally, I would like to draw the reader's attention to observations, based on experiences in the game reserve Ujung Kulon (established in 1921), at Java's southwestern point. Here it was discovered that *banteng* and deer were doing badly after hunters and agriculturalist had been kept out for some ten years, because the fields covered with *alang-alang* (a coarse grass (*Imperata cylindrica*)) that used to be burned annually, were now slowly reverting to forest, thereby depriving the ungulates of their feeding grounds.[4]

Generally speaking, population growth was low in Java prior to the late eighteenth century, and in most other areas of Indonesia before 1850 or even 1900. Increased population growth after these dates certainly influenced the presence of game, not only through increased hunting pressure, but perhaps even more through the loss of habitat.

Java was already densely populated around 1800, with some 55 to 60 persons per km^2. European visitors to Java in the fifteenth and sixteenth centuries also commented upon the high population density they encountered there. It is not a coincidence that Central and East Java had a typical monsoon climate, with a pronounced dry season. Therefore, it cannot be regarded as part of the 'humid tropics', and it was probably never covered with humid tropical forests (Bellwood 1985:9). It is usually assumed that areas with a typical monsoon climate are more attractive to humans, and historically they have sustained fairly high population densities at an early stage of societal development. To this more favourable climate, Java added its fertile volcanic soils. It may come as a surprise, therefore, that such a populous island as was Java around 1800, was nevertheless still rich in game around that time. Even forty years later, when the German naturalist Franz Junghuhn was exploring Java's flora and fauna, this situation had not changed much.[5]

Part of the explanation may be the mechanism, described above, by which in humid tropical forest areas – such as western Java – the number of game increases when agriculture has created favourable niches for terrestrial mammals. An equally important factor seems to have been that Java's population, using less arable per capita then for instance contemporary

4 Wallace 1869:362; Hornaday 1885:108-9; Morin 1909:117; Brasser 1926:139; *Verslag NIVN* 1929-30:83; Heynsius-Viruly and Van Heurn 1935:48, 52, 63; *Verslag NIVN* 1936-38:172-213, 359, 385-7; Schilling 1952:100, 114-5.

5 KITLV Manuscript collection H 277 (J.A. Stützer, *Journal von die Reise nach Cheribon* (1786-7)); Radermacher and Van Hogendorp 1779:23-4; Leschenault 1811:434; Pfyffer 1838:22; Junghuhn 1853-54; De Haan 1911-12, II:808-13, III:152.

Europeans, was living concentrated in the valleys of southern Central Java and along the northcoast, thereby leaving large, often interconnected tracts of forest and grassland/savannah to game animals. It was only when this chain of 'wastelands' was increasingly broken by clearing for subsistence and plantation agriculture that wildlife became scarce.

Subsistence hunting

Early nomadic hunters hunted because they had to feed themselves and their families. We do not know what happened when the Southern Mongoloid agriculturists came to Indonesia, where nomadic foraging was then dominant. Did they transplant their agricultural customs lock, stock, and barrel? Or did they revert, partly or entirely, to a hunter-gatherer existence? Whatever happened during the earlier stages of their presence in the archipelago, by the seventeenth century the large majority had probably settled down as agriculturists, although many did not cultivate permanent fields.

In the eighteenth and nineteenth centuries, when we start to get more detailed information on the daily activities of sedentary peasants and semi-sedentary slash-and-burn agriculturists, hunting and trapping do not seem to have been major subsistence activities, although they were certainly not absent. There were, of course, exceptions, such as the nomadic Kubu and the sedentary Batak and Mentawai who were still spending quite some time hunting and trapping in the nineteenth century. This applies also to the nomadic Punan in Kalimantan, to a lesser extent to some of the sedentary Dayak tribes, and possibly to the sedentary Toraja in Sulawesi. Turning to the Australoid groups we can mention the Alifuru/Galela of Halmahera and the Alifuru of Seram as avid hunters, at least during the first half of the nineteenth century. Papuans and the people of Aru were also described as good hunters, but I have no idea how much of their time was taken up with it.

For the present discussion, the nomadic Kubu and Punan are the most interesting groups. If it is true that they arrived a few thousand years ago as agriculturists, with perhaps some hunting on the side, we must assume that at one time they reverted to full-time foraging and gave up agriculture almost entirely. One reason for this move out of agriculture could have been the possibility, mentioned earlier, that the very arrival of these agriculturists stimulated the growth of the numbers of game animals through the creation of species-rich border areas between ecological zones. In due time, this would have made hunting more rewarding. I fear that it will not be easy to test this hypothesis, given the probability that these 'reversals' may have taken place quite early.[6]

6 For a different, but somewhat comparable approach, see Hoffman 1984.

However, in terms of impact these small groups could not turn the tide of ever-increasing numbers of sedentary and semi-sedentary peasants. It may be assumed that agriculture, fishing, and livestock rearing, in addition to non-farm sidelines that provided the peasantry with some cash or with products to be bartered for food, made hunting less important. Moreover, peasants growing wet rice did not have much time for hunting. Nevertheless, hunting and trapping certainly did not disappear. In the first place, some hunting was undertaken to rid the villages of 'pests', such as wild pigs, elephants who trampled the crops, and even tigers, provided they had 'sinned' (killed humans, pets, or cattle). This form of hunting was in fact a prerequisite for survival as agriculturists.

In the second place, we should not underestimate the hunting and trapping that were undertaken in peasant societies for the provision of food, perhaps partly for family consumption, but also for the – supra-local – market. Throughout the archipelago the main victims of this form of hunting were deer, of which the large majority were Java deer or *rusa* (*Cervus timorensis*). Locally, other species were hunted, such as the barking deer or *kijang* (*Muntiacus muntjac*), and the *sambar* (*Cervus unicolor*). Only among some Dayak groups were there taboos against eating deer. My impression is that, at least in Sumatra, Java, Bali, Timor, and Southeast Sulawesi, the indigenous deer hunts were often led by professionals. Most deer hunting was actually trapping, the deer being caught in snares, traps, and pits. The Dutch called these subsistence hunters *dendeng* (dried meat) hunters, as opposed to those who hunted as a pastime. Deer hides were also an important commercial product.[7]

Local overhunting of deer occurred already in the nineteenth century. Around 1790, the indigenous companions of Clemens de Harris, Commander of Banyuwangi (East Java), rounded up 900 deer in two days. In the same region, around 1845, Junghuhn counted some 50,000 deer in eight hours. When Junghuhn reported his findings, professional indigenous hunters soon went after the deer, and at the end of the century both *rusa* and *kijang* had almost disappeared. However, when planter-cum-hunter Ledeboer turned part of his enormous estate in this area into a wildlife reserve, the few deer that had been left started to multiply quickly, and already around 1915 there were again large numbers of them.[8] In many other areas, deer were not so lucky, and their numbers went down steadily.

In eastern Indonesia, another animal much hunted for its meat (and its fur) was the cuscus (*Phalanger* spp.), a marsupial typical for the fauna to the east of the Wallace Line. According to Valentijn (ca 1700), the cuscus

[7] Valentijn 1724-26, III/2:136; Van Hasselt 1882:414; *Catalogus* 1883:93; Helfrich 1889:585; Hagen 1890:108-9; Snouck Hurgronje 1893-94, I:426; Klerks 1897:101-2; Groneman 1902:19; Brasser 1926:222; Van der Paardt 1929:55; *Verslag NIVN* 1936-38:147; Sellato 1994:65.

[8] *Bataviasche Courant* 21- and 28-9-1820; Junghuhn 1853-54, I:634; Groneman 1902:19; Bijl de Vroe 1980:116.

was already rare in Ambon. It was hunted by Christians and 'Heathens', but not by Muslims, who regarded it as *haram*. Wallace was amazed that the various species of cuscus still survived in the Moluccas, where they were a very popular item on the menu. The naturalist Henry Forbes (ca 1880) mentioned that they were caught in nooses, or tricked into coming down from the trees at night. Another naturalist, Kopstein (ca 1925) also reported that they were relentlessly hunted down by Ambonese, Alifuru and Papuans, but that their numbers did not seem to drop. They have survived until today, and they are now officially protected. However, they are still being hunted.[9]

Remaining for the moment in eastern Indonesia, a similar story could be told about turtles and megapodes, although it must be admitted that by doing so I am stretching the notion of hunting and trapping a bit. I am referring to the fact that for ages the people of Sulawesi and the Moluccas have been collecting the highly appreciated eggs of large (sea) turtles and megapodes, apparently without damaging these populations beyond repair, at least until the end of the colonial period. Several turtle species, such as the green turtle (*Chelonia mydas*), lay their eggs in pits they have dug on a beach. As they often return to the same beach every year, the eggs are not difficult to find once such a patch of beach has been discovered by humans. The people of the island of Buton (near Sulawesi) were experts in catching the female turtles just before they deposited their eggs, as such females fetched high prices on the markets of Ambon and Makassar. All species are now seriously endangered. The megapodes or mound-builders, such as the *maleo* (*Macrocephalon maleo*) of Sulawesi, are birds that also bury their large eggs, either in hot sand or in self-constructed mounds of vegetable matter. The eggs were – and are – considered as delicacies and people collected the eggs as soon as they were buried; they also caught the birds for consumption. In some areas the collection of these eggs was farmed out (*sasi*), and large numbers of them appeared each year on the market. Many megapodes are now seriously endangered, and some are locally extinct.[10]

Returning now to western Indonesia, we encounter, apart from the deer hunters, small numbers of specialized professionals. The sources on this topic mention elephant hunters, rhinoceros hunters, and even tiger and crocodile hunters. In Sumatra, the general term for such specialized indigenous hunters was *pawang* (*pawang gajah*, *pawang badak*, etc.). These people were regarded as spiritual specialists, capable of dealing with animal and forest spirits which threatened lesser mortals who would

9 Valentijn 1724-26, III/1:272-6; Wallace 1869:399; Forbes 1885:291; Kopstein 1930:66-7; K. MacKinnon 1992:116, 123-5, 156-9.

10 Valentijn 1724-26, III/1:320; Wallace 1869:270-2, 402-3; Kopstein 1930:30, 36-7; K. MacKinnon 1992:120-2, 165-6, 237; Jones et al. 1995. The turtles also laid their eggs on the beaches of Sumatra, Java, and Kalimantan (Bock 1881:26; Mohnike 1883:410).

venture into the realm of these animals. The tiger and crocodile hunters/ catchers were at the same time tiger or crocodile charmers, who were supposed to be able to let these animals do their bidding.[11]

Not only were specialized hunters also ritual specialists, hunting was also sometimes undertaken for spiritual (or ritual) purposes. This is documented for the Mentawai (off Sumatra's westcoast), the Baduy (western Java), for Bali, the Minahasa (Sulawesi), and for the Nuaulu (Seram). In some societies it even seems that all hunting is ritual in nature (Huaulu, Seram; Tanimbar Islands). This is – among other things – related to the fact that hunting, apart from being an economic activity of vital importance for many Indonesian societies past and present, also confers male identity and status. Hence the many taboos regarding sexual intercourse and contact with (pregnant, menstruating) women in general with which the hunter is confronted.[12] To the environmental historian this context opens hitherto uncharted fields of investigation, encompassing topics such as hunting versus procreation, and hunting as a status marker versus sustainable hunting. It also explains why it is very difficult to keep certain groups from hunting protected species.

The state

The early Javanese state may have influenced the presence of game in many ways. In the first place it stimulated settled agriculture, probably showing a certain preference for wet rice production. Wet rice agriculture is a labour-intensive activity, which must have kept the peasantry from spending much time on hunting and trapping. As wet rice agriculture and the use of water-buffaloes or cattle go hand in hand, the need for meat from wild animals was probably also slightly reduced.

Java was not the only island where rice-growing on permanent fields and the presence of a court went hand in hand. We can think of Aceh and the Minangkabau area in Sumatra, Banjarmasin in Kalimantan, Goa and Bone in Southeast Sulawesi, Bali, Lombok, and Bima (Sumbawa).

In its turn, Java's wet-rice-producing peasantry has been instrumental in maintaining a considerable state apparatus or, in other words, a large ruling élite. It is probably a universal phenomenon that kings and aristocrats are hunters, and the Javanese élite was no exception to this rule. The Old Javanese epic *Negarakertagama*, dated circa 1360, and reporting on the activities of the court of Majapahit, mentions a large-scale 'royal

11 Mohnike 1874:178-9, 201; Van Hasselt 1882:414; Hagen 1890:108-9; Helfrich 1889:585; Brasser 1926:88-90; Westenenk 1962:107.

12 Padt-Brugge 1866:324; Jacobs and Meijer 1891:21-2; Van der Paardt 1929:56; Kent 1989:6-7; Schefold 1991:18; Valeri 1994; Shepherd 1995:44; Ellen 1996; S. McKinnon 1996. I am indebted to James Scott, who suggested to look into this topic when commenting upon an earlier version of this paper.

chase'. Game consisted predominantly of wild boar, wild cattle (*banteng*) and deer. Early sixteenth-century Portuguese writers Tomé Pires and Duarte Barbosa praised the Javanese kings as great sportsmen and hunters and skilful horsemen, riding to hounds and using excellent birds of prey. Most of their time was said to be spent hunting.[13]

Around 1600, the princes of Banten, Jakatra, Cirebon, Gabang (western Java), Mataram, and Tuban (Central Java) were all mentioned as hunters. In the seventeenth century, the rulers of Mataram had deer parks (*krapyak*) and other game reserves (*larangan*) laid out for hunting purposes. Such reserves could also be found in the seventeenth and eighteenth centuries in western Java (Priangan, Cirebon). Many of these reserves survived into the nineteenth century. Particularly in the Priangan, deer hunts became a favoured pastime of the local aristocracy as well as a status marker. Junior members of the nobility had to show their prowess during these dangerous undertakings. In 1867, there were still five large game reserves in the Priangan, together taking up some 12,000 hectares. Here, the aristocracy went after deer, wild boar, rhinoceros, and, occasionally, tiger. Sometimes, hundreds of deer were killed during such a hunt, but there seems to have been an unlimited supply of them. Around 1870, however, rhinos were becoming scarce. Between the 1870s and the 1910s these game reserves disappeared, largely because the Priangan aristocracy had lost its special status by then, but also because the pressure of alternative indigenous and European claims to these areas was mounting. From that time onwards game was no longer abundant in these areas.[14]

Finally, the tiger rituals at the courts of Central Java in the seventeenth to nineteenth centuries should be mentioned. The Central Javanese rulers did not personally hunt tiger, but they employed professional tiger catchers in order to provide them with tigers for their rituals. This was an important element in the gradual disappearance of tigers in Central Java in the nineteenth century (Boomgaard 1994a).

Royal hunts also took place in seventeenth-century Sumatra, where we encounter the princes of Aceh, Pedir (North Sumatra), and Banten (hunting in the Lampung area of South Sumatra), predominantly in pursuit of elephants and tigers.[15] The elephants were not killed but trapped in order to have them tamed. They could then join the ranks of the royal

[13] Dames 1921, II:194; Cortesão 1944:168-76; Pigeaud 1960-63, III:57-62, IV:145-9.

[14] Junghuhn 1853-54, I:304; Andrásy 1859:57-65; Jagor 1866:162-3; Groneman 1874:13, 28-9, 170-2, 339; Couperus 1887:318; De Bie 1888; De Haan 1910-12, IV:379-80, 422-3; Keuning 1942:35-6; Foster 1943:161; Boomgaard 1993:312-3; Hoadley 1994:15, 29.

[15] Whether all these princes really took an active part in the pursuit of dangerous game remains to be seen. Malay sultans were supposed to love hunting, and therefore indigenous sources often depict them as great hunters (Andaya 1979:158, 196-7). And what to think of the female sultan of Aceh – the only example I have seen of a female ruler hunting – reported to be on an elephant hunt in 1656? (*Generale Missiven* III:92 (4-12-1656)).

elephants, of which many South and Southeast Asian rulers had large collections, even in places such as Java, where elephants had to be imported. I have seen no evidence of separate royal hunting grounds in Sumatra.[16]

Deer hunting was also an aristocratic occupation in Southwest Sulawesi (Celebes) in the seventeenth century, and this time we do find reports on special hunting reserves (*ongko*). A bag of 2,000 to 3,000 deer during a royal hunt was mentioned. The hunting grounds were still there in the nineteenth and twentieth centuries, as were the deer hunts. It is possible that both the deer and the hunting technique – although in a slightly different form – were imported from Java at an early stage. In the 1930s it was suggested that the existence of these reserves may have kept the local deer population from becoming extinct.[17]

In the nineteenth and twentieth centuries, similar deer hunts in large game reserves could be found in Banjarmasin, a coastal sultanate in Kalimantan, perhaps on Halmahera for the sultan of Tidore, and in the states of Lombok and Timor. Again, the deer and the deer hunt may both have been introduced from outside. In the case of Timor this may have been a late introduction, because two eighteenth-century reports dealing with game and hunting in Timor do not mention deer.[18]

The Dutch colonial state was, unlike the British Raj, not a hunting state. Neither was the Dutch East Indies Company (VOC) that preceded it. This does not mean, however, that VOC officials did not hunt at all. Local VOC representatives hunted from time to time, sometimes in the company of the indigenous nobility, as did soldiers employed by the VOC, because they had easy access to and were familiar with 'modern' weapons. There is even the case of a governor of Java's northeast coast, J.F. van Reede tot de Parkeler (ca 1800), who imitated the local nobility by laying out a hunting reserve, but that seems to have been exceptional. But then, Van Rede was a Dutch nobleman, and therefore more inclined to take hunting seriously than the large majority of his commoner compatriots.[19]

This situation did not change much when after 1800 the Dutch colonial state took over from the VOC, at least not for the first three quarters of the century. 'We did not hear of any Dutch gentlemen who are sportsmen [hunters]; but had Java continued in the English possession, I have no doubt it would ere this have been celebrated for its field sports', as the Englishman Jukes, visiting Java in 1844, has it. Or, shorter but equally

16 De Haan 1910-12, IV:379; Foster 1934:157; Keuning 1942:176; Cortesão 1944:167; Unger 1948:88; De Graaf 1954:58-77; Lombard 1967:143; Boomgaard 1993:314.

17 *Generale Missiven* III:215 (14-12-1658) Valentijn 1724-26, III/2:131-6; Roorda van Eysinga 1830-2, I:156; Francis 1856-59, II:8-9; Buddingh 1867, II:10; Ligtvoet 1880:140; Bock 1881:19; Van Weede 1908:292-4; *Verslag NIVN* 1936-8:147; Schilling 1952:33.

18 Van Hogendorp 1779:297; Wallace 1869:180-2, 216, 282, 319; Jacobs 1883:158-9; *Verslag NIVN* 1936-8:151; Dampier 1939:168-70.

19 KITLV Manuscript collection H 277; *Daghregister* 1644:4 and 8/2, 23/4; Heydt 1744:77-119; De Haan 1910-12, IV:244, 380.

dismissive: 'The Dutch do not hunt', to quote Whitney, who hunted in Sumatra some sixty years later. Although seen from today's perspective this would seem to be admirable behaviour on the part of the Dutch, the statements just quoted somewhat exaggerated Dutch abstinence from hunting, particularly around 1900. Shooting game by Europeans in Indonesia was not negligible, and may have increased slightly between 1800 and 1870, probably to the same degree as the number of Europeans and the influence of the Dutch colonial state expanded. The higher echelons of the colonial bureaucracy – governor-general, governor, resident – are not on record as great hunters. Lower officials did hunt from time to time, even in the line of duty, for instance when they were called upon to put a stop to the activities of man-eating or cattle-killing tigers.[20]

Most hunters, however, seem to have been planters or military men. That the military hunted is more or less self-evident. That owners or administrators of estates or plantations were often avid hunters is not hard to understand either. As their lands were frequently established in 'wild' areas, they started their hunting careers when they attempted to rid their grounds of 'pests'. Some planters had also links to the local indigenous aristocracy, were sometimes related to them, and copied their hunting behaviour. In other areas they had indigenous concubines and started local Indo-European dynasties, in which the hunting traditions were transmitted from father to son.

It is my impression that after circa 1870 hunting became much more fashionable among the Europeans in Indonesia. This was partly, and perhaps largely, a question of numbers. The Dutch empire expanded considerably but not always peacefully between 1870 and 1910, and the enlarged military presence made for more hunting. With the Agrarian Law of 1870 it had become much easier for Europeans to obtain long *erfpacht* (land lease), and many Dutchmen availed themselves of this opportunity. At the same time there was an increasing interest in obtaining land concessions for tobacco planting in Sumatra (Deli). The new estates in Java and Sumatra often had to be cleared in or near heavily forested and other uncultivated areas. This attracted a lot of 'noxious' game, and planters thus turned hunters. It seems that the new estates, particularly those in Java, attracted quite a few members of the Dutch aristocracy, who already had a hunting background before they came to the archipelago.[21]

With the general increase in the number of Dutch people who came to the Indonesian archipelago (this time also of women!) we see the development of a new 'white', imperial morality. It was directed against the slack morals of the old European colonists (drinking, concubinage, whoring, and laziness), and espoused sobriety, self-control, and physical exercise,

20 Van Rees 1863-65, II:94-7; Buddingh 1867, I:393; De Beauvoir 1873:280.

21 Hagen 1890:66-72; Van Weede 1908:136-62; Boissevain 1909:333; Morin 1909:67-9, 117; Brasser 1926:5-17; Feith 1940:76-7; Wormser 1941:190-205; Bijl de Vroe 1980:122-34.

or, in other words, 'clean living'. Hunting, a 'manly sport' par excellence, became a recognized means to stay fit and healthy, which, moreover, kept men out of mischief. The 'sunday-hunter' was born.[22] Finally, it should be mentioned that after 1900 a small number of Europeans, predominantly Indo-Europeans, turned professional elephant and tiger hunters. Summing up, it can be said that although hunting by Europeans did not reach the levels of British India, it was not absent either, and it seems to have increased considerably after 1870.

By 1910 the period of growth seems to have been over. I can think of four elements that may have been responsible for this process. In the first place, military intervention was much less important after this year, and it may be assumed that hunting by the military decreased. Secondly, we also witness the first attempt by the authorities to limit hunting by law: the first ordonnance for the protection of wild animals was published in 1909. Although the ordonnance was hardly effective outside of Java, it recognized that the state had a role to play in the protection of endangered species. Another step in this direction was the establishment of the first nature reserves in 1919. More rigorous game laws were published in 1924 (introducing hunting permits) and 1931 (forbidding the export of most wild animals or their products). Finally, in 1932 an ordonnance for the creation of wildlife reserves was put on the Statute Book (Boomgaard 1993).

What, in the third place, also slowed down some time after 1910 was the creation of European-run estates. In Java, the creation of forest reserves and the growth of indigenous agriculture slowly but surely had claimed most areas where new estates could have been staked out. In the Outer Islands similar processes operated, to which can be added that many estates had gone bankrupt because operating them on an economically sound basis had proved more difficult than had been anticipated. Finally, it should not be overlooked that, at least in Java, large game had become rare in many areas. My impression is that by the 1920s and 1930s, only wild boar was still abundantly available in most areas. European hunting had not stopped entirely at the eve of World War II, but life had been made much more difficult for serious hunters.

Both the VOC (and the EIC) and the Netherlands Indies government influenced the numbers of game in another way, namely by the promise of bounties for animals captured or killed. This was mostly directed against tigers, leopards, and crocodiles, but other animals, such as rhinos, were occasionally included. Nineteenth-century sources often suggested that the bounties did not stimulate the indigenous population to kill more tigers, but my impression is that this may not be entirely true, and that, moreover, some (Indo-)European hunters made a living of killing tigers for bounties.

22 Van Weede 1908:126; Boissevain 1909:57-8; Brasser 1926:59; Van der Paardt 1929:55.

Finally, I should mention another instance of Dutch government policy that may have influenced hunting. As we have seen earlier, Indonesian states and densely settled rice-growing areas went hand in hand. This was often at least partly the result of conscious state policies. The Dutch were no different in this respect. Throughout the existence of the VOC and that of the Dutch colonial state, the Dutch attempted to settle dispersed groups and make them grow crops on permanent fields. This was sometimes done for security reasons, and often directed at populations living in upland areas, who then had to come down to the coast. It is possible that these attempts influenced hunting behaviour, in the sense that it may have made the people concerned less dependent on hunting.

Cultural change

Cultural change is linked to indigenous state formation and the arrival of the European merchant companies. Nevertheless it rates a separate section, as it was often not the state as such that instituted these changes. Over the last 2,000 years, many Indonesians underwent the influence of Hinduism, Buddhism, Islam, and Christianity. Now it seems that Indonesians are and always have been gifted syncretists, so old beliefs were not thrown out radically in favour of the new religion.

Nevertheless, the conversion to Islam was a major watershed as regards eating pork. As I have shown above, hunting wild boar was a favourite pastime of the nobility. Anthropological and historical research has shown that the 'heathen' peoples of Sumatra, Kalimantan, the Lesser Sundas, and Irian Barat also hunted wild boar on a large scale and often continue to do so to this very day. So one may expect that boar hunting stopped in many areas of Java and Sumatra and in the coastal areas of Kalimantan, Sulawesi, and the Moluccas between, let us say, 1500 and 1600. There are numerous examples that this was indeed what happened.[23] I have to admit that it is unlikely that all recently converted Muslims all of a sudden stopped hunting and eating wild boar. It may have taken some time and probably some supportive action from orthodox Muslim rulers before the converted peoples had 'internalized' that pigs were *haram*. Sometimes the ruling group itself was far from orthodox, as witness the Priangan nobility who went on hunting – though perhaps not eating – wild boar. As late as 1917, the chief *penghulu* (highest Muslim official) of Cianjur was leading a boar hunt of a group of Priangan nobles and their Dutch guests.[24] Nevertheless, I think it is fair to say that the

23 Valentijn 1724-26, III/1:269; Heydt 1744:120; Junghuhn 1853-54, I:305; Wallace 1869:224; Couperus 1887:318; Fayle 1929:76-7.

24 Bijl de Vroe 1980:134. The *penghulu* were appointed by the Dutch colonial government; in the Priangan they were members of the Priangan nobility.

majority of the Indonesian Muslims, living in areas where conversion had taken place some time ago, abstained from hunting and eating pig.[25]

The mind boggles at the possible implications of the changes in eating habits. Historical sources should be screened carefully in order to find out whether there is any evidence of increasing numbers of pigs and their main enemies, the big cats, or whether people just killed more deer, thereby depriving tigers and leopards of their other favoured prey.

Similar stories could be told about other animals, such as monkeys and snakes. In the nineteenth century the Batak (North Sumatra), the Kubu, and the Mentawai (Siberut, off Sumatra's west coast), all non-Muslim groups, were said to eat snakes and monkeys. The Punan (Kalimantan) eat monkeys and the Papuans (Irian) eat snakes (there are no monkeys in Irian). If we look at the Javanese, however, we encounter many nineteenth-century reports stating emphatically that they would not kill snakes and monkeys if they could avoid it, let alone eat them. Yet it seems likely that, let us say before 1500, they did eat snakes. At least, this is suggested by a Chinese source, dating from the early fifteenth century, where it is said that the Javanese ate disgusting things, such as snakes, ants, insects, and worms. This tallies nicely with an Old Javanese text, the *Niti Sastra,* that urges people of rank not to eat dogs, rats, snakes, lizards, and caterpillars. I am inclined to believe that this means that such animals were eaten, because if not, the injunction would have been superfluous. The two texts do not mention monkeys, which might imply that these were not (any more?) eaten by the Javanese.[26]

In one respect, the coming of Islam did not influence local beliefs concerning animals very much. I am referring to the often-made observation that the people from Sumatra, Java, and Bali would not kill a tiger (or leopard), unless the tiger had slain a relative or killed a water-buffalo. This refusal was connected to the animist belief that people and tigers where somehow related, or the Hindu notion of reincarnation (Boomgaard 1994b). As we have seen, this belief did not influence the behaviour of the Javanese rulers.

What is not easily explained is that the, equally animist, tribal population of Kalimantan does not seem to have shared this attitude towards big cats. The Punan and other Dayak tribes hunted the clouded leopard for its skin, which was used as a war-cloak by the aristocrats. The skin was also to be found in the market-place. According to the naturalist Mjöberg, writing in 1930, 'the Borneo native has great reverence for the

25 I can think of only one major exception, namely the wild-boar hunts organized by Dutch civil servants when wild boar became a 'plague' to the peasantry. In Java such hunts were undertaken regularly since the 1850s.

26 Crawfurd 1820, II:233; Boers 1838:290; Roorda 1841-50, III:103; Andrásy 1859:70; De Beauvoir 1873:243; Mohnike 1874:198; Bock 1881:70; Van Hasselt 1882:414; Kopstein 1930:21; Van der Valk 1940:181; Mills 1970:93.

tree-tigers' soul, and has to go through a special purification rite if he, with the help of his dogs, has caused the death of one of these clouded leopards'. Such 'great reverence' did, of course, not do much for the clouded leopard.[27]

Changes in demand

Although the meat of elephants, crocodiles, and rhinos could be eaten, and was eaten by certain groups, it is highly unlikely that fulfilling the subsistence needs of the local people was the main motive of the *pawang* we encountered earlier. It is quite clear that several products of these animals – other than meat – were items of trade in the regional, supra-regional, and even international markets.

Ivory, of course, had been exported from Sumatra for ages. The Chinese, Indians, and Arabs had been buying it prior to the arrival of the Europeans. Elephants themselves were also traded, as witness the many royal elephants in Java before 1600. Ivory also played a role in the trade between Sumatra and eastern Indonesia. Here, ivory was one of the prestige goods that were – and sometimes still are – used, among other things, for bride prices and fines. This is attested from the seventeenth century onwards for the Aru, Kai, and Tanimbar Islands, Solor, and Flores. In the nineteenth century, international and regional demand was sustained. However, judging from the – admittedly somewhat abstruse – trade statistics, the trade in ivory was never important during the period 1825 to 1940. The average annual value of the ivory traded during the nineteenth century was not much more then 10,000 guilders, or probably some 100 tusks. Trade figures of 20,000 guilders or more were very rare. After 1875 the annual trade figures dropped.[28]

There was also a steady demand, dating from an early period, for rhinoceros horn, that was (and is) particularly popular with the Chinese (as an aphrodisiac and as an antidote against poison). A similar story could be told about deer antlers and deer hides.

In all these cases we were dealing with products that had been exported for ages. From time to time one could observe ups and downs in the export figures, related to price fluctuations and/or the lack of alternative sources of income. What should be regarded as a novel phenomenon – from the late nineteenth century onwards – was the enormous influence of fashion in Europe and America on hunting and trapping of certain animals in the archipelago. The most notorious examples are birds of paradise, the feathers of which started to adorn women's hats in the 1860s, with a peak

27 Bock 1881:152-3; *Catalogus* 1883:90-1; Shelford 1916:27-8; Brasser 1926:223; Mjöberg 1930:22.

28 Valentijn 1724-26, III/2:42; Roorda 1841-50, II:50, 57; Van Doren 1864:68, 83; Forbes 1885:315; Kohl 1989; Boomgaard 1996:42-77.

in demand around 1910 (see also Cribb, this volume), snakes and crocodiles, because their skins became quite fashionable for shoes and ladies' handbags in the 1920s. Feathers of the birds of paradise had been a trade item for many centuries, but it had always been a mere trickle until the end of the nineteenth century. It seems that the influence of these demand booms was limited as regards the bird of paradise, but quite disastrous for the Java snakes.[29]

Increasing numbers of Europeans came to the archipelago in the nineteenth century as collectors of specimens of what was then called natural history. This was not an entirely new phenomenon. It had started in the seventeenth century with shells, of which many rich Europeans had extensive collections in their curiosity cabinets. In the eighteenth century learned societies started to collect all sorts of preserved animals, a task that was taken over by museums in the nineteenth century. What was new in the nineteenth century was the number of people involved and the numbers of specimens collected.

In the late nineteenth and early twentieth centuries we witness other sources of demand, namely animal shows, circuses, zoological gardens, pet shops, and medical laboratories. This demand created a new group of specialists, namely professional European animal catchers, who came to Indonesia, and often remained there for one or two decades. They created local networks of indigenous people who would catch animals for them. They were interested in live animals, to which hunting methods had to be adapted, but local environmental impact was the same, as the animals concerned left the area. Tens of thousands of monkeys (specifically rhesus monkeys), snakes, and birds, and hundreds and perhaps even thousands of big cats were shipped to the United States of America and Europe. The new demand was on a much larger scale and came in addition to the already existing pet-keeping habits of the local population.

Technological change

Hunting by indigenous people before the arrival of the Europeans was done with the aid of spears, long hunting knives, blowpipes, bows and arrows, and dogs. The latter were already present in Java when the Portuguese arrived there for the first time, but it seems that in Kalimantan their introduction came much later. Bows and arrows were, one or two exceptions apart, only used by the eastern Indonesian Australoid groups, whereas blowpipes were (almost?) entirely restricted to the Indo-Malaysian Mongoloids. One wonders whether this rather rigid border between bows and arrows on the one hand and blowpipes on the other, so neatly coin-

[29] *Catalogus* 1883:96; Van Weede 1908:240-1; Kopstein 1930:79-80, 150; Boomgaard 1993:331; Swadling 1996:83-108.

ciding with the dividing line between the two Indonesian 'races', was purely a matter of cultural heritage. An alternative explanation could be that differences in fauna made for differences in weaponry. There is no historical record for a transition in the Indonesian archipelago from bows and arrows to blowpipes, as there is for the Malaysian Semang. In their case, it seems likely that the introduction of firearms had made the bow and arrow redundant.[30]

Apart from these weapons (and dogs), there were various contraptions in use to catch animals, such as nooses, snares, traps, pits, and cages with live bait. It was also quite common, mainly in the case of 'pests' (wild boar, elephants, tigers), to use poison. In fact, it is my impression that trapping (poison included) may have been more important than hunting, which may be one of the reasons why the importance of animal catching for peasant populations may have been underestimated in the literature. Trapping is neither a glamorous nor a conspicuous activity, and it does not have the cultural connotations of hunting. Nevertheless, it is my contention that it has been quite influential, both in an ecological and in a dietary sense.

When the Portuguese gave their first descriptions of Java, muskets were already in evidence, presumably matchlocks. Soon, Indonesians started to produce matchlocks themselves. In the seventeenth century, the Dutch introduced flintlocks, which must have been quite some improvement if one wanted to hunt tiger. In the nineteenth century, technological change went into higher gear, and soon rifles (with spiral grooves) were introduced, percussion locks, and finally breechloaders. However, the indigenous population, having adopted matchlocks at a very early stage, did not take easily to the later firearms, although by the middle of the nineteenth century flintlocks had become somewhat more general. A problem with breech-loading and percussion-type rifles was that the indigenous people could no longer make their own shot or bullets, but had to buy cartridges. It is therefore questionable whether nineteenth-century European hunting equipment did influence indigenous hunting practices all that much. Of course it did influence European hunting in Indonesia.

The Dutch were not always happy with the fact that Indonesians had firearms at their disposal, even the less efficient ones. Therefore, possession of firearms by Indonesians was forbidden, unless a permit had been granted. Nevertheless, during times of war or unrest, the Dutch often rounded up the guns of the indigenous population. This happened several times in Sumatra during the nineteenth and twentieth centuries, and reports sometimes stated that the number of tigers or elephants had increased because of these measures. So even the flintlocks may have made some impact.

There have been various debates surrounding the transition from bows

30 Pleyte 1891; Schebesta 1928:77; Rambo 1978; Harrison 1984:318.

and arrows, blowpipes, and spears to firearms, questioning the received wisdom that firearms had increased the productivity of the hunter, and, of course, the environmental damage done. Although those who attacked the existing orthodoxy came up with some interesting points (firearms are noisy and scare away game!), there is no doubt that the amount of meat acquired per hour spent on hunting increased when modern shotguns – from the 1970s – were used. It has been argued that this also applied to firearms dating from the nineteenth century, but that is a more difficult point to prove.[31] What I found a particularly interesting point is the fact that, although the efficacy of the modern shotgun is beyond dispute, it is also much more expensive to acquire. This means that people have to work harder in order to buy one. Therefore, it could be argued that, if the extra hours worked were added to the hunting hours, the amount of meat produced per hour remained the same. People just put in more hours and therefore had more food. If the work done in order to obtain the gun entailed environmentally damaging activities, the introduction of firearms may have had a double environmental effect.

Environmental impact

Around 1900 – but locally much earlier – large forest areas in Indonesia had been turned into *alang-alang* fields. I cannot share Michael Dove's opinion that *alang-alang* – in more than small amounts – was a product of colonial systems of agriculture. We find large expanses of *alang-alang* in the nineteenth century in locations where Europeans had seldom visited, let alone introduced new agricultural systems, such as the Batak area circa 1835, Halmahera circa 1860, Central Sulawesi circa 1900, and Irian Barat circa 1925.[32] *Alang-alang* was a product of repeated burning, and we know that hunters often burned certain areas in order to attract game with the young shoots of *alang-alang* that would spring up. These fires made inroads in the adjacent forests and prevented the regrowth of trees. I am not prepared to argue that all *alang-alang* was produced by hunters, but there is sufficient evidence for Sumatra, Java, and Sulawesi that hunters annually burned large, and expanding, areas of grassland for the hunt. Even here, however, some fields may have been created for livestock grazing, and it will not be easy for the historian to separate the two. Nevertheless, it is clear that hunting was responsible for a considerable loss of forest cover in many areas.

There were other, less visible effects. Much more research will have to be done in order to establish what the (near) disappearance of certain

[31] Rambo 1978; Hames 1979; Yost and Kelley 1983.
[32] Müller 1855:119; Wallace 1869:319-22; Adriani and Kruyt 1900; Salverda 1911; Kopstein 1930:18; Endert 1933; Dove 1986:169.

animal species did to the environment. We do know, for instance, that the increased hunting pressure on tigers led to an increase in the number of wild pig 'plagues'. The corollary of this was that overhunting of pigs made for more maneating tigers, and, eventually, drove the tigers away. There is also evidence that the massive killing of snakes for their skins led to increased problems with rats and mice. If bats – eaten by many Indonesian groups – and birds were killed on sufficiently large a scale, it might cause an increase in the number of – noxious – insects. Bats, birds, and various mammals are also responsible for the dispersal of seeds, and in that sense overhunting may threaten the very existence of the forests. I know of only one example – from Malaysia – were the disappearance of mammals did, indeed, lead to a forest breaking down, but it is unlikely that it is the only one (Whitmore 1990:78). But for most species we cannot even begin to guess what a sizeable reduction of their numbers or disappearance meant.

The Javanese literati must have been aware of the connection between the preservation of forested areas and the survival of wild animals, as witness the well-known fable, to be found in the Old Javanese *Niti Sastra* and quoted by Crawfurd, of the pact between the forest and the tiger to protect each other. When they fell out, the forest, no longer protected by the tiger, was cut down, and the tiger, no longer able to hide himself in the forest, was killed (Crawfurd 1820, II:33). It is, indeed, plausible to assume that the removal of large and potentially dangerous game (tiger, *banteng*, rhino, elephant, bear) made it easier to cut down a forest. At the same time ongoing deforestation facilitated killing these animals.

Conclusions

What was the upshot of all this? There is no doubt that, in the long run, hunting pressure increased with the population. However, this was, as we have seen, not the only factor to be taken into consideration. State formation, pressure from indigenous and colonial states on nomadic and semi-nomadic people (in order to get them settled), protection of nature, conversion to another religion, demand from outside of Indonesia, increased importance of plantations, technological change – they all seem to have influenced hunting behaviour.

There were some surprises, such as the fact that around 1800 densely settled Java was still rather rich in game. On the other hand, the situation in Kalimantan around 1880 and 1930 was less rosy than could have been expected from such a sparsely settled area, without plantations or European hunters. Both phenomena may be partly explained by the garden-hunting hypothesis.

Another surprise may have been to many that so much hunting, and particularly trapping, went on in basically sedentary peasant societies. Some of this was done by specialists, but a lot of trapping was probably done during a lull in the agricultural cycle.

It is less surprising that several species of game became increasingly endangered after circa 1870, through the combined effects of more and better firearms, higher population growth rates, growing numbers of plantations, and increased European hunting pressure. The last two factors slowed down after circa 1910, but clearings for subsistence agriculture continued at a fairly high pace. In the 1920s and 1930s, conservation of nature and a much more restrictive policy regarding deforestation, including the prohibition of burning grasslands for hunting, came to the aid of the beleaguered game animals, sometimes (rhino and *banteng* in Java) just in time to save them from extinction.

Finally, the resilience of certain species, such as the relentlessly hunted deer and cuscus, was highly impressive.

Generally speaking, hunting (and trapping) was seldom the only factor responsible for the (near) extinction of a species. It operated in combination with loss of habitat, natural phenomena such as long droughts, epizootics, and, finally, when populations became very small, inbreeding. Nevertheless, it is hard to imagine that these animals would have fared equally badly if there had been no hunting at all.

References

Adriani, N. and A.C. Kruyt
1900 'Van Posso naar Mori', *Mededeelingen van wege het Nederlandsche Zendelinggenootschap* 44:135-214.

Andaya, B.W.
1979 *Perak, the abode of grace; A study of an eighteenth-century Malay state.* Kuala Lumpur: Oxford University Press.

Andrásy, Graf E.
1859 *Reise des Grafen Emanuel Andrásy in Ostindien: Ceylon, Java, China, Bengalen.* Pest: Geibel.

Bataviasche Courant
1816-27 *Bataviasche Courant.* Batavia: Landsdrukkerij.

Beauvoir, Comte de
1873 *Voyage autour du monde: Australie, Java, Siam, Pékin, Yeddo, San Francisco.* Paris: Plon.

Bellwood, P.
1985 *Prehistory of the Indo-Malaysian archipelago.* Sydney: Academic Press.

Bie, H. de
1888 'Hertenjachten in de Preanger', *Tijdschrift voor het Binnenlandsch Bestuur* 1:159-79.

Bock, C.
1881 *The head-hunters of Borneo; A narrative of travel up the Mahakkam and down the Barito [...].* London: Sampson Low, Marston, Searle and Rivington.

Boers, J.W.
1838 'De Koeboes', *Tijdschrift voor Neêrland's Indië* 1, II:286-95.

Boissevain, Ch.
1909 *Tropisch Nederland; Indrukken eener reis door Nederlandsch-Indië.* Haarlem: Tjeenk Willink.

Boomgaard, P.
1993 'Protection de la nature en Indonésie pendant la fin de la période coloniale (1889-1949)', in: J. Pouchepadass (ed.), *Colonisations et environnement*, pp. 307-44. Paris: Société Française d'Histoire d'Outre-mer.
1994a 'Death to the tiger! The development of tiger and leopard rituals in Java, 1605-1906', *South East Asia Research* 2:141-75.
1994b 'Man-eating; The deadly encounter between people and tigers/leopards in Indonesia, 1600-1950'. Paper prepared for the ASEASUK annual conference, London.
1996 (with the assistance of R. de Bakker), *Forests and forestry 1823-1941.* Amsterdam: Royal Tropical Institute. [Changing Economy in Indonesia 16.]

Brasser, J.C.
1926 *Jacht op groot wild in Nederlandsch Oost-Indië; Verhalen uit verre wildernissen.* Zutphen: Thieme.

Buddingh, S.A.
1867 *Neêrlands Oost-Indië; Reizen over Java, Madura [...] gedaan gedurende het tijdvak van 1852-1857.* Amsterdam: Van Kesteren. 3 vols.

Burnet, M. and D.O. White
1972 *Natural history of infectious disease.* Cambridge: Cambridge University Press.

Bijl de Vroe, C.L.M.
1980 *Rondom de Buitenzorgse troon; Indisch dagboek, 1914-1919.* Edited by M. Schouten. Haarlem: Fibula van Dishoeck.

Catalogus
1883 *Catalogus der afdeeling Nederlandsche koloniën van de Internationale Koloniale en Uitvoerhandel Tentoonstelling 1883 te Amsterdam; Groep II.* Leiden: Brill.

Cohen, M.N.
1989 *Health and the rise of civilization.* New Haven/London: Yale University Press.

Cordes, J.W.H.
1874-5 'Herinneringen aan Sumatra's Westkust; Bijdrage tot de kennis van de boschstreken over een gedeelte van Sumatra', *Tijdschrift Nederlandsche Maatschappij ter Bevordering van Nijverheid,* 15(derde serie):289-396, 16:1-48.

Cortesão, A. (ed.)
1944 *The Suma Oriental of Tomé Pires; An account of the East, from the Red Sea to Japan, written in Malacca and India in 1512-1515 [...].* London: Hakluyt Society. 2 vols. [Works Hakluyt Society, second series 89-90.]

Couperus, G.W.
1887 'Le gibier et la chasse à Java', *Revue Coloniale Internationale* 2:299-334.

Crawfurd, J.
1820 *History of the Indian archipelago [...]*. Edinburgh: Constable. 3 vols.

Daghregister
1888-1931 *Daghregister gehouden int Casteel Batavia vant passerende daer ter plaetse als over geheel Nederlandts-India, 1624-1682*. Edited by J.A. van der Chijs et al. Batavia: Landsdrukkerij, 's-Gravenhage: Nijhoff. 31 vols.

Dames, M.L. (ed.)
1918-21 *The book of Duarte Barbosa; An account of the countries bordering on the Indian Ocean and their inhabitants, written by Duarte Barbosa and completed about the year 1518 A.D.* London: Hakluyt Society. 2 vols. [Works Hakluyt Society, second series 44 and 49.]

Dampier, W.
1939 *A voyage to New Holland*. Edited by J.A. Williamson. London: Argonaut Press.

Doren, J.B.J. van
1864 'De Tenimber-eilanden, ten Zuid-Westen van de Keij-eilanden', *Bijdragen tot de Taal-, Land- en Volkenkunde* 11:67-101.

Dove, M.R.
1986 'The practical reason of weeds in Indonesia; Peasant vs. state views of Imperata and Chromolaena', *Human Ecology* 14:163-90.

Ellen, R.
1996 'Cuscus and cockerels; Killing rituals and ritual killings among the Nuaulu of Seram', in: S. Howell (ed.), *For the sake of our future; Sacrificing in Eastern Indonesia*, pp. 263-81. Leiden: Research School CNWS.

Endert, F.H.
1933 'Eenige resultaten van de boschverkenning in de Buitengewesten', *Tectona* 26:391-422.

Fayle, C.E. (ed.)
1929 *Voyages to the East Indies; Christopher Fryke and Christopher Schweitzer*. London: Cassell.

Feith, Jhr. J.
1940 *Sport in Indië*. Deventer: Van Hoeve.

Forbes, H.O.
1885 *A naturalist's wanderings in the Eastern archipelago*. New York: Harper.

Foster, W. (ed.)
1934 *The voyage of Thomas Best to the East Indies 1612-14*. London: Hakluyt Society. [Works Hakluyt Society, second series 75.]
1943 *The voyage of Sir Henry Middleton to the Moluccas 1604-1606*. London: Hakluyt Society. [Works Hakluyt Society, second series 88.]

Francis, E.
1856-59 *Herinneringen uit den levensloop van een 'Indisch' ambtenaar van 1815 tot 1851*. Batavia: Van Dorp. 3 vols.

Generale Missiven
1960-88 *Generale Missiven van Gouverneurs-Generaal en Raden aan Heren XVII der Verenigde Oostindische Compagnie*. Edited by W.P. Coolhaas [vol. 1-8] and J. van Goor [vol. 9]. 's-Gravenhage: Nijhoff. 9 vols. [RGP, Grote Serie, 9 vols.]

Graaf, H.J. de
1954 *De regering van panembahan Sénapati Ingalaga*. 's-Gravenhage: Nijhoff. [KITLV, Verhandelingen 13.]

Groneman, J.
1874 *Bladen uit het dagboek van een Indisch geneesheer*. Groningen: Wolters.
1902 *Op het Jang-gebergte in Oost-Java*. Zutphen: Thieme.

Haan, F. de
1910-12 *Priangan; De Preanger-Regentschappen onder het Nederlandsch bestuur tot 1811*. Batavia: Bataviaasch Genootschap van Kunsten en Wetenschappen. 4 vols.

Hagen, B.
1890 'Die Pflanzen- und Thierwelt von Deli auf der Ostküste Sumatra's; Naturwissenschaftliche Skizzen und Beiträge', *Tijdschrift Koninklijk Nederlandsch Aardrijkskundig Genootschap* 7(tweede serie):1-280.

Hames, R.B.
1979 'A comparison of the efficiency of the shotgun and the bow in neotropical forest hunting', *Human Ecology* 7:219-52.

Harris, M. and E.B. Ross
1987 *Death, sex, and fertility; Population regulation in preindustrial and developing societies*. New York: Columbia University Press.

Harrison, T.
1984 'The prehistory of Borneo', in: P. van de Velde (ed.), *Prehistoric Indonesia; A reader*, pp. 297-326. Dordrecht/Cinnaminson: Foris. [KITLV, Verhandelingen 104.]

Hasselt, A.L. van
1882 *Volksbeschrijving van Midden-Sumatra*. Leiden: Brill. [Vol. 3 of P.J. Veth (ed.), *Midden-Sumatra; Reizen en onderzoekingen der Sumatra-Expeditie 1877-1879*.]

Headland, T.N.
1987 'The wild yam question; How well could independent hunter-gatherers live in a tropical rain-forest ecosystem?', *Human Ecology* 15:463-91.

Headland, T.N. and R.C. Bailey
1991 'Introduction; Have hunter-gatherers ever lived in tropical rain forest independently of agriculture?', *Human Ecology* 19:115-22.

Helfrich, O.L.
1889 'Bijdrage tot de geographische, geologische en ethnographische kennis der Afdeeling Kroë', *Bijdragen tot de Taal-, Land- en Volkenkunde* 38:517-632.

Heydt, J.W.
1744 *Allerneuester geographisch- und topographischer Schau-Platz von Africa und Ost-Indien [...]*. Willhermsdorff: Tetschner.

Heynsius-Viruly, A. and Jhr. F.C. van Heurn
1935 'Overzicht van de uit Nederlandsch-Indië ontvangen gegevens [...], met biologische aanteekeningen omtrent de betreffende diersoorten', *Mededeelingen Nederlandsche Commissie voor Internationale Natuurbescherming* 10(supplement):25-77.

Hoadley, M.C.
1994 *Towards a feudal mode of production; West Java, 1680-1800.* Singapore: ISEAS.

Hoffman, C.L.
1984 'Punan foragers in the trading networks of Southeast Asia', in: C. Shrire (ed.), *Past and present in hunter-gatherer studies*, pp. 123-49. Orlando: Academic Press.

Hogendorp, W. van
1779 'Beschryving van het eiland Timor, voor zoo verre het tot nog toe bekend is', *Verhandelingen van het Bataviaasch Genootschap der Konsten en Wetenschappen* 1:273-306.

Hornaday, W.T.
1885 *Two years in the jungle; The experiences of a hunter and naturalist in the Malay peninsula and Borneo.* New York: Sribner's.

Hutterer, K.L.
1983 'The natural and cultural history of Southeast Asian agriculture; Ecological and evolutionary considerations', *Anthropos* 78:169-212.

Jacobs, J.
1883 *Eenigen tijd onder de Baliërs; Eene reisbeschrijving met aanteekeningen betreffende hygiène, land- en volkenkunde van de eilanden Bali en Lombok.* Batavia: Kolff.

Jacobs, J. and J.J. Meijer
1891 *De Badoej's.* 's-Gravenhage: Nijhoff.

Jagor, F.
1866 *Singapore-Malacca-Java; Reiseskizzen.* Berlin: Springer.

Jones, D., R. Dekker and C. Roselaar
1995 *The Megapodes, Megapodiidae.* Oxford: Oxford University Press.

Jones, E.L.
1981 *The European miracle; Environments, economies, and geopolitics in the history of Europe and Asia.* Cambridge: Cambridge University Press.

Jukes, J.B.
1847 *Narrative of the surveying voyage of H.M.S. Fly [...] during the years 1842-1846 [...].* London: Boone. 2 vols.

Junghuhn, F.W.
1853-54 *Java, zijne gedaante, zijn plantentooi en inwendige bouw.* Tweede verbeterde uitgave. 's-Gravenhage: Mieling. 3 vols.

Kent, S.
1989 'Cross-cultural perceptions of farmers as hunters and the value of meat', in: S. Kent (ed.), *Farmers as hunters; The implications of sedentism*, pp. 1-18. Cambridge: Cambridge University Press.

Keuning, J. (ed.)
1942 *De tweede schipvaart der Nederlanders naar Oost-Indië onder Jacob Cornelisz. van Neck en Wijbrant Warwijck 1598-1600.* Vol. 3. 's-Gravenhage: Nijhoff. [Werken Linschoten-Vereeniging 44.]

Klerks, E.A.
1897 'Geographisch en ethnographisch opstel over de landschappen Korintji, Serampas en Soengai Tenang', *Tijdschrift voor Indische Taal-, Land- en Volkenkunde (TBG)* 39:1-114.

Kohl, K.-H.
1989 'Der Elefant mit den sieben Stosszähnen; Heiratsregeln im Lewolema-Gebiet und ihre Begründung im Mythos', in: H. Zinser et al. (eds.), *Foedera Naturai; Klaus Heinrich zum 60. Geburtstag,* pp. 157-68. N.p.: Königshausen und Neumann.

Kopstein, F.
1930 *Zoologische Tropenreise; Mit Kamera und Feldstecher durch die indo-australische Tierwelt.* Batavia: Kolff.

Leschenault [de la Tour]
1811 'Notice sur un lac d'acide sulfurique [...] dans la province de Bagnia-Vangi [sic] [...]', *Annales du Muséum d'Histoire Naturelle* 18:425-46.

Ligtvoet, A.
1880 'Transcriptie van het dagboek der vorsten van Gowa en Tello met vertaling en aanteekeningen', *Bijdragen tot de Taal-, Land- en Volkenkunde* 28:1-259.

Linares, O.F.
1976 ' "Garden Hunting" in the American tropics', *Human Ecology* 4:331-49.

Lombard, D.
1967 *Le Sultanat d'Atjéh au temps d'Iskandar Muda 1607-1636.* Paris: EFEO.

MacKinnon, K.
1992 *Nature's treasurehouse; The wildlife of Indonesia.* Jakarta: Gramedia.

McKinnon, S.
1996 'Hot death and the spirit of pigs; The sacrificial form of the hunt in the Tanimbar Islands', in: S. Howell (ed.), *For the sake of our future; Sacrificing in Eastern Indonesia,* pp. 337-50. Leiden: Research School CNWS.

Mills, J.V.G. (ed.)
1970 *Ma Huan; Ying-yai sheng-lan; 'The overall survey of the ocean's shores' [1433].* Cambridge: Cambridge University Press. [Works Hakluyt Society, extra series 42.]

Mjöberg, E.
1930 *Forest life and adventures in the Malay archipelago.* London: Allen and Unwin.

Mohnike, O.
1874 *Banka und Palemban; Nebst Mittheilungen über Sumatra im Allgemeinen.* Münster: Aschendorff.
1883 *Blicke auf das Pflanzen- und Thierleben in den Niederländischen Malaienländern.* Münster: Aschendorff.

Morin, H.
(ca 1909) *Unter der Tropensonne; Streifzüge aus Java, Sumatra und Ceylon.* München: Ifaria.

Müller, S.
1839 *Zoölogie.* Leiden: La Lau. [Verhandelingen over de natuurlijke geschiedenis der Nederlandsche overzeesche bezittingen 2.]
1855 *Reizen en onderzoekingen in Sumatra tusschen de jaren 1833 tot 1838.* 's-Gravenhage: Fuhri.

Otto, E.
1903 *Pflanzer- und Jägerleben auf Sumatra.* Berlin: Süsserott.

Paardt, Th. van der
1929 'Onbewoond Noord-West Bali', *Tijdschrift Koninklijk Nederlandsch Aardrijkskundig Genootschap* 46(tweede serie):45-77.

Padt-Brugge, R.
1866 'Beschrijving der zeden en gewoonten van de bewoners der Minahassa [...] 1679', *Bijdragen tot de Taal-, Land- en Volkenkunde* 13:304-31.

Persoon, G.A.
1994 *Vluchten of veranderen; Processen van verandering en ontwikkeling bij tribale groepen in Indonesië.* [PhD thesis, Rijksuniversiteit Leiden.]

Pfyffer zu Neueck, J.J.X.
1838 *Schetsen van het eiland Java en deszelfs onderscheidenen bewoners.* Amsterdam: Van Kesteren.

Pigeaud, Th.G.Th.
1960-63 *Java in the 14th century; A study in cultural history.* The Hague: Nijhoff. 5 vols. [KITLV, Translation Series 4.]

Pleyte, C.M.
1891 'Sumpitan and bow in Indonesia', *Internationales Archiv für Ethnographie* 4:265-81.

Radermacher, J.C.M. and W. van Hogendorp
1779 'Korte schets van de bezittingen der Nederlandsche Oost-Indische Maatschappye', *Verhandelingen van het Bataviaasch Genootschap der Konsten en Wetenschappen* 1:1-70.

Rambo, A.T.
1978 'Bows, blowpipes and blunderbusses; Ecological implications of weapons change among the Malaysian Negritos', *Malayan Nature Journal* 32:209-16.
1985 *Primitive polluters; Semang impact on the Malaysian tropical rainforest ecosystem.* Ann Arbor: Museum of Anthropology, University of Michigan. [Anthropological Papers Museum of Anthropology, University of Michigan 76.]

Rees, W.A. van
1863-65 *Herinneringen uit de loopbaan van een Indisch officier.* 's-Gravenhage: Visser, Leiden: Van den Heuvell en Van Santen. 4 vols. in 2 parts.

Roorda van Eysinga, P.P.
1830-32 *Verschillende reizen en lotgevallen van S. Roorda van Eysinga.* Amsterdam: Van der Heij. 4 vols.

1841-50 *Handboek der land- en volkenkunde, geschied-, taal-, aardrijks- en staatkunde van Nederlandsch Indië.* Amsterdam: Van Bakkenes. 3 vols. in 5 parts.

Salverda, A.Th.L.
1911 'Boschbeheer in de Buitenbezittingen', *Tectona* 4:667-86.

Schebesta, P.
1928 *Among the forest dwarfs of Malaya.* London: Hutchinson.

Schefold, R.
1991 'Harmonie und Luftgewehre', *Entwicklung/Développement* 34:17-9.

Schilling, T.
1952 *Tijgermensen van Anai; Jacht op grof wild.* Amsterdam: Meulenhoff.

Seidensticker, J. and Suyono
1980 *The Javan tiger and the Meru Betiri reserve; A plan for management.* Gland, Switzerland: World Wildlife Fund.

Sellato, B.
1994 *Nomads of the Borneo rainforest; The economics, politics, and ideology of settling down.* Honolulu: University of Hawaii Press.

Shelford, R.W.C.
1916 *A naturalist in Borneo.* London: Fisher Unwin.

Shepherd, J.R.
1995 *Marriage and mandatory abortion among the 17th-century Siraya.* Arlington: American Anthropological Association. [American Ethnological Society Monograph Series 6.]

Simmons, I.G.
1989 *Changing the face of the earth; Culture, environment, history.* Oxford: Blackwell.

Snouck Hurgronje, C.
1893-94 *De Atjéhers.* Batavia: Landsdrukkerij, Leiden: Brill. 2 vols.

Swadling, P.
1996 *Plumes from paradise; Trade cycles in outer Southeast Asia and their impact on New Guinea and nearby islands until 1920.* Boroko: Papua New Guinea National Museum.

Unger, W.S. (ed.)
1948 *De oudste reizen van de Zeeuwen naar Oost-Indië, 1598-1604.* 's-Gravenhage: Nijhoff. [Werken Linschoten-Vereeniging 51.]

Valentijn, F.
1724-26 *Oud en nieuw Oost-Indiën.* Dordrecht: Van Braam, Amsterdam: Onder de Linden. 5 vols.

Valeri, V.
1994 'Wild victims; Hunting as sacrifice and sacrifice as hunting in Huaulu', *History of Religions* 34:101-31.

Valk, A.C. van der
1940 *Vangen en jagen in Sumatra's wildernis.* Amsterdam: Meulenhoff.

Verslag NIVN
1912-38 *Verslag Nederlandsch-Indische Vereeniging tot Natuurbescherming.* [11 issues.]

Volz, W.
1912 *Nord-Sumatra; Bericht über eine [...] in den Jahren 1904-1906 ausgeführte Forschungsreise. Band II: Die Gajoländer*. Berlin: Reimer.

Wallace, A.R.
1869 *The Malay archipelago; The land of the orang-utan, and the bird of paradise; A narrative of travel, with studies of man and nature*. New York: Harper.

Weede, Jhr. H.M. van
1908 *Indische reisherinneringen*. Haarlem: Tjeenk Willink.

Westenenk, L.C.
1962 *Waar mens en tijger buren zijn*. Den Haag: Leopold.

Whitmore, T.C.
1990 *An introduction to tropical rain forests*. Oxford: Clarendon Press.

Whitney, C.
1905 *Jungle trails and jungle people; Travel, adventure and observation in the Far East*. New York: Scribner's.

Wormser, C.W.
[1941] *Drie en dertig jaren op Java. Deel II: In de bergen*. Amsterdam: Ten Have, Bandoeng: Vorkink. [First edition 1928.]

Yost, J. and P.M. Kelley
1983 'Shotguns, blowguns and spears; The analysis of technological efficiency', in: R.B. Hames and W.T. Vickers (eds), *The adaptive responses of native Amazonians*, pp. 189-224. New York: Academic Press.

JEYA KATHIRITHAMBY-WELLS

Human impact on large mammal populations in Peninsular Malaysia from the nineteenth to the mid-twentieth century

Introduction

Statistical information on the status of the elephant *(Elephas maximus)*, tiger (*Panthera tigris*), *seladang* or wild cattle (*Bos gaurus*) and the two species of rhinoceros (*Rhinoceros sondaicus* and *Dicerorhinus sumatrensis*) in Peninsular Malaya during the colonial period is virtually absent. A mammal survey, in emulation of one covering India in the 1920s, was proposed in 1927 by H.C. Robinson, Oxford-trained zoologist and Director of the Federated Malay States Museums.[1] The task was never undertaken due to lack of government interest, necessary financial provisions and technical staff. Even data for the 1960s, the earliest available (see Table 2), are based on estimates.

The density of the rain forest reduces visibility such that estimating populations has constituted a major problem for naturalists and hunters. A further factor contributing to the rarity of sighting big mammals in the Peninsula is the rain forest's smaller carrying capacity for large herbivores compared to the savannahs of mainland Southeast Asia. Like the flora, the fauna of the rain forest is rich in species but with relatively low numbers in each species. It was believed among the aboriginal Senoi Semai, for example, that there was never more than one pair of tigers on one hill (Dentan 1968a:25). Elephant and deer, for example, have been found in far greater abundance in the grasslands of northern Thailand and Cambodia than in the forests of Malaya.[2]

The diversity of mammal species of equatorial and island Southeast Asia results partly from vegetational changes during cold phases (Ice

* I would like to thank Dr D.R. Wells and Lord Cranbrook for comments and references. The Netherlands Institute for Advanced Study, Wassenaar, provided facilities and research support, including the assistance of Drs Yves de Roo who helped in the preparation of maps and tables.

1 T. Hubback, 'Game Protection in Malaya', Undated Memo in no. 29, CO 717/ 62/4

2 On the grasslands of Thailand and Cambodia see Wharton 1968:132-3, 138-9; McNeely and Wachtel 1988:27-30.

Seladang herd with a young bull as sentinel (Foenander 1952:opposite p. 64).

Ages) of the Pleistocene period, reaching back some two million years ago. Such periods exposed much of the Sunda Shelf and generated a continental type of climate with stretches of savannah, interspersed by forest refugia. It is during such a period of eustatic recession of the sea that continental mammals such as the tiger, rhinoceros, deer, and wild cattle are believed to have invaded the archipelago. Conversely, the restoration of high forest cover after 12,000 BP, by reducing grass and browse, eliminated most of these large ungulates from all or parts of the surviving Sunda region. It resulted, for example, in the Holocene extinction in Borneo of the one-horned Javan rhinoceros (*Rhinoceros sondaicus*) (Cranbrook 1986; Wells 1988:170).

Shifting cultivation

This article is a tentative study of how aboriginal and early pioneering activity in the Peninsula boosted secondary vegetation and grassland simulating Pleistocene cold periods, to the advantage of large mammal populations. By opening up the forest canopy on a rotational cycle and increasing the regeneration of secondary vegetation, shifting cultivation helped to expand the food resources of herbivores. It added to the natural browse and graze in forest gaps, in areas of periodic flooding of lowland river banks, wind-throw sites, and landslips. In contrast, it can be argued that during the twentieth century new patterns of land use reversed the process of assisting large mammal population growth.

Shifting or *ladang* cultivation has been generally associated in modern times with the Proto-Malay and Senoi (Temiar and Semai) Aborigines in the foothills and valleys of the main range (Wyatt-Smith 1958:140; see also Map 2). The *belukar* (secondary forest) surrounding such clearings (important for their survival) further increased food and browse for big mammals (Rambo 1982:269-70). Shifting cultivation was also practised by the Malays and became particularly widespread during the early phase of nineteenth-century settlement and political instability preceding colonial rule. A Malaysian historical geographer, Zaharah Haji Mahmud, goes further to say that even during stable conditions, shifting dry rice cultivation coexisted with irrigated (*sawah*) rice as 'a supplementary and complementary adjunct' within a compatible environment. The Muda and Merbok valleys, the upper valleys of Kedah, the Berang valley of Terengganu, the Krian valley of Perak and the Upper Tembeling of Pahang are named as important examples (Zaharah 1992:313-4). Though shifting cultivation was a widespread neolithic practice, it would appear that some aboriginal groups acquired it from the Malays. Among the Temiar of Ulu Kelantan, for example, shifting cultivation is described as *jerami*, a word borrowed from the Malays, meaning straw or stubble (Cole 1959:263).

With opportunities available for cash cropping, wage labour, and forest-produce collection, many Malays were disinclined towards settled agriculture. Even with the introduction of land titles, Malays found loopholes in the law for continuing shifting cultivation. It was often practised under Temporary Occupation Licences (TOL) which under the Land Code Enactment of the FMS of 1937 imposed no restriction on agricultural land use (Wyatt-Smith 1958:148). In Negeri Sembilan, Malays and Javanese permitted to occupy the Labu Reserve were found, in 1932, to be engaged in *jagung* (maize), *ubi kayu* (cassava), *keladi* (taro), and *nanas* (pineapple) growing, with no sign of cultivating wet rice.[3] In Pahang, apart from restricted areas of wet rice in densely populated areas and *paya* or swamp rice planted behind river levees in the Temerloh District, dry rice cultivation was universal (Cant 1972:34-6). The TOL is observed to have wrought havoc in the Tembeling valley. Large stretches of virgin forest, particularly along the river banks up to the Kuala Tahan, are reported to have been cleared by Malay settlers, leaving behind a trail of *belukar*. Around Kuala Tahan, the Malays evidently preferred to live like Senoi, clearing a fresh *ladang* each year, earning supplementary income from selling rattan collected by the Orang Asli (Aborigines) to the Chinese and tapping *jelutong* (*Dyera costulata*). In the event of the Land Office refusing to issue new TOL, they had the old licences renewed annually while effectively occupying new blocks of land each year.[4] In addition to *ladang*, *tenggala* or ploughed permanent 'dry' fields, involving cultivation of river terraces every 4-5 years, between fallow periods, was popular in Pahang. Practised in Ulu Jelai and downstream from Kuala Tembeling to the north of Pekan, it added extensive areas of grassland and scrub, providing excellent graze (Hill 1977:158).[5]

Malay and Orang Asli shifting cultivation continued to pose a serious problem well into the period preceding World War II and had significant implications for mammal populations. Abandoned clearings (*ladang*) produced ideal browse in the form of young shoots of ruderal plants, attracting concentrations of herbivores (Medway 1965:223-34). Large-scale immigration and settlement during the nineteenth and early twentieth centuries also brought widespread subsistence agriculture, involving a large variety of crops attractive to big mammals. As an editorial of the English Press, *The Straits Times* observed in 1936:

> 'It is forgotten or conveniently overlooked that these animals dislike the smell of man and normally prefer to give his habitation a wide berth. It is because they come under influences that are not normal that they leave the jungle for the paddy field, *kampung* garden and rubber estate. In their seasonal round of the jungle

3 State Forest Officer, Negeri Sembilan and Melaka to Adviser, Forestry, CF 287/32, Hutan Malaya (Arkib Negara, Kuala Lumpur).

4 T. Hubback to H.P. Mead, 29-7-1935, no. 1, Ad. F. 294/35, Hutan Malaya (Arkib Negara, Kuala Lumpur).

5 In 1901 paddy acreage in Pahang constituted 27,400 with over 11,500 acres under *ladang* and *tenggala* cultivation. See Lim 1977:61.

> elephants are influenced to a great extent by fruit-bearing trees, especially the durian and frequently they find a clearing established where a year before there was only jungle. In these clearings they strike food which is new to them but much to their liking – sweet potato for example – and in a short time the crop has disappeared.'[6]

The attraction of animals to clearings is, in fact, advanced as one of the reasons for frequent shifting, under ideal conditions, among aboriginal populations engaged in cultivation. In the absence of a specialized pastoral culture in Peninsular Malaya, the aboriginal and Malay populace relied for supplementary protein on *seladang*, deer (*rusa*/sambar or *Cervus unicolor* and barking deer *Muntiacus muntjac* or *kijang*). The Orang Asli consumed also wild pig (*Sus scrofa* and *Sus barbatus*) and other mammals. The relation between mammals and man under these conditions was symbiotic and the constant threat of wildlife to crops was accepted as a normal hazard which merely enhanced the incentives for hunting.

In addition, the pioneer phase of settlement brought the widespread practice of shifting cultivation to other communities. The abundance of land and the lack of administrative constraints during the early phase of development facilitated the rapid spread of shifting cultivation among Chinese and Indonesian immigrants who cultivated pepper, gambier, and cassava in Johore, Negeri Sembilan, Selangor, and Perak. After the soil had become exhausted they moved on to fresh plots, leaving a trail of *lalang* (*Imperata cylindrica*) and *belukar* (Jackson 1965:80; Sadka 1968:345-7).

Tigers and early cash-cropping

Preceding British intervention during the last quarter of the nineteenth century, more than 90% of the peninsula is believed to have been covered by forest. Direct confrontation between tiger and man was precipitated by widespread clearance and shifting cultivation in areas of pioneer settlement. By the middle decades of the nineteenth century, Singapore, Johore, and Negeri Sembilan saw forest clearance for the shifting cultivation of cassava, pepper, and gambier by Chinese. Some of the crops cultivated, and especially the clearings abandoned, which brought the extension of scrub and herbivorous plants, had the effect of increasing the food resources for wild hog and deer, attractive to tigers (Oxley 1849:594-5). According to experienced hunter H.H. Banks, 'a stretch of *lalang* grass country, broken up with ravines and dotted with patches of secondary jungle and backed with forest covered hills' was where the wild pig afforded the tiger plentiful food (Banks 1923:255).

In Singapore, where previously the tiger is thought to have been

6 'Wild Life', Editorial, *The Straits Times*, 22-5-1936.

absent, or at best scarce, settlement following British occupation in 1819 appears to have attracted a fair number from the mainland (Earl 1837: 358). Tigers are believed to have turned man-eaters only when provoked or when deprived of traditional sources of food. Such a condition soon prevailed with the combined effect probably of the loss of habitat to cultivation and the hunting of wild pig and deer by new settlers. Though pushing back the forest was seen as a means of reducing the threat posed by wildlife, there was a reverse trend, at least during the early period. The open country, with patches of *lalang* or abandoned clearings, appears to have provided the ideal conditions for their predatory activities. By the mid-century the island is believed to have harboured about twenty pairs of tigers which took a regular toll on lives, particularly the Chinese gambier and pepper planters who lived in isolated pockets of clearings in the woods. By an 1840 estimate, the mortality from attack by tigers averaged one per month (Turnbull 1972:153), rising towards the end of the decade to an average of 200 per annum. Twenty years later, the estimate of one per day, even allowing for a margin of exaggeration, suggests a dramatic increase. The problem was sufficiently serious to favour complete felling of the jungle as a means of flushing out the tigers (Cameron 1965:90-102). Encouraging cultivation was seen as a practical means of achieving this. It was the combined effect of hunting and the systematic clearance of scrub for settlement and permanent cultivation which brought the final disappearance of the tiger menace in Singapore.

In the Peninsula, tigers were common in areas around abandoned *ladang* where deer were present, and posed a serious danger to Chinese shifting cultivators. According to one mid-nineteenth-century observer, upwards of fifty Chinese in a single village in Johore were carried off by tigers (Oliphant 1969:28). Even during the 1930s, tigers were reported to be on the increase in Ulu Temengor, Perak, where hill-rice cultivation attracted *rusa* (Wild Life Commission of Malaya 1932, I:219). Besides the 'game killer' and 'man-eater' associated with human habitation, there was also the 'cattle killer', living in jungles close to settlements (Mayer 1922:173-4). The reliance of tigers on cattle raised by the Malay peasantry is attested to by the larger number of tigers found in Kelantan and Terengganu (Burgess 1954:57-8). Besides open areas around settlements, tigers were also known to have frequented pathways and jungle-fringed roadsides where cattle grazed. Disappearance of travellers along roads was not uncommon in Negeri Sembilan where tigers and leopards are described at the turn of the century as being 'numerous'. Bullock-cart drivers sought safety by travelling in convoys while passing through tiger-infested areas (Rathborne 1898:44-5, 93).

The seladang

In the case of wild cattle (*gaur* or *seladang*), Charles Wharton has linked its occurrence in Southeast Asia with the widespread activities of slash-and-burn which helped maintain, if not create, the vegetation on which it was dependent (see Wharton 1968:107-10; Dentan 1968b:89-92). According to Malaya's first game warden and an expert on the *seladang*, Theodore Hubback, the Peninsular *seladang* (*Bos gaurus*) had followed the 'cleared path of settlers from the mainland' (Hubback 1938). The incidence of *seladang* has been related to abandoned aboriginal clearings (Wharton 1968:147-8; Banks 1934:163-4; Kitchner 1961:198) which provided their favourite browse in the form of *lalang* shoots and grass, in addition to young bamboo and ferns (Pelandok 1938:63). The adjoining forest at the same time constituted an essential part of the habitat for providing cool shade away from the sun and the essential retreat. 'It is unlikely', wrote Medway, 'that any significant portion of the *seladang* population of Malaya exists far beyond the limits of peripheral human settlement' (Medway 1965:224-5).

Preceding the spread of large-scale settlement in the lowlands, it would appear that the *seladang* took advantage of natural conditions of lowland grass. In 1874 there was reported to be a large herd of as many as 150 *seladang* at Kampung Gajah, north of Telok Anson, on the banks of the Perak River, within the flood plain (Burns 1976:101). Sir George Maxwell noted during his service in Pahang at the beginning of this century that *seladang* were comparatively common in the downriver districts (Maxwell 1960:71). But pressure in the lowlands contributed to enhancing the importance of sub-montane slopes as the *seladang*'s most favoured habitat,[7] though it seldom occurred above circa 300 metres elevation. Its distribution in the present century has been along the foothills of the main range and the upper course of rivers in Pahang, Terengganu, and Kelantan (Medway 1965:224, 229 Figure 1). These areas, between about 150 metres and 450 metres where the Orang Asli practised shifting cultivation, proved particularly congenial to the *seladang* (see Hubback 1905:48)[8] which probably derived its name from the word *ladang*. The low density of human population (65-130 per km^2) in such areas (see Map 3) allowed for shifting cultivation which, when abandoned, provided a constant supply of food and suited the shy habits of the *seladang*. Howard Banks, a planter and hunter who later turned conservationist and Honorary Game Warden for Negeri Sembilan, located *seladang* in the hills leading to Kuala Pilah

7 'The Late Capt. Syers', *The Selangor Journal* 5-23(23-6-1897):372.

8 There were many reasons why hillsides were favoured for dry rice. Hillsides afforded easier clearance whereby the weight of the felled trees on the upper slopes brought down vegetation on the lower slopes. It was also conducive to the upward draft necessary when the hillside was fired before planting. After planting, the slopes provided better drainage and light for the crops.

and Jelebu and noted *padang* (open grasslands) and *belukar* to be their favourite haunts (Banks 1934:163-5). Conditions were particularly ideal, such as in Ulu Temengor at Sungai Piah and Grik, where mineral licks occurred in areas of aboriginal shifting cultivation.[9] Another area of Orang Asli shifting cultivators and salt licks, important for *seladang*, was Ulu Serting in Negeri Sembilan (Hubback 1905:265). Keen of scent, sight, and hearing, the *seladang* has been described as 'by far the most difficult to shoot of all the big game in the Peninsula'.[10] When provoked, however, the *seladang* could prove dangerous. As George Maxwell, a keen hunter, observed: 'A wounded seladang was probably the most dangerous of all big game. Not only will it charge, but it will hunt down a man with the utmost vindictiveness and tenacity of purpose' (Maxwell 1960:70). In 1897, the Police Commissioner of Selangor, H.C. Syers, probably the most skilled Peninsular big-game hunter at the time, died from wounds sustained after being gored by a *seladang* in Pahang.[11]

The elephant

Though basically a forest animal and of wider distribution than the *seladang*, Asian elephant herds are 'best adapted to alluvial floodplain habitats, of which grasslands are an important component' (Olivier 1978a: 146). Like the *seladang*, elephants are attracted to human habitation which when reduced in the Peninsula, resulted in their restriction largely to hill areas (Olivier 1980:315). Found in herds up to fifteen, they flourished within 'disturbed and successional vegetation types' produced either by natural events or human activity (Olivier 1980:316), feeding on grass, roots, creepers, bamboo shoots, wild ginger, wild banana, *mengkuang* (*Pandanus* spp.), and rattan. They occurred in open country (*padang*) and in *belukar*, as well as in swamps thick with the edible food-producing *kelubi* palm (*Eleiodoxa conferta*) (Burgess 1954:67). The attraction of elephants to secondary growth was noted by forester and hunter E.C. Foenander,[12] in the territory around Kampung Cenor, in the Temerloh area on the banks of the Pahang River. The primary forest in the area had been destroyed by the Great Flood of 1926, and was replaced by 'a mass of creepers, vines and rattans' which became a popular feeding ground for elephants where they could be hunted down (Foenander 1939:59). Within the wide area of secondary forest, elephant populations were concentrated around salt licks essential for their survival (Mohd. Khan 1967:21).

9 Wild Life Commission of Malaya 1932, II:218. For more on mineral licks and mammals see p. 15.
10 'Some notes about Seladang', *The Selangor Journal* 1-24(11-8-1893):383.
11 'The Late Capt. Syers', *The Selangor Journal* 5-23(1897):372.
12 Foenander joined the Forest Service in Malaya in 1920 as a ranger. He worked largely in Pahang and, in 1946, rose to the position of Assistant-Conservator of Forests.

Human activity enhanced natural conditions and elephants habitually invaded cultivation associated with Malay *kampung* and Orang Asli settlements, feeding on coconuts and the fronds of palm, green paddy, and cassava.[13] They relished especially durians, found in Orang Asli orchards within the forest.[14] Their partiality to a whole range of cultivated foods, including sweet potato, fruit, young coconut palm, and rubber, meant their greater proliferation on the forest edge with, as of the mid-nineteenth century, the extension of Malay village and plantation settlements.[15] Both Jeram and Sungai Gumut (a tributary of the Bernam) in Selangor were described as areas of elephant concentration because of the presence of durian and other fruit, associated with human settlement (Gullick 1975: 39, 45).

The rhinoceroses

Drawn as much as the elephant to secondary lowland forest and salt licks, the rhinoceros often shares the same wallow with the elephant (Talbot 1969:206). A field study on the two-horned Sumatran rhinoceros (*Dicerorhinus sumatrensis*) conducted in the Sungai Dusun Reserve in Selangor during 1965-1966 identified more than forty species of plants on which the rhino fed, out of which over half the number were noted to be secondary forest or fringe vegetation found in clearings, land-slips, river banks, and wind-throws (Strickland 1967:11). The rhino thrived in disturbed habitat as much as the *seladang*, tiger, tapir, deer, and wild pig. Their occurrence in marginal areas may account for their early depletion through being hunted for their prized horns. Believed to provide a remedy for almost any ailment, the horns were carefully graded for the market by the Malays into four types: the *sumbu lilin* (wax-coloured horn), the *sumbu nila* (the blue horn), the *sumbu api* (the flame-coloured horn), and the ordinary *sumbu hitam* (black horn) (Maxwell 1969:11). The commercial incentives were such that rhinos continued to be hunted down despite the belief prevalent amongst the indigenous people that they were *kramat* or holy. By the early decades of the century they were confined to isolated fringe locations. Thus, in 1928-1929 Siamese are reported to have come over the border to shoot rhino in Belum, Upper Perak (Wild Life Commission of Malaya 1932, II:219).

Unlike the two-horned rhino which occurs at all altitudes, its one-horned counterpart, the Javan rhinoceros (*Rhinoceros sondaicus*) was confined to lower elevations which made it even more vulnerable to hunters, resulting in its rapid extermination in the Peninsula. Rare on the west

13 V.G. Bell, Conservator of Forests, Sungai Patani, Kedah, 21-1-1827, Hutan Malaya, CF 172/27:4.

14 They are said to have eaten durians by stamping them and picking out the seeds with their trunk. According to Malay reports, they are said to have eaten them whole as well, though this is an unlikely story (Hubback 1905:63, 72-3).

15 *What Malaysia is saying* 1936:73.

coast, there is no definite evidence of its occurrence on the east coast of the Peninsula (Loch 1937:185-6). Despite its rarity by 1899, its alleged danger led to the killing by Sir George Maxwell of the notorious Pinji rhino of Kinta, Perak (Maxwell 1960:10-1; Loch 1937:137-8). By 1925 they were described as verging on extinction throughout the Peninsula.[16] Ironically, probably the last one-horned rhino in the Peninsula, the 'Sungai Lampang Rhino' in Ulu Bernam, Perak, was shot in 1932 for the British Museum, by pioneer conservationist and Honorary Game Warden Theodore Hubback (Stevens 1968:116).

Pioneer settlement and large mammals in Selangor

The effect of human habitation, especially communities practising shifting cultivation, on large mammal populations is demonstrated in the case of Selangor where, at the turn of the century, development and deforestation took place more rapidly than in any other state. With the rush of immigrants, its population more than doubled between 1891 and 1901. In the latter year a total population of 168,700, including some 40,000 Malay-Indonesians and 110,000 Chinese (Dodge 1980:463, Table 5) were engaged predominantly in mining and shifting cultivation of cassava, gambier, pepper, and coffee. Mining activity is reported to have left large areas of wasteland covered in *lalang*, prone to fires during the dry season and subject to a continuous process of regeneration.[17] Chinese gambier and pepper plantations once abandoned or left untended are described as 'almost choking with lalang'.[18]

During the nineteenth-century pioneer phase, except in Kedah, Negeri Sembilan, and Kelantan where *sawah* was predominant, there was extensive clearance for *ladang* cultivation. In Selangor, although many Minangkabaus from Negeri Sembilan took up permanent rice cultivation in swamp lands, large numbers of other Sumatrans (from Mandailing and Kampar especially) cultivated subsistence hill rice, in combination with earning cash income from coffee, tobacco, and gambier cultivation.[19] The clearance of forests to provide the growing population and the mining industry with wood and fuel was also significant. With the export of forest produce almost negligible, revenues from forest administration were drawn largely from the issue of passes to Chinese for woodcutting and charcoal burning.[20] Large tracks of grassland which from 'the distance looked like

16 See Table 1, 'Sightings of Elephants, *Seladang* and Rhinoceros in various parts of the Peninsula'.
17 Memo W.H. Treacher, Acting Resident General, 30-5-1900, CO 273, 263/40314, 14-11-1900.
18 'Through "No Man's Land" to Kuala Lumpur', *The Selangor Journal* 2-10 (26-1-1894):152.
19 *The Selangor Journal* 2-10 (26-1-1894):149.
20 Memo W.H. Treacher, Acting Resident, 30-5-1900, CO 273, 263/40314, 14-11-1900.

close-cropped meadows'[21] were noted, particularly around Kuala Lumpur and the state was the first to face a fuel crisis. In 1887 shifting cultivation was prohibited in Selangor (Lim 1977:66) but there was no way of either extending the law to cover the aboriginal communities, or of apprehending others engaged in the practice. By 1886, what forest remained was described as predominantly secondary (Ridley 1896:445-6). Bamboo was also prevalent in old mining sites.[22] These habitats, generated by deforestation during the early phase of development, provided rich feeding ground especially for elephants, *seladang* and deer and would appear to have boosted the number of big mammals.

In contrast to primary forest where hunting was generally difficult, the grassland and open terrain of Selangor became one of the prime hunting grounds in the Peninsula. We are informed that at the turn of the century the open country around Batu Caves in Selangor was a special haunt of the *seladang* (Rathborne 1898:154). Despite the elusive habits of the *seladang*, up to 1893 as many as twenty-two had been shot by European hunters.[23] American naturalist William Hornaday, collecting zoological specimens in Selangor in the 1870s, noted the high grass on the banks of Sungai Buloh trampled by elephants and was informed that wild cattle also were plentiful in some of the surrounding areas. Similarly, in Batu, to the north of Kuala Lumpur, both *seladang* and elephants, 'browsing off the tufts of grass', were found to be abundant (Hornaday 1879:125, 127-8; Rathborne 1898:154-5).

In the wake of settlement and development in the nineteenth century, the Orang Asli populations in the Peninsula as a whole are believed to have moved further away from the coast into the surrounding hills (Evans 1915:10).[24] In the case of Selangor, the interface between Orang Asli and non-aboriginal populations proved, up till about the beginning of the present century, mutually beneficial. Aborigines of Proto-Malay and Senoi origins, particularly those of the hill country of Ulu Langat and Ulu Selangor, cleared the forest edge, whether for timber or the cultivation of

21 'Verandah in Kuala Lumpur', *The Selangor Journal* 3-23 (26-6-1895).

22 It was reported in 1894 that in 2,000 acres of previously abandoned mining land in Rasa valley, some 600 acres were reworked by Lok Yew and others who were able to make a saving of some 40% on capital invested through putting the bamboo growing on it to good use. 'Notes of the Resident's visit to the Districts in Selangor', *The Selangor Journal* 2-13 (9-3-1894):205. W.W. Skeat, travelling in the northeast of the Peninsula during the turn of the century, noted that extensive areas of bamboo were favourable to wild *seladang*.

23 'Some notes about seladang', *The Selangor Journal* 1-24 (11-8-1893):383. This figure presumably covers the period from British intervention in 1874 to 1893. According to Hubback *seladang* were never heard of in Selangor. It could be that on the disappearance of the *seladang* in Selangor by the end of the century hunters resident in the state shot animals on the borders of Perak and Pahang (Hubback 1905:48).

24 In 1964 there were 4,560 Orang Asli in Selangor, comprising 3,290 Proto-Malays and 1,270 Senoi out of a total of 53,000 for the peninsula (Carey 1976:11).

tapioca, *taro* and bananas.[25] Such activity left abandoned sites of grass and secondary vegetation which attracted deer, wild pig, elephants, and *seladang*. Elephants were also reputedly partial to durians from the much valued Orang Asli orchards. The boost in the mammal population, in turn, substantially increased game for the Orang Asli.

Before the advent of guns, the Malays and Orang Asli hunted big game by digging pitfalls and suspending a wooden spear or *penerut* over a game track (Hubback 1905:20). The indigenous spring trap (*pelantek* or *belantek*) of wood and bamboo used by the aborigines of Selangor could also take animals of a considerable size.[26] By the turn of the century, however, with increased cash earnings, the Selangor Aborigines took advantage of the greater availability of guns. Often borrowing guns from each other, they hunted valuable big game such as elephant, *seladang*, and rhinoceros, reserving the use of blowpipes, traps, and spears for smaller mammals such as the monkey, deer, and pig. According to the following song of the Senoi of Ulu Langat, deer were killed with a throwing stick.

Rise up, oho! and take your squailer
Take your squailer and stab the [...] deer[27]
Bear him now homewards and cook my [...]deer,
And when you have cooked him, quarter my[...]deer
And give unto each an equal portion (Skeat and Blagden 1906, II:150).

Another song suggests that gun and shot were reserved for the greater profit and excitement of hunting down the few remaining rhinos:

The Rhinoceros that roams and climbs the mountains
What skill have I to strive with the Rhinoceros? [...]
The Rhinoceros follows me.
I then take a gun and shoot the Rhinoceros.
The bullet has hit him. The Rhinoceros has fallen.
See that yea singe [...] and quarter the Rhinoceros,
And give to eat a little to every one;
But sell the horn to the Chinese foreigners.

Skeat and Blagden note that the Negrito had traditionally hunted rhino by piling leaves on the animal and setting fire to it while it wallowed in the mud (Skeat and Blagden 1906, II:202-3); but this cannot be corroborated. In the case of the *seladang* and other big game, the Orang Asli of Selangor, like their counterparts in the other states, were often engaged as trackers by white hunters. The prospect of a *seladang* hunt normally

25 See, for example, Report by C.N. Maxwell, Acting District Officer, Ulu Langat, *The Selangor Journal* 5-25 (6-8-1896):398.

26 'The Sakais of Selangor', *The Selangor Journal* 1-6 (2-12-1892):100.

27 Skeat and Blagden (1906) incorrectly describe the deer as 'roe-deer', found only in the temperate regions.

created excitement among the Orang Asli as it left them with plenty of meat after the European hunters claimed the heads as trophies.

Human-elephant interactions

Competition between man and big mammals within settled habitats is particularly evident in the case of elephants. According to Olivier who made a special study of elephants in the Peninsula in the 1970s, 'Because of the strong ecological relationship between elephants and floodplains, the development activities of man have had an adverse impact on elephants from the very start' (Olivier 1980:316). Elephants require a larger natural range than other mammals, calculated at between 40 and 80 km^2 per average herd of 8-10 animals and are therefore the first animals to be affected by development (Mohd. Khan 1967:21; Stevens 1968:118; Olivier 1980:317). During the eighteenth century, Kedah and Perak were major exporters of elephants to the Indian subcontinent. The disturbed habitat of early settlement in the floodplains would have provided ideal conditions allowing for their easy capture for export. The combination of trapping and economic expansion during this period which in the case of Kedah is described to have been 'at its most intense' (Zaharah 1970:97), would have resulted in the long-term loss of habitat for elephants. They were subsequently confined – particularly in Kedah which was given to extensive rice cultivation – to the hill regions of the interior, but still in large concentrations, assisted by Orang Asli shifting cultivation.[28]

Though highlands under natural conditions offer climax forest with a low elephant-carrying capacity, the activities of shifting cultivation by both Malay and aboriginal populations which resulted in disturbed habitat, greatly enhanced elephant densities (Olivier 1980:317). In Ulu Muda, Kedah, where there was a high concentration of elephants, the area was occupied by a small population of Negrito (Semang) Kensiu and Kenta (Schebesta 1928:15; Carey 1976:30). Though originally hunter-gatherers, their contact with settled populations on the coast led to their gradual adoption of supplementary *ladang* (Endicott 1979:170-1). When Schebesta visited the Negrito on the Siong River, north of Baling in the 1920s, he found them at the edge of the great stretch of Malay ricefields and rubber plantations stretching into the interior (Schebesta 1928:204-5). In Upper Perak, supplementary shifting cultivation by hunter-gatherer Negrito, sometimes shared with the Malays, would also have provided suitable country for elephants. The exemption of Orang Asli from government land rent encouraged Malays in these parts to continue cultivating hill *padi* by employing Negritos (Evans 1915:188).

Unlike the Negrito who lived within ecotones of forest areas fringing

[28] This is borne out in their distribution in modern times. See Map 4.

on Malay *kampung* and Chinese settlements, the Senoi were engaged in more extensive swidden agriculture on hill slopes. The Temiar among the Senoi, in particular, are described as 'sophisticated and highly successful' shifting cultivators. Living often in isolated highland clearings, at altitudes between 150 metres and 600 metres (Hill 1977:167) and within proximity of primary forest, they cultivated hill paddy, some every year, others every two years, using abandoned secondary forest serving as sources of important plant and animal resources (Rambo 1982:269; Noone 1949:1). Ulu Plus, calculated to have had a population of about 2,000 Temiar in the 1930s (Noone 1936:41),[29] and Cemor-Korbu (north of Kuala Kangsar) with about 800 Temiar, were important elephant country. The Korbu, Plus, and Temengor areas were also rich in bamboo, an important food source for elephants, in addition to the *buloh sewor* (*Bambusa wrayi*), growing on higher altitudes, utilized by the Temiar for the manufacture of blowpipes (Noone 1955:1-2).

Temiar familiarity with elephants, arising from their shared habitat, made them expert *gembala* or elephant drivers. From the large number of wild elephant herds which occurred in the region, many were trapped and tamed. Owned widely by Malays, particularly by the ruler and chiefs (MacNair 1878:16,18; Kelham 1928:49), they constituted the main beasts of burden and were commonly used in Perak for the conveyance of tin from the mines. In the Plus area where the elephants caused heavy damage to rubber, the Temiar assisted in a successful programme of elephant control initiated by planter F.J. Davy (Noone 1936:49). In Grik and Temengor in Upper Perak, also inhabited by Temiar, elephants were used for conveying timber from the forests.

As in Perak, Malay and Senoi cultivation resulting in *belukar* may be associated with the large concentrations of big game, including elephants, found in the upper reaches of the Tembeling, Lower Serting, Jelai, and Tekai in Pahang. Here, human activity in the form of *jelutong* tapping may also have added to disturbed habitats favourable to big mammals. In Serting, for example, extensive *jelutong* tapping, especially by Chinese, was accompanied by the felling of trees and the clearing of *jelutong* tracks (Browne 1932:234-9).

Areas of aboriginal shifting cultivation often overlapped with mineral licks which elephants shared with other mammals. There were the hot springs in the Temiar country of Ulu Nenggiri, across the Main Range, in Kelantan, and salt licks in Sungai Rening and Sungai Misong, down the Telom, which contributed to the abundance of game, including the elephant, *seladang* and tiger (Noone 1936:49; Clifford 1898:5-6) The Grik-

29 With the aboriginal population at the beginning of the century estimated at around 20,000, compared to 16,700 in 1969, the Temiar population would have been proportionately larger as suggested by the figures of Skeat and Blagden. See Annadale and Robinson 1903:29; Skeat and Blagden 1906, I:577; Carey 1976:11.

Temengor area boasted the Peninsula's largest concentration of salt licks and drew elephants and other mammals such as *seladang*, *rusa*, tapir, and wild pig which frequented the old cultivation plots along river banks (Stevens 1968:62).

Development and the mammal 'nuisance'

A significant boost to the mammal population by the turn of the century is suggested by the sharp rise in big-game hunting in the Peninsula among an expanding European community. Opportunities for hunting were evidently enhanced by the unprecedented extension of secondary forest and herbaceous vegetation resulting from the early phase of forest clearance following British intervention. At the time of the early European explorations of the Peninsula forest, it was deemed a veritable hunting ground for the 'sportsman who has exhausted every variety of game to be found in the jungles of India' (Oliphant 1969, I:24). European planters, civil servants, and foresters in the Peninsula soon generated a counterpart imperial tradition nurtured on the subcontinent. Hunting skills were put on a par with character building associated with sport and considered an honourable diversion from the drudgery of professional business. H.C. Syers was much admired by contemporaries for his hunting prowess. Apart from Hubback, there were others such as Howard Banks, planter-turned-game warden of Negeri Sembilan. The cult of trophy hunting was developed into a fine art. The size of the largest *seladang* horns taken by Syers of Selangor and J.B.M. Leech of Perak at the end of the century matched the record of G.P. Sanderson, a leading big-game hunter in India.[30] In elephant shooting, the 41.25 kg pair of tusks bagged by William Hay, admired for his 'cool courage, daring and resource', was superseded by the Pahang Tungku Kudin's all-time record of a 44.5 kg pair, not far off the mark of the African record of 98 pounds (Foenander 1939:59, 1952:114). These tusks suggest the presence of mature stocks of big mammals.

From early times, the exorbitant prices paid in China for the horns and tusks of elephants, rhinoceros, and *seladang* had created within the indigenous population, both aboriginal and Malay, a strong tradition of hunting big game for profit, in addition to subsistence.[31] The valuable horns, tusks, and bones of the rhinoceros, elephant, and tiger were sold for their presumed medical efficacy, while sambar meat was widely relished by the Malays. They are also known to have consumed the meat of the *seladang*, reputedly more succulent than domestic beef (Syers 1933:201-2). Big game which could be trapped only with great skill and difficulty was taken with greater ease once guns became more widespread with improved cash earnings. Among the Malay the shooting, especially of *rusa* deer for

30 'Some notes about Seladang', *The Selangor Journal* (11-8-1893):382.
31 'Shooting in Selangor', *The Selangor Journal* (2-12-1892):90.

consumption, was popular. As among the Orang Asli, guns were often shared. Malay forest rangers, equipped with guns, were some of the keenest hunters. Malays usually went out in hunting parties of 20-30 and divided the spoils. So widespread was *rusa* hunting that by 1931 there was fear of extermination in some parts of Perak. In Sitiawan, Perak, the rhinos had all been shot and trapped and in Kuala Selinsing the *seladang* population of 80-90 was greatly reduced by hunting and trapping by Senoi (Wild Life Commission of Malaya 1932, II:211-2).

The earliest Peninsular legislation for the protection of wildlife was passed in 1896 in Pahang, motivated partly by the state government's emphasis on forest revenues.[32] Because of their economic value, the shooting of elephant, *seladang*, and rhinoceros, like the extraction of stipulated mineral and vegetable products, required a licence. The enactment also prohibited the shooting of female elephants and deer and all immature big game. With the development of hunting as a sport among the European community, interest in ensuring sustainable stocks provided a strong incentive for the 1911 Federal Enactment for the Protection of certain kinds of Wild Animals and Birds. It prohibited shooting without licence, shooting of female elephants and deer, and shooting of immature big game of all kinds. It provided also for the creation of sanctuaries and reserves with the consent of the Ruler-in-Council.[33] The lack of an administrative organization in the form of an independently constituted game department meant, however, that licences were freely and, often, indiscriminately issued by District Officers and *penghulu* (village chief). The sharp rise in economic prosperity during the first two decades which contributed to the increased circulation of guns was directly related to land clearance and concomitant pressure on the mammal population. Even earlier, during the turn of the century in Selangor, coffee plantations and macadamized roads had pushed their way into the hinterland, driving the mammals away from their favourite haunts. As one observer recorded in 1898:

> 'the *sladang* has been driven from its haunts. The grunt of the frightened wild pig disturbed at its meal is no longer heard, and the rhinoceros has abandoned its wallow. The large shady trees are gone beneath which the elephants used to sway their trunks.' (Rathborne 1898:157.)

Once introduced at the beginning of the century, rubber rapidly extended into the foothills, impinging on the existing habitat of big mammals. The total acreage of rubber in the Peninsula doubled between 1915 and 1922 from over a million to 2.3 million (see Drabble 1973:215, Appendix C) and reached 3.45 million by 1940, occupying some 11% of the land. Rubber saplings and the small kitchen gardens of plantation workers were foraged

32 'Products on State Land', Enactment No. IV of 1896, *Pahang laws*, Kuala Lumpur 1898:64-7.

33 Enactment XI of 1911, *F.M.S. Enactments 1911*, Kuala Lumpur 1912:31-44.

by elephants as well as by deer and pigs which, in turn, attracted the tiger (Aikin and Leigh 1992:55). In Pahang, even the *seladang* which occasionally damaged rice allegedly attacked rubber by nibbling top shoots. Though young rubber offered ready and concentrated acreage of food, their availability was both patchy and sporadic so that they would appear to have merely compensated to some extent for the permanent loss of richer secondary forest and grassland to monocultivation. The introduction of oil palm in the 1930s proved a special boon to the *seladang*, able to feed both on the *Centrosema* cover crop and the grasses and ferns struggling through them. In 1938 Temerloh alone is reported to have harboured a few hundred *seladang* (Pelandok 1938:63). It is unlikely, however, that the ground vegetation could support anything like the variety of flora characteristic of mixed grassland and *belukar*.

Official denunciation in 1904 of the uneconomic use of land dedicated to shifting cultivation was followed by a tightening of land laws to discourage the practice. Though shifting cultivation proved difficult to curb in the remoter parts such as in Pahang and Terengganu and was continued especially by the aboriginal populations, it gradually became less prevalent on the west coast (Lim 1976:49-65). The introduction of land tenure, the creation of Malay Reservations in 1923, and government irrigation schemes served as inducements for Malays to accept permanent settlement less compatible with *belukar*- and grassland-dependent mammals. The combined effects of what would appear to be an early boost to the mammal population and the subsequent loss of habitat to settled agriculture brought an unprecedented confrontation between large mammals and humans. The attack of rubber tappers and peasants by tigers making their forays into villages was not uncommon and accounted for constant efforts to eliminate them. It was reported in 1931 that the tiger population had 'suffered very badly through shooting'. Whether the killings were in excess of the birth rate, however, is not known.[34] Elephants which were in the habit of returning to familiar haunts impinged on plantations when these were opened up in proximity or across their regular beats near salt licks. The impact of cultivation on elephant herds was best demonstrated in Carey Island, off Kelang. By the mid-1920s cultivation of rubber and coconut had edged out the herd which was reduced to near-starvation. In 1927 Sir George Maxwell pleaded for putting an end to the plight of small and isolated herds of 6-15 animals through their destruction, though any initiative in this direction was left to the planters (Maxwell 1927:197-201). In Perak, damage to rubber holdings was estimated at $200,000. (Wild Life Commission of Malaya 1932, I:9). In 1927, the Jenderak Planting Syndicate in Pahang lost some 12,000 trees – about a quarter of its holdings – to destruction by elephants (Wild Life Commission of Malaya 1932, III:330-1). The damage done by deer which fed on the bark of rubber

34 'Notes of the day', *The Malay Mail*, 5-1-1931.

trees was also extensive, causing a plantation in Mersing, Johore, a loss estimated at 9,000 trees between 1919 and 1923 (Wild Life Commission of Malaya 1932, I:67).

The influential Planter's Association pressed for the elephant and sambar to be removed from the scheduled list, claiming that fencing – recommended by Hubback as a means of keeping out the animals – was ineffective. Particularly during the 1920s and 1930s, raids on plantations escalated and by the 1925 Wild Animals and Bird Protection Enactment planters could, with the written permission of the District Officer, shoot any big game posing a threat within a mile of a cultivated area. Dozens of elephants were also killed in Pahang and Negeri Sembilan under exemptions granted by game wardens.[35] Pressed by protests from agriculturists the game wardens themselves were brought to act in the interests of agriculture rather than wildlife. In Johore, Captain Ahmad bin Muhamedun claimed that during his service as game warden (1925-1929), he had killed 30-40 elephants in the interests of agriculture (Wild Life Commission of Malaya 1932, II:76). With many elephants wounded by buckshot and turned rogue, a vicious circle was established with increased destruction and killings. Thus, in addition to hunting for pure sport, there was the elimination, by the same planters, of elephants and sambar regarded as 'vermin'. In Tekam, Pahang, a man is reported to have received $40 for every head of elephant killed.[36] The profit motive was apparent in cases where bull elephants, carrying valuable ivory, were allegedly shot in preference to cows, even though cows were claimed to be equally if not more destructive.[37] It was also reported that many among the local populace planted cassava or paddy in remote areas with the aim specifically of attracting game which they then shot as predators.[38]

The conflict of opinion between those who perpetuated the extermination of mammals infringing on plantations and those who opposed it, became a matter of serious public debate in the 1920s and 1930s (see Kathirithamby-Wells 1996). Concern over the fate of the big mammals in particular led to the creation of the first generation of wildlife reserves and parks. The most important were the Kerau (1923) and Gunung Tahan (1925) Reserves in Pahang; the Serting Reserve in Negeri Sembilan (1923); the Sungkai Reserve in Perak (1928) and Endau-Kota Tinggi (1933), Endau-Kluang (1933) and Segamat (1937) Reserves in Johore.[39] Overlapping areas of Orang Asli activity with areas of grasslands and secondary forest, they

35 Hubback, Royal Society's Club, London, to W.D. Ellis, Colonial Office, 30-12-1929, no. 8, CO 717/ 62/4.

36 Hubback to A. Caldecott, Bournemouth, 12-4-1933, CO 717/96/1.

37 I.H.N. Evans, Acting Director of Museums to Under Secretary, FMS with a copy to the Under Secretary, Colonial Office, London, 26-7-1929, in Hubback to Caldecott, CO 717/96/1.

38 Hubback, 'Sunlaws', Pahang to E. Campbell, MP, London, 10-11-1932, no. 9238, CO 717/89/7.

39 The only sanctuary preceding these was the Coir Game Reserve (Perak) of 4,955 ha with a salt-lick, created to protect a herd of *seladang* but this area was gradually excised for agriculture.

represented natural sanctuaries of big mammals. The wildlife controversy which was given wide publicity in the press led to the appointment of a Wild Life Commission which, with Hubback as Chairman, held 64 sessions during 1930-1931 throughout the Peninsula. Its investigations, published in 1932 in a three-volume report (Wild Life Commission of Malaya 1932), culminated in the creation in 1938 of a National Park and was considered a practical means of conserving wildlife within restricted boundaries while allowing for the protection of agriculture at large. Yielding to arguments that in addition to the National Park some 70% of the land then under forest reservation would afford protection to endangered species, the importance of protecting habitat-related mammal communities was set aside. Thus, the Serting Game Reserve in Negeri Sembilan, established in 1923, was revoked in 1929. It was argued that the protection of a dozen or more *seladang* did not warrant setting aside some 30,000 ha of land valuable for agriculture, particularly as an estimated 500 *seladang* still remained in the Peninsula.[40]

The 1932 decentralization undermined the Wild Life Commission's recommendations for creating an integrated federal wildlife department aimed at greater efficiency in protecting big mammals. Further, the 1933 retrenchments, following the Great Depression, with a cut in budget for Wild Life Preservation, did not bode well for mammal protection. Involving up to a 25% reduction in the case of Pahang – the most important state for wildlife – it resulted in the curtailment of an already skeletal staff of game wardens and rangers.[41] In contrast, the number of guns licenced steadily increased and the sale of ammunition rose from about 1 million in 1929 to about 1.7 million in 1938, bringing a considerable revenue for the government.[42] Provision was lacking for the proper supervision of the game laws so that poaching was on the increase.[43] In the case of valuable big game, the profits from spoils far exceeded the fines which were consequently ineffective. In the 1930s the fine for killing a rhino was only $250 as compared to its value at $2,500.[44] In 1935, on the basis of two stray feathers discovered, the magistrate's court in Kuala Lipis, Pahang, imposed a fine of $15 or two weeks 'rigorous imprisonment' on a Malay for having shot and killed five *berkok* or Large Green pigeons (*Treron capellei*). Yet, there was difficulty in tracking down persons responsible for the killing in Ulu Tembeling of a *seladang* and an elephant, the latter carrying a pair of 18-23 kg tusks.[45]

40 Memo J.A.W. Simmons, British Resident, Negeri Sembilan, 16-12-1929, Enclosure in H.C. Scott to Lord Passfield, 5-1-1930, no. 8, CO 717/69/3.

41 Hubback to Campbell, 1-2-1933, CO 717/96/1.

42 Hubback, Royal Society Club, St. James Street, London to T. Drummond Shields, MP, 21-12-1929, CO 717/62/4, no. 62310; 'Game in the F.M.S.: Chief Warden's Report', *The Malay Mail*, 14-2-1938.

43 'Wild life', *The Straits Times*, 11-3-1933.

44 'Question of heavy fines for infringement', *The Straits Times*, 31-1-1935.

45 'Game protection in Pahang', *The Malaya Tribune*, 27-10-1934.

That the ambiguous attitude of humans to mammals became more pronounced with the hurried pace of development and deforestation is apparent in the evidence given by local witnesses to the Wild Life Commission. Despite the threat of big mammals to life and agriculture, rural Malays were deeply aware of the role of fauna in the composite ecology of the Peninsula. Following the probable increase in the mammal population, the rhino excepted, during the nineteenth century, the loss of habitat to monoculture in the early decades of this century increased their vulnerability. Yet, there was anxiety, particularly among the *kampung* inhabitants, (Wild Life Commission of Malaya 1932, II:211) that the decline in numbers within individual species would seriously affect the ecological balance. Of the big mammals, tigers were considered important for keeping down wild pigs destructive to agriculture, although they posed a threat to humans and cattle (Wild Life Commission of Malaya 1932, II:219, 250-1). The serious decrease of sambar, it was feared, 'may upset the balance to such an extent that tigers may take to cattle killing'. Even the elephant, widely acknowledged to be destructive, was considered useful for keeping down the *bertam* palm (*Eugeissona tristis*) and bamboo and preventing forest regeneration (Wild Life Commission of Malaya 1932, II:83).

It is likely that the ban against arms during the Japanese occupation (1941-1945), coupled with widespread cultivation in jungle clearings to overcome food shortage, resulted in gains for the mammal population, even compensating for declines in the period preceding the war. In Terengganu the tiger population had increased to such a level that they posed a serious problem (Locke 1954:2). Independence in 1957 inaugurated a new phase of development based on the Ford Foundation Paarlberg Report of 1963, stressing the need for developing forest resources and giving agriculture priority over forest conservation and protection of species (Stevens 1968:19). The period 1960-1963 saw a large number of elephants and tigers killed for agricultural protection (Medway 1965:227, Table 2). Between 1965 and 1968 there was an alarming fall in the number of big mammals particularly in states such as Pahang, Johore, and Terengganu, affected by new plantations and large government-sponsored rural development and resettlement schemes. As game warden Mohammed Khan commented in 1965 with reference to the destruction of elephants in Perak: 'It is very doubtful if the yearly toll of destruction is being balanced by the yearly production of calves' (Mohd. Khan 1965:139). In Perak, the building of large hydroelectric dams, as in the case of the Temengor, destroyed prime sites favoured by large mammals, including two Sumatran rhinos, 50 *seladang*, and 40 elephants and the Kenering Dam flooded several mineral licks (Stevens 1968:61-2). In fact, the wildlife fund recommended by the Wild Life Commission of 1930-1931 was never implemented; game preservation became a non-government affair and largely a matter of public concern (Kathirithamby-Wells 1996).

Conclusion

This survey, though lacking in hard evidence in the form of statistical data, suggests that human activity at a non-intensive level is not incompatible with the survival of mammals and, indeed, can increase the range and carrying capacity of their habitats. The diachronic pattern of development in the Malay states brought by pioneer activity meant that big mammal populations peaked earlier in Johore and the westcoast states. Selangor and Negeri Sembilan, where large mammals had been plentiful as of the 1870s, saw their sharp decline by the 1920s (see Table 1). In Pahang, development came much later and proceeded at a slower pace over a relatively large terrain. Here the type of clearance favourable to big mammals appears to have occurred only during the early decades of the century, with a significant decline in populations not until the post-colonial era.

In the case of the Malay Peninsula, the much-maligned *belukar* or secondary jungle and *lalang* – the product largely of human activity – would seem to have played a significant role in supporting species not indigenous to the equatorial rain forest habitat. This period of pioneer development, with humans as the main agents for generating secondary forest, represents a unique phase in the environmental history of the post-Ice Age Peninsula, with big mammal populations probably reaching maximum levels by the early decades of this century. Some of the rarer mammals, like the rhinoceros, may have been brought to near extinction through killing for sport in the days preceding the institution of game laws and subsequently through poaching. Notwithstanding the effect of habitat loss and disturbed conditions which were less than conducive to breeding on population levels, sharp declines were partly the result of killings in the interest of agriculture, however short-term these may have been in ecological terms. It has been argued that secondary forest growth which can be expected to increase with development, could well boost the population of mammals like pigs and deer (Stevens 1968:32).[46] Such niches, however, without sufficiently extensive forest retreats, cannot support either the shy *seladang* or the elephant operating within a wide radius.

[46] It is reported that in 1971 alone an estimated 6 million kg of wild pig were taken from the forests of the Peninsula (Rambo 1982:253).

Abbreviations

CF	Conservator of Forests
CO	Colonial Office Records, Public Records Office, London
FMS	The Federated Malay States
TOL	Temporary Occupation Licences

References

Aiken, S.R. and C.H. Leigh
1992 *Vanishing rain forest; The ecological transition in Malaysia*. Oxford: Clarendon Press.

Annandale, N. and Robinson, F.C.
1903 *Fasciculi Malayenses*. London: University Press of Liverpool.

Banks, H.H.
1923 'The tiger' in: *Illustrated guide to the Federated Malay States 1923*. London: Malay State Information Agency.
1934 'Hunting seladang in Malaya', *British Malaya* 9-4:163-5, 195-6.

Bird, I.
1967 *The golden Chersonese, and the way thither*. Kuala Lumpur: Oxford University Press. [Reprint of London: Murray, 1883.]

Browne, F.G.
1932 'Pahang river trips', *Malayan Forester* 2:233-40.

Burgess, P.F.
1954 'Some elephant and tiger shooting experiences', *Malayan Forester* 17:63-9.

Burns, P.L. (ed.)
1976 *The journals of J.W.W. Birch, first British Resident to Perak, 1874-1875*. Kuala Lumpur: Oxford University Press.

Cameron, J.
1965 *Our tropical possessions in Malayan India*. Kuala Lumpur: Oxford University Press. [Reprint of London: Smith, Elder, 1865.]

Cant, R.G.
1972 *An historical geography of Pahang*. Singapore: Malaysian Branch of the Royal Asiatic Society. [Monograph 4.]

Carey, Iskandar
1976 *Orang Asli; The aboriginal tribes of peninsular Malaya*. Kuala Lumpur: Oxford University Press.

Clifford, H.
1898 *Studies in brown humanity, being scrawls and smudges in sepia, white, and yellow*. London: Grant Richards.

Cole, R.
1959 'Temiar Senoi agriculture; A note on aboriginal shifting cultivation in Ulu Kelantan, Malaya', *Malayan Forester* 22:191-201, 260-71.

Cranbrook, Earl of
1986 'A review of fossil and prehistoric remains of rhinoceroses of Borneo', *Sabah Museum and Archives Journal* 1-1:50-110.

Dentan, R.K.
1968a *The Semai; A nonviolent people of Malaya*. New York: Holt, Rinehart and Winston.

1968b 'Some problems in determining the conservation needs of the hill peoples of Southeast Asia', in: L.M. Talbot and M.H. Talbot (eds), *Conservation in tropical Southeast Asia; Proceedings of the conference on conservation of nature and natural resources in tropical Southeast Asia, Nov. 1965*, pp. 89-92. Morges, Switzerland: International Union for the Conservation of Nature.

Dodge, N.
1980 'Population estimates for the Malay Peninsula in the nineteenth century, with special reference to the East Coast States', *Population Studies* 34-3.

Drabble, J.
1973 *Rubber in Malaya 1876-1922; The genesis of the industry.* Kuala Lumpur: Oxford University Press.

Earl, G.W.
1837 *The eastern seas or voyages and adventures in the Indian Archipelago [...].* London: Allen.

Endicott, K.
1979 'The impact of economic modernisation on the orang Asli (Aborigines) of northern peninsular Malaysia', in: J.C. Jackson and M. Rudner (eds), *Issues in Malaysian development*, pp. 167-204. Singapore: Heinemann.

Evans, I.H.N.
1915 'Notes on the Aboriginal inhabitants of Ijok in the district of Selama, Perak', *Journal of the Federated Malay States Museums* 5-4:176-86.
1937 *The Negritos of Malaya; Their types, distribution and habits.* Cambridge: Cambridge University Press.

Foenander, E.C.
1939 'Elephant hunting in Pahang', *Malayan Forester* 8-1:57-66.
1952 *Big game of Malaya; Their types, distribution and habits.* London: Batchworth Press.

Gullick, J.M.
1975 'Selangor 1876-1882, The Bloomfield Douglas diary', *Journal of the Malaysian Branch of the Royal Asiatic Society* 48-2:1-51.

Hill, R.D.
1977 *Rice in Malaya; A study in historical geography.* Kuala Lumpur: Oxford University Press.

Hornaday, W.T.
1879 'Account of a naturalist's visit to the territory of Selangor', *Journal of the Straits Branch of the Royal Asiatic Society* 3:124-31.

Hubback, T.R.
1905 *Elephant and seladang hunting in the Federated Malay States.* London: Ward.
1923 'Big game shooting', in: *Illustrated guide to the Federated Malay States 1923.* London: Malay State Information Agency.
1938 'The Malayan gaur or seladang', *Journal of the Bombay Natural History Society* 40-1:8-19.

Jackson, J. C.
1965 'Chinese agricultural pioneering in Singapore and Johore, 1800-1917', *Journal of the Malaysian Branch of the Royal Asiatic Society* 38-1:77-105.

Kathirithamby-Wells, J.
1996 'Nature and nation in colonial Malaya'. Paper presented at the 14th conference of the International Association of Historians of Asia, Chulalangkorn University, Bangkok, 20-24 May 1996.

Kelham, H.R.
1928 'Malaya in olden days', *British Malaya* 3-2:47-50.

Kitchner, H.J.
1961 'The bleak prospect for the seladang in Malaya', in: J. Wyatt-Smith and P. Wycherley (eds), *Nature conservation in western Malaysia, 1961*, pp. 197-201. Kuala Lumpur: Malayan Nature Society.

Lake, H.W. and H.J. Kelsall
1894 (with the assistance of H.N. Ridley) 'Journey to the Sembrong River; From Kuala Endau to Batu Pahat', *Journal of the Straits Branch of the Royal Asiatic Society* 26:1-33.

Lim Teck Ghee
1976 *Origins of a colonial economy; Land and agriculture in Perak, 1874-1897*. Penang: Penerbit Universiti Sains Malaysia.
1977 *Peasants and their agricultural economy in colonial Malaya, 1874-1941*. Kuala Lumpur: Oxford University Press.

Loch, C.W.
1937 'Rhinoceros Sondaicus; The Javan or Lesser One-horned Rhinoceros and its geographical distribution', *Journal of the Malayan Branch of the Royal Asiatic Society* 15-2:130-49.

Locke, A.
1954 *The tigers of Terengganu*. London: Museum Press.

MacNair, F.
1878 *Perak and the Malays; 'Sarong' and 'kris'*. London: Trinsley.

McNeely, J.A. and P.S. Wachtel
1988 *Soul of the tiger; Searching for nature's answers in Southeast Asia*. Singapore: Oxford University Press.

Maxwell, George
1927 'Big game and planters', *British Malaya* 2-8:187-201.
1960 *In Malay forests*. Singapore: Eastern Universities Press. [Reprint; First edition London: Blackwood, 1907.]

Mayer, C.
1922 *Trapping wild animals in Malay jungles*. London: Unwin.

Medway, Lord
1965 'Game statistics for Malaya, 1960-63', *Malayan Nature Journal* 19:223-34.

Mohd. Khan bin Momin Khan
1965 'Problems with the elephant, and where it is in danger in the State of Perak', *Malayan Nature Journal* 19:138-40.
1967 'Movements of a herd of elephants in the Upper Perak area', *Malayan Nature Journal* 20:18-23.

Newbold, T.J.
1837 'Sketch of the State of Muar', in: J.M. Moor, *Notices of the Indian archipelago [...]*. Singapore: Cass.

Noone, H.D.
1936 'Report on the settlements and welfare of the Ple-Temiar Senoi of the Perak-Kelantan watershed', *Journal of the Federated Malay States Museum* 19-1:1-85.
1949 'The first fruits of the hillrice harvest among the Temiar Senoi of the Plus-Temiar Aboriginal area', *Bulletin of the Raffles Museum* 4(Series B):5-7.
1955 'Notes on the trade in blowpipes and blowpipe bamboo in north Malaya', *Federation Museums Journal* 1-2:1-8.

Oliphant, L.
1969 *Narrative of the Earl of Elgin's mission to China and Japan in the years 1857, 1858 and 1859*. New York: Kelley. 2 vols. [Reprint; First edition Edinburgh: Blackwood, 1859.]

Olivier, R.C.D.
1978a 'Conservation of the Asian elephant', *Environmental Conservation* 5-1:146.
1978b *On the ecology of the Asian elephant: With particular reference to Malaya and Sri Lanka*. [PhD thesis, Cambridge University.]
1980 'Reconciling elephant conservation and development in Asia; Ecological bases and possible approaches', in: J.I. Furtato (ed.), *Proceedings of the Vth International Symposium of Tropical Ecology 16-21 April 1979, Kuala Lumpur, Malaysia*, pp. 315-22. Kuala Lumpur: International Society of Tropical Ecology.

Oxley, T.
1849 'The zoology of Singapore', *Journal of the Indian Archipelago and Eastern Asia* 3:594-8.

Pelandok
1938 'Seladang hunting in Pahang', *Malayan Forester* 7-2:61-71.

Rambo, A.T.
1982 'Malaysia; Orang Asli adaptive strategies; Implications for Malaysian natural resource development planning', in: C. MacAndrews and Chia Lin Sien (eds), *Too rapid rural development; Perceptions and perspectives from Southeast Asia*, pp. 251-99. Athens/Ohio: Ohio University Press.

Rathborne, A.
1898 *Camping and tramping in Malaya; Fifteen years' pioneering in the native states of the Malay Peninsula*. London: Sonnenschein.

Ridley, H.N.
1896 'The forests of Selangor', *Selangor Journal* 4-26 (4-9-1896):444-7.

Sadka, E.
1968 *The protected Malay states, 1874-1895*. Kuala Lumpur: University of Malaya Press.

Schebesta, P.
1928 *Among the forest dwarfs of Malaya*. London: Hutchinson.

Skeat, W.W.
1953 'Reminiscences of the Cambridge University expedition to the North-Eastern Malay states, 1899-1900', *Journal of the Malayan Branch of the Royal Asiatic Society* 26-4:9-147.

Skeat, W.W., and C.O. Blagden
1906 *Pagan races of the Malay Peninsula*. London: Macmillan. 2 vols.

Stevens, W.E.
1968 'The conservation of wild life in Malaysia'. Seremban, Malaysia: Office of the Chief Game Warden, Federal Game Department. [Unpublished Report.]

Strickland, D.L.
1967 'Ecology of the rhinoceros in Malaya', *Malayan Nature Society* 20:1-17.

Syers, H.C.
1933 'Shooting in Selangor', *British Malaya* 7-9:186-8.

Talbot, L.M.
1969 'Javan rhinoceros; Lesser one-horned Rhinoceros (*Rhinoceros sondaicus Desmarest*)', *Oryx* 5:204-15.

Turnbull, C.M.
1972 *The Straits Settlements 1826-67; Indian presidency to crown colony*. London: Athlone Press.

Wells, D.R.
1988 'Birds', in: Earl of Cranbrook (ed.), *Key environments: Malaysia*, pp. 167-95. Oxford: Pergamon Press.

Wharton, C.H.
1968 'Man, fire and wild cattle in Southeast Asia', in: *Proceedings; Tall timbers fire ecology annual conference*, pp. 107-67. Tallahassee, Florida: Tall Timbers Research Station.

What Malaysia is saying
1936 'What Malaysia is saying', *British Malaya* 11-3:73.

Wild Life Commission of Malaya
1932 *Report of the Wild Life Commission of Malaya, 1930*. Singapore: Government Press. 3 vols.

Wyatt-Smith, J.
1958 'Shifting cultivation in Malaya', *Malayan Forester* 21-3:139-90.

Zaharah binti Hj. Mahmud
1970 'The period and the nature of "traditional" settlement in the Malay Peninsula', *Journal of the Malaysian Branch of the Royal Asiatic Society* 43-2:81-113.
1992 'The traditional Malay ecumene of the Peninsula; A historical-geographical essay on the position of wet rice', in: Voon Phin Keong and Tunku Shamsul Bahrin (eds), *The view from within; Geographical essays on Malaysia and Southeast Asia*, pp. 297-321. Kuala Lumpur: Department of Geography, University of Malaya.

Table 1. Sightings of elephants, seladang, and rhinoceroses in various parts of the peninsula.

State	Elephant	Seladang	Sumatran rhinoceros	Javan rhinoceros
Johore				
Gunung Pulai 1859[1]	–	+++	++	–
Sembrong 1894[2]	+	–	–	–
Throughout 1925[3]	+	–	–	–
Selangor				
S. Buloh/Batu 1878[4]	+++	+++	+	–
Throughout 1923[5]	+	+	–	–
Negeri Sembilan				
Muar 1836[6]	+++	–	++	–
Throughout 1925[7]	++	+	–	–
Perak				
Lower Perak 1883[8]	+++	–	++	–
Throughout 1947[9]	++	+	–	–
Peninsular North				
Upper Perak-East Pahang 1923[10]	+++	+++	++	+
Upper Perak-Kelantan 1936[11]	+++	–	–	–
Pahang				
K. Tahan 1899-90[12]	+++	+++	+++	–
Temerloh 1938[13]	+++	–	–	–
Throughout 1947[14]	+++	++	+	–

\+ *rare*
++ *some*
+++ *plentiful*
– *no data available or species extinct or absent*

Sources:

1 Oliphant 1969, I:24-5.
2 Lake and Kelsall 1894:13.
3 Hubback, 8 May 1925, No. 20854, CO 717.
4 Hornaday 1879:127.
5 Hornaday 1879:128; Hubback 1923:237.
6 Newbold 1837:73.
7 Hubback, 8 May 1925, No. 20854, CO 717.
8 Bird 1967:336-7.
9 Hubback 1923:237.
10 Noone 1936:44.
11 Skeat 1953:103.
12 Pelandok 1938:63.
13 *Annual Report*, Game Department, 1947.
14 *Annual Report*, Game Department, 1947.

Table 2. Estimated numbers of elephants, *seladang* and rhinoceroses in West Malaysia compared with previous reports.

State	Elephant[1]		Seladang[2]		Rhinoceros[3]	
	1965	1968	1965	1968	1963	1968
Johor	43	53	50	13	10	5
Pahang	229	127	405	121	8	3
Negeri Sembilan	39	8	26	0	0	0
Selangor	11	5	0	0	5	3
Perak	82	97	45	91	10	5
Terengganu	119	111	36	10	5	0
Kelantan	57	36	74	57	5	1
Kedah	90	23	70	17	4	0
Perlis	12	3	0	0	0	0
National Park	n.a.*	23	n.a.	37	n.a.	3
Total	682	486	706	346	47	20

[1] Medway 1965

[2] Milton 1963

* Data not available

Source: W.E. Stevens, 'The concentration of wild life in West Malaysia', Ministry of Lands and Mines, Malaysia, October 1968.

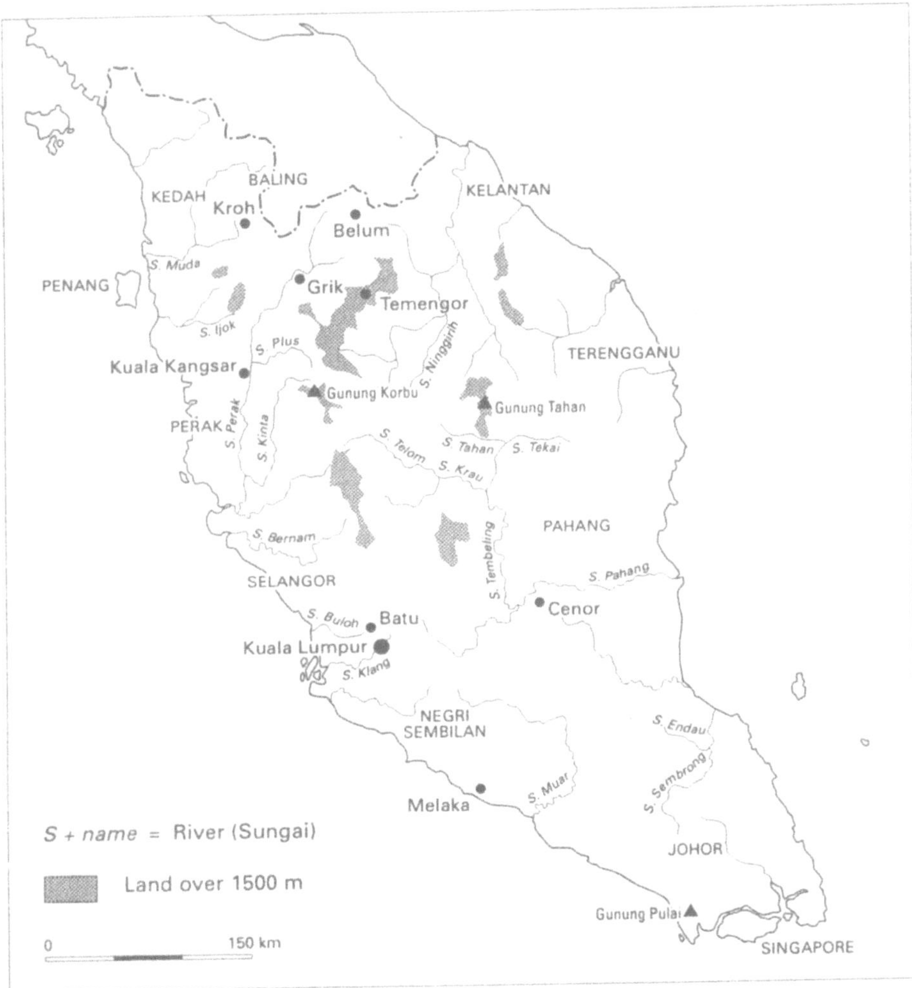

Map 1. The Malay Peninsula

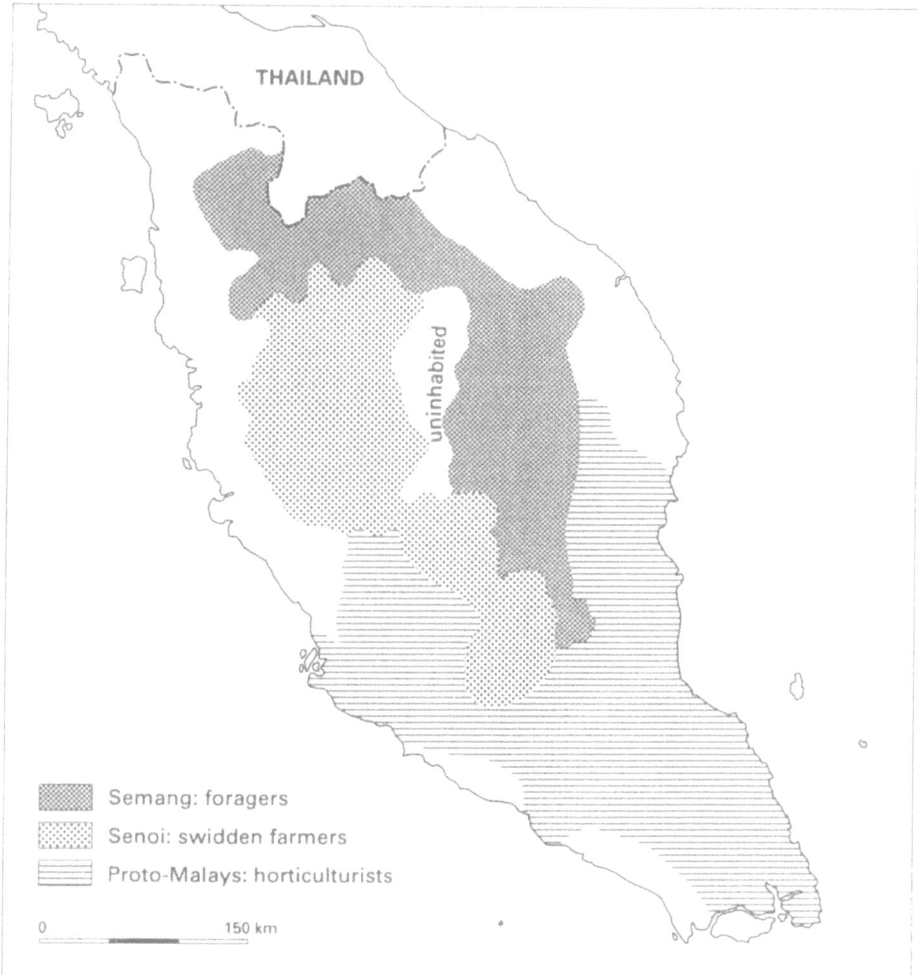

Map 2. Distribution of West Malaysian aborigines

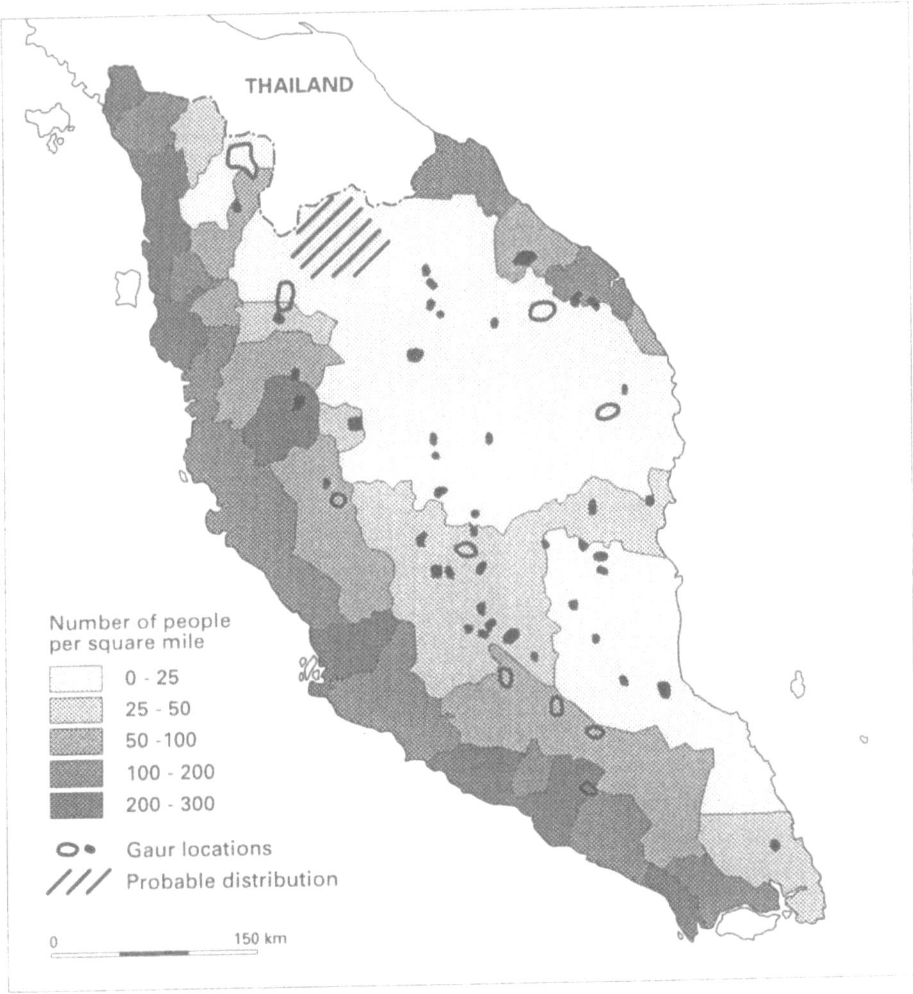

Map 3. Human population and gaur distribution in West Malaysia

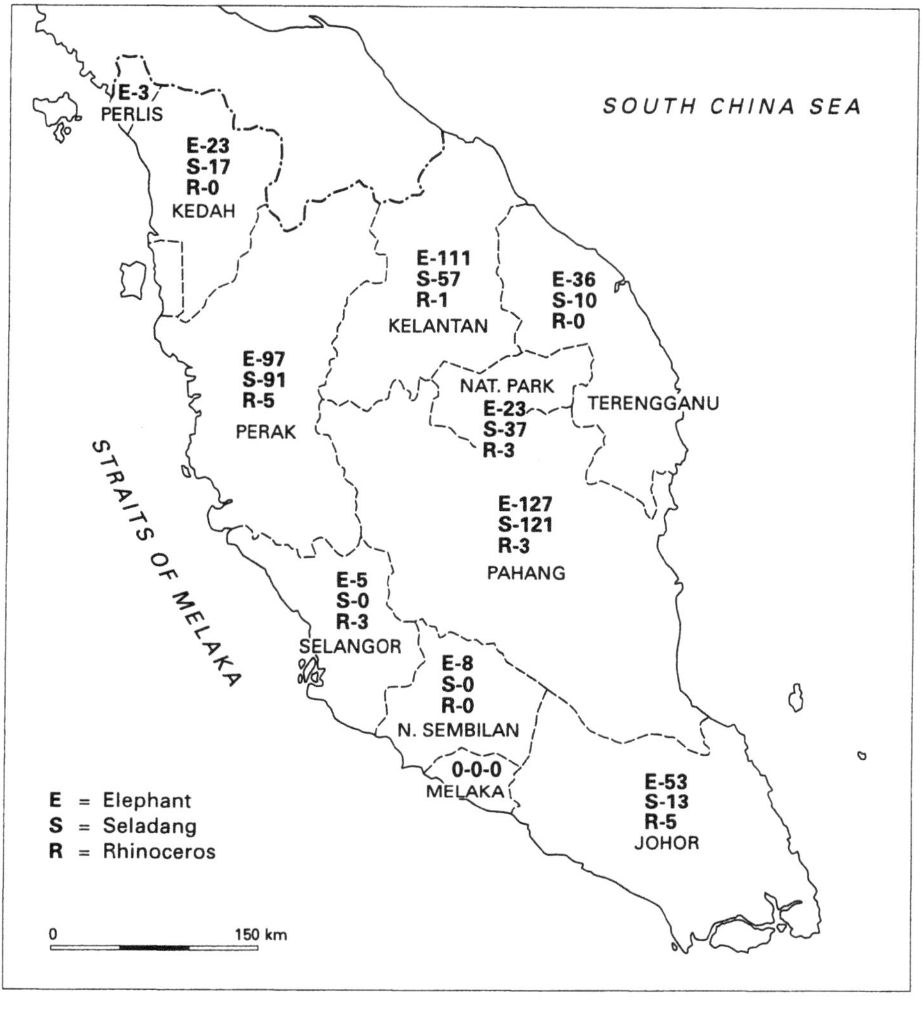

Map 4. Distribution of rare large mammals in West Malaysia, 1968

MASYHURI

Fishing industry and environment off the north coast of Java, 1850-1900

Introduction

Up to the 1870s, the sea fishing sector in the waters around Java and Madura experienced a period of growth. Local fish production was more or less sufficient to meet the demands of local consumption, and import of fish from abroad was not necessary. Java and Madura were the most important centres of fishing activity in the archipelago. Fish-salting industries also thrived in this period, while offshore fishing far outweighed inshore fishing in importance (Masyhuri 1995:65-76).

The growth of the fishing industry in Java and Madura was made possible by the fact that the sea fishing sector was integrated into a wider trading system. Up to the 1860s the colonial government operated a revenue-farming system in the fishing sector. The tax 'farmers', usually rich Chinese merchants, dominated the industry and were extremely successful in building up business networks based on fishing and expanding them into other sectors. Under their aegis the fishing industry was drawn into the commercial economy.

At the end of the nineteenth century the fishing industry in Java and Madura declined and there was a shift from large-scale offshore fishing to much smaller-scale inshore fishing, until ultimately the Java fishery became an inshore industry. The major factor behind this shift was an alteration in the role of the revenue farmer which triggered off drastic changes throughout the entire industry, including its processing and marketing branches (Masyhuri 1995:90-105).

The aim of this article is to examine the interplay between the fishing industry and the marine environment in the Java Sea during the period 1850-1900, and in particular to assess the environmental consequences of the shift from offshore to inshore fishing.

* I would like to thank Professor Peter Boomgaard, Dr David Henley, Dr Freek Colombijn and M.A.S. Hilam for their comments on the draft version of this paper.

Mayang boat; photograph by the Regent of Demak (KITLV photo collection 3686).

Marine ecology and the fishing industry

The Java Sea covers some 100,000 km².[1] Its floor is flat and muddy and the surrounding coasts are mostly low-lying (Bottemanne 1946b:13-4). As part of the Sunda Shelf the Java Sea is shallow, ranging in depth from 20 metres off the coast of South Sumatra to just over 60 metres in its easternmost parts (Hardenberg 1931:81). A great variety of fish species are found in its waters. It is supposed that this fish fauna is generally similar to that of other Sunda Shelf waters (Delsman 1927b:3-4). According to Bleeker there were around 500 fish species in the Java Sea. Weber and De Beaufort, however, discussed some 850 species and estimated that the total number must be twice as large, while Delsman and Hardenberg assumed that the Java Sea contained around 1,500 fish species.[2] Unfortunately we do not know how many species can be caught in economically viable quantities either, although Maxwell notes that there are around 250 species of high economic value. The waters around the Malay Peninsula, by way of comparison, contain around 1,000 fish species of which 50 can be caught in quantities of 100 tonnes per year and 20 in quantities of 1,000 tonnes per year.[3] Research conducted in 1930 indicated that the average fish stock of the Java Sea was around 4 tonnes per km^2; for the North Sea off the Netherlands in 1910, the equivalent figure was 2.3 tonnes per km^2 (Bottemanne 1946b:15).

In short, the Java Sea is a productive area with a high density of fish stock (*Vischrijkdom* 1933). One reason for this is that it is also rich in plankton.[4] Quantitative research on zooplankton carried out at the beginning of 1970s by the Lembaga Oseanologi Nasional (Indonesian Institute of Oceanology) showed that the concentration of zooplankton in the Java Sea was higher than that found in the South China Sea or in the seas of eastern Indonesia. The highest densities of all were found in the inshore waters of Bangka Island and off the Sumatran mainland near Bengkalis and Bagansiapiapi (Soegiarto and Birowo 1975, II:27-9, map 456).[5] Plankton is the productive base of marine ecosystems, providing food for larger animals including fish (*Zeeën* 1922:215; Delsman and Hardenberg 1934:20-2).

The economically most significant fish species in the Java Sea are *lemuru* (*Clupea leiogaster*), *kembung laki-laki* (*Scomber kanagurta*), *kembung perempuan* (*Scomber neglectus*), *tengiri* (*Scomberomorus commerson*), *layang* (*Decapterus* spp.), *teri* (*Stolephorus baganensis*), and *bandeng*

1 Topographically, Indonesian waters can be divided into three zones – the Sunda Shelf, the Sahul Shelf, and the deep-sea areas. The Sunda Shelf connects the three islands of Sumatra, Kalimantan, and Java with mainland Asia. It includes the southern parts of the China Sea, the Gulf of Thailand, and the Java Sea.

2 Delsman 1927a:24; Delsman and Hardenberg 1934; ARA MvK, verbaal 2-7-1909, no. 6.

3 Kesteven 1949:36; Firth 1966:6; see also Fowler 1938.

4 *Java Bode*, 8-9-1916; Allen and Cupp 1935:1; Schippers 1928:19-21; Delsman 1939a.

5 See also Arinardi 1980.

(*Chanos chanos*) (Delsman and Hardenberg 1934:129). *Lemuru* is caught in large quantities around Anyer and Banyuwangi (Palm 1962:221). *Kembung* and *tengiri* can be caught almost everywhere off the north coast of Java and Madura. *Kembung perempuan* lives mainly in inshore waters, while *kembung laki-laki* prefers offshore waters (Delsman and Hardenberg 1934:330-2). What is called *layang* by local people is not one single species but a mixture of several, including *Decapterus Kurra, Decapterus Russelli* and *Decapterus Macrosoma* (Delsman and Hardenberg 1934:306). The *layang* are pelagic fish which live in clear offshore waters of high salinity (Hardenberg 1937:295-7; Sunier 1914:4). *Teri* is a small fish found only in inshore areas, while *bandeng* is the most important food fish obtained from brackish water.

The Java fisheries can be divided into three categories: offshore, inshore, and onshore (the last consisting of aquaculture in brackish-water fishponds). Offshore fishing, as already mentioned, was the most prominent up to the middle of the nineteenth century, and within this category the most important form was *mayang* fishing. *Mayang* is the name of the type of boat used to fish for the most important offshore species, the *layang*. The *mayang* is the largest traditional fishing boat of Java and Madura, capable of working safely at more than 40 kilometres offshore (Van Kampen 1922:22). The net which is used to catch *layang*, known as the *payang* net, is also distinctive. According to Javanese folk etymology, the words *mayang* and *payang* are both derived from the name of the *layang* fish. The *mayang* fishery represents artisanal fishing on quite a large scale. Until the end of the 1870s it was the most important type of fishery in the Java Sea, and Java was the most important *mayang* fishing centre in the archipelago. *Mayang* fishing developed particularly in Tegal, Pekalongan, Jepara, Rembang, Pasuruan, and several centres on the north coast of Madura (*Zeevisscherijen* 1882:117). *Mayang* fisheries were also found in Banten, in Jakarta, and around Cirebon.

Paradoxically, the Java Sea is basically not a suitable habitat for the *layang*. For much of the year its waters have too high a turbidity and too low a salinity for this group of species. However, an annual migration of *layang* takes place to the Java Sea from the eastern part of Indonesia. The steady monsoon winds which prevail for about six months of the year cause a drift or current along the long axis of the Java Sea which is directed westward in the east monsoon (June-August) and reverses to flow eastward during the west monsoon (December-February). Calculations based on meteorological observations at Jakarta indicate that twice a year the Java Sea is completely swept clean (Delsman 1939b:143). The east monsoon winds bring clear water of high salinity into the Java Sea from the Flores Sea, and partly also from the Pacific Ocean, via the Makasar Straits.[6] This is the time when the waters of the Java Sea, especially at a distance

6 Delsman 1939b:143; Hardenberg 1937:297; Soeriaatmadja 1956; Van Weel 1923:27.

of 40 or 50 kilometres offshore, become suitable for the *layang* and the habitat of the fish extends temporarily in a westward direction. During this period, *layang* fish from deep eastern waters such as the Flores Sea follow the westward current and migrate into the Java Sea. The high concentration of plankton in the Java Sea is probably an important reason for this migration, since the *layang* feed on plankton. Of the 1,365,747 kilogrammes of fresh fish landed at Ketapang (Madura) in 1910, 1,240,667 kilogrammes were produced during the *layang* fishing season of September to December and almost all of this quantity consisted of *layang* (Van Dort 1936:36; Vink 1912:24). The *layang* has the commercial advantage that it can easily be preserved by salting. Most of the fish consumed on Java is purchased in salted form.

The aquaculture or 'onshore fisheries' of the north coast of Java have a long history. The civil code of Majapahit already lists penalties for theft of fish from fishponds (Jonker 1885). In the nineteenth century, coastal fishponds were found particularly in the areas from Jepara eastward to Pasuruan, and also in Banten, although here they suffered much damage from the Krakatau eruption of 1883. At the beginning of the 1860s the reported total area of the fishponds was 46.068 *bahu* (1 *bahu* = 7096.5 m^2).[7] As mentioned, the most important species reared in fishponds was the *bandeng*.

The yield from inshore fishing, by contrast, consisted of mixed species in which no single type or group dominated. While offshore fishing was carried out mainly by *mayang* boats, inshore fishing included various types of boat and tackle. *Teri* and shrimp were two important products. Inshore fishing took a small-scale artisanal form and was found everywhere on the north coast of Java and Madura. The inshore waters of the Java Sea tend to be very turbid due to the sediment carried by the many rivers of Sumatra, Java, and Kalimantan. Nearly all the major rivers of Indonesia are located in this area. Their existence also influences the salinity of the coastal waters. Near the mouth of a river, turbidity is generally higher and salinity lower than offshore (Soegiarto and Birowo 1975, I:61). Turbidity also varies over time, being higher during the west monsoon than in the east monsoon (Soegiarto 1977:4). Shallow, turbid water and low salinity are important reasons why *layang* are hardly ever caught from inshore waters in the Java Sea.

In the 1870s there were some 250,000 professional fishermen living on the coasts of Java and Madura.[8] If the size of their families averaged five persons, then the number of people dependent on the fishing sector can be estimated at 1,250,000 persons (Boomgaard 1989:172).

7 Nieuwenhuis 1915:1-6; Van Spall 1865:5-11; ARA MvK, verbaal 16-9-1864, no. 2 (bijlage).
8 ANRI, missive gouvernements-secretaris 7-12-1905, no. 4099 (bijlage).

Changes in the industry and their environmental impact

Beginning in the 1870s there was a recession in the fishing industry. The number of fishing boats shrank drastically, from some 15,000 in 1870 to as few as 4,257 in 1903.[9] Investment also dwindled until there were scarcely any new boats built, even though the bulk of the boats launched in the 1860s were old, had suffered damage, or had disappeared beneath the waves.

In 1864 the colonial government abolished the tax-farming system in the fishing sector. The farmers were stripped of their rights, including the former privilege of buying cheap salt for the preservation of fish. Business costs were pushed up enormously and the fishing industry no longer yielded a profitable return. The loss of subsidized salt meant that fish salters were now forced to use salt manufactured for the consumer market, which as a government monopoly was much more expensive. Under these circumstances the former revenue farmers abandoned their roles in financing the fishing industry and marketing its products. Because they no longer needed to bind the fishermen to them in order to control the purchase of fish, investment in new boats ground to a halt and the industry suffered from capital erosion. Only a small quantity of fish could still be processed, and fish marketing also suffered. This was a time in which the fishing sector was isolated from the wider economic system (Masyhuri 1995:125).

As the *mayang* fisheries declined, inshore fishing become more important. In Cirebon, for instance, coastal fishing intensified and the fishermen took all fish, including young fish, without distinction. The Dutch authorities referred to this as *roofvisscherij* or 'robber fishing'.[10] By 1890 it was reported that adult fish were hardly to be seen any more in the Cirebon markets.[11] Moreover, several of the types of tackle used, including the *jala hela* and the *jaring krikit* (pull nets), had the effect of destroying fish eggs. Unregulated fishing using destructive methods caused serious damage to fish habitats and reduction of fish stocks. Similar phenomena were also noted in other areas of West, Central, and East Java. At the end of the nineteenth century overfishing in inshore areas became a hot issue among colonial officials, while offshore waters were fished only to a fraction of their potential.[12]

At the same time, coastal ecosystems were also coming under pressure from an unwanted side-effect of agricultural expansion in the interior. Soil erosion inland meant that the rivers of Java carried large quantities of silt to the sea (Donner 1987:308). In some places this resulted in the development of mudbanks along the coast, with catastrophic consequences for the fishermen living and working nearby. The most obvious environmental

9 ARA MvK, verbaal 28-10-1905, no. 10 (bijlagen); Hasselman 1914:Bijlage J.
10 ARA MvK, verbaal 15-18-2-1895, no. 16; 23-11-1903, no. 103.
11 ANRI, Besluit 3-4-1892, no. 7 (bijlagen).
12 ANRI, Besluit 3-4-1892, no. 7 (bijlagen); Bottemanne 1946a:179.

change along the coastline, however, was the expansion of brackish-water fishponds. In part this expansion had to do with increasing private investment in aquaculture following the economic liberalization of 1870.[13] But it may also have been connected with the contraction in the fishing sector. Van Spall notes that the reclamation of land for fishponds was carried out by groups of fishermen.[14] In Surabaya, these cooperative groups consisted of between 10 and 20 fishermen under a single leader.[15] Unfortunately we do not know how many offshore fisherman left their work and moved into the fishpond sector.

As mentioned above, the total area of brackish-water fishponds in Java in 1862 was 46,068 *bahu* spread over Besuki, Probolinggo, Pasuruan, Surabaya, Rembang, Jepara, and Semarang. Surabaya, with 35,211 *bahu*, had the largest area, followed by Pasuruan and Jepara. The coastline around Surabaya features tidal mudflats, scoured channels, and estuarine inlets, making it particularly suited for fishpond construction. At the western end of the island, brackish-water fishponds could be found around Jakarta and in Banten. Not all north coast areas, however, had extensive fishponds. Tegal and Pekalongan, for instance, had just 14 and 3 *bahu* respectively.[16]

An investigation held in 1886 showed that by that time the area of brackish-water fishponds in Java had increased to 59,703 *bahu*. In Semarang, Jepara, and Rembang the number of ponds had grown many times over, and growth in Madura was also considerable. Again, the pattern was uneven: in Pasuruan the fishpond area grew only slightly, and in Probolinggo and Besuki not at all.[17] By 1915, however, the total area of fishponds had increased again to around 78,000 *bahu* (Gerdes Oosterbeek 1921:580). From that time onward, at least until the 1930s, it did not grow much further.[18] Over the whole period from the 1860s until the 1930s, the area of brackish-water fishponds in Java and Madura increased by a reported 31,932 *bahu* or 69%.

The extension of brackish-water fishponds on the north coast of Java led to increased production of *bandeng* fish. But it also reduced the area of mangrove swamps, with significant environmental consequences. Fishponds are usually excavated on the mangrove fringe of low-lying coasts, especially on delta margins. Mangrove forests help to stabilize coastal terrain because of the sheltering effect of their canopy and the binding effect of their roots. They provide good protection against wave attack. The diminution of mangrove swamp often means disruption to the coastline. In many areas erosion began after the clearance of a mangrove fringe. On the

13 ARA MvK, verbaal 24-9-1866, no. 7 (bijlagen); verbaal 18-2-1874, no. 44.

14 ARA MvK, verbaal 16-9-1864, no. 2. The results of this research were published in the *Tijdschrift voor Nijverheid en Landbouw in Nederlandsch-Indië* (Van Spall 1865).

15 ARA MvK, verbaal 11-4-1904, no. 17 (bijlage).

16 ARA MvK, verbaal 16-9-1864, no. 2 (bijlage).

17 ARA MvK, verbaal 11-4-1904, no. 17. See also De Jaager and Lawick van Pabst (1903).

18 ARA MvK, verbaal 16-11-1905, no. 8 (bijlage); Burger 1930:403.

eastern shore of Jakarta Bay, for instance, the shoreline retreated by up to 145 metres after the construction of fishponds. The shoreline at Sungaibuntu also experienced erosion after fishponds were built on Cilincing beach. In the 65 years between 1873 and 1938 the shoreline of this area retreated about 50 metres (Bird and Ongkosongo 1980:19-23). The full extent of the effects of fishpond construction on the Java shoreline awaits further study.

Besides helping to protect the shoreline against erosion, mangroves also provide a productive habitat for fish, crustaceans, and the like. The organic matter which they produce, particularly in the form of fallen leaves, is transformed by bacteria into detritic particles which are eventually transported to surrounding waters by tidal flushing. These particles, including various life forms found within them, serve as food for larger organisms such as fish, shrimp, and crabs (White, Martosubroto and Sadorra 1989:38-40). Mangroves provide a breeding ground and nursery for many forms of marine life, and the conversion of mangrove swamp to fishponds disrupts the life cycle of fish and shrimp.

This factor on the one hand and overfishing on the other led to declining sea-fish production. The catch from the waters around Pekalongan, for instance, fell steadily until many fishermen had to leave for other areas. Fishermen in Probolinggo also complained of decreasing catches.[19] Worse, because of strong competition from producers outside Indonesia, declining production was not compensated by rising prices. The decline of offshore fishing was accompanied by a sharp rise of the amount of fish imported into Java, particularly salted fish from Siam. While fish imports in 1875 were still limited, 28,029 tonnes of salted fish were imported in 1895 and 33,553 tonnes in 1904.[20] Competition in the marketing of fish intensified, and the price of the imported fish was far lower than that of local production. One ton of salted fish imported from Siam in 1900, for instance, cost around *f* 150, whereas one ton of locally-produced salted fish cost *f* 335.[21]

Up to the 1860s, when the fishing industry in Java was still flourishing, the annual income per capita of the population of Java and Madura from the fishing sector was around *f* 3. In the period after 1870 this figure decreased steadily, not only because of stagnant production but also because the population increased continuously. By the turn of the century, annual income per capita of the population of Java and Madura from the fishing sector had fallen to *f* 0.60 (Verloop 1904:320).

The rates of return for the fishermen themselves decreased too. During the period when offshore fishing was dominant, the average daily income

19 ANRI, Besluit 3-4-1892, no. 7 (bijlagen).

20 *Onderzoek* 1905:64 (note). See also *Koloniaal Verslag* 1902 Bijlage BB No. 4; ANRI, missive gouvernements-secretaris 7-12-1905, no. 4099 (bijlage).

21 ARA MvK, verbaal 16-11-1905, no. 13 (bijlagen); *Zeevischvangst* 1905:339.

per capita of fisherman in Java and Madura reached a reported 29.60 cents. This was considerably better than the income of the average Javanese farmer, which in the 1880s was not more than *f* 5.80 per month or 19.33 cents per day (Boomgaard 1989:176-7; *Zeevisscherijen* 1882:81), and nearly the same as that of a sugar factory labourer, which was around 30 cents per day in the 1870s (Dros 1992, Table 5.2). During the period when the fishing industry in Java had experienced deterioration and shifted to small-scale inshore fishing, however, the income of the fishermen fell sharply and many sought alternative ways of making a living.

Conclusion

The development of the fishing industry was strongly influenced by the marine environment. Different marine habitats support different fish species, and the fisherman who wants to catch a certain species needs to employ a particular tackle and method. *Layang*, for instance, are pelagic fish which live in offshore waters, and special equipment – the large *mayang* boat and the *payang* net – was developed to catch them.

Conversely, fishing also affected the marine environment. Environmental change on the north coast of Java was partly a consequence of changes in the fishing industry. When the *mayang* fisheries underwent a period of decline caused by abolition of the salt subsidy and the farming system in the fishing sector, fishing activities began to move from large-scale offshore fishing into smaller-scale inshore fishing. This shift, political in origin, had ecological consequences. The burden borne by the coastal area became heavier, and this seems to have accelerated environmental change. Overfishing took place in some inshore waters, while the offshore waters of the Java Sea were underexploited in relation to their full biological potential.

In the same period there was an expansion in the area of brackish-water fishponds. This too had ecological consequences. In some areas the conversion of mangrove swamp into fishponds led to coastal erosion and accelerated shifting of the shoreline.

Abbreviations

ANRI	Arsip Nasional Republik Indonesia, Jakarta
ARA	Algemeen Rijksarchief, Den Haag
MvK	Archive of the Ministerie van Koloniën, ARA

References

Allen, W.E. and E.E. Cupp
1935 'Plankton diatoms of the Java Sea', *Annales du Jardin Botanique de Buitenzorg* 44:101-224.

Arinardi, O.H.
1980 'Perbandingan kandungan dan komponen zooplankton di perairan Teluk Jakarta dan sekitarnya antara musim barat dan musim timur', in: A. Nontji and Asikin Djamali (eds), *Teluk Jakarta; Pengkajian fisika, kimia, biologi dan geologi tahun 1975-1979*, pp. 99-106. Jakarta: Lembaga Oseanologi Nasional, LIPI.

Bird, E.C.F. and Otto S.R. Ongkosongo
1980 *Environmental changes on the coasts of Indonesia*. Tokyo: United Nations University.

Boomgaard, P.
1989 *Children of the colonial state; Population growth and economic development in Java, 1795-1880*. Amsterdam: Free University Press. [CASA Monographs 1.]

Bottemanne, C.J.
1946a 'De Indische zeevisscherij', *Economisch Weekblad voor Nederlandsch-Indië* 12-23:196-7.
1946b *Het Indische zeevisscherijprobleem*. Batavia: Departement van Economische Zaken. [Mededeelingen van het Departement van Economische Zaken in Nederlandsch-Indië 3.]

Burger, D.H.
1930 'De zoutwatervischvijvers in het regentschap Pati', *Koloniaal Tijdschrift* 19:402-34.

Delsman, H.C.
1927a 'Fauna en flora', in: A.D. Rinkes et al., *Het Indische boek der zee*, pp. 18-28. Batavia: Kolff.
1927b 'Het zee-onderzoek en zijn resultaten', in: A.D. Rinkes et al., *Het Indische boek der zee*, pp. 1-9. Batavia: Kolff.
1939a 'Fishing and fish-culture in the Netherlands Indies', *Bulletin of the Colonial Institute of Amsterdam* 2:92-105.
1939b 'Preliminary plankton investigation in the Java Sea', *Treubia* 17:139-84.

Delsman H.C. and J.D.F. Hardenberg
1934 *De Indische zeevisschen en zeevisscherij*. Batavia: Visser.

Donner, W.
1987 *Land use and environment in Indonesia*. London: Hurst.

Dort, Th.K.L. van
1936 *Richtlijnen voor een zeevischindustrie op Java*. Den Haag: Van Stockum.

Dros, N.
1992 *Wages 1820-1940*. Amsterdam: Royal Tropical Institute. [Changing Economy in Indonesia 13.]

Firth, R.
1966 *Malay fishermen; Their peasant economy*. London: Routledge and Kegan Paul.

Fowler, H.W.
1938 *A list of fishes known from Malaya*. Singapore: Government Printing Office.

Gerdes Oosterbeek, W.F.
1921 'Visscherij', in: *Encyclopaedie van Nederlandsch-Indië*, vol. IV, pp. 573-86. 's-Gravenhage: Nijhoff, Leiden: Brill.

Hardenberg, J.D.F.
1931 'The fish fauna of the Rokan mouth', *Treubia* 13:81-168.
1937 'Preliminary report on a migration of fish in the Java Sea', *Treubia* 16:295-300.

Hasselman, C. J.
1914 *Algemeen overzicht van de uitkomsten van het welvaart-onderzoek, gehouden op Java en Madoera in 1904-1905*. 's-Gravenhage: Nijhoff.

Jaager, C.J. de, and H.J.W. Lawick van Pabst
1903 *Rapport nopens de vischvijvers op Java en Madoera naar aanleiding van het onderzoek ingesteld door de controleurs der 1e klasse bij het Binnenlandsch Bestuur [...] op genoemde eilanden*. Batavia: Landsdrukkerij.

Jonker, J.C.G.
1885 *Een oud-Javaansch wetboek vergeleken met Indische rechtsbronnen*. Leiden: Brill. [PhD thesis, Rijksuniversiteit Leiden.]

Kampen, P.N. van
1922 *Visscherij en vischteelt in Nederlandsch-Indië*. Haarlem: Tjeenk Willink.

Kesteven, G.L. (ed.)
1949 *Malayan fisheries; A handbook prepared for the inaugural meeting of the Indo-Pacific Council*. Singapore: Malaya Publising House.

Masyhuri
1995 *Pasang surut usaha perikanan laut; Tinjauan sosial-ekonomi kenelayanan di Jawa dan Madura, 1850-1940*. [PhD thesis, Vrije Universiteit Amsterdam.]

Nieuwenhuis, J.H.
1915 'Kort bericht omtrent de zeevisscherij in de afdeeling Pandeglang (Banten)', *Bestuur van de Centrale Kas te Batavia* 1:1-6.

Onderzoek
1905 *Onderzoek naar de mindere welvaart der Inlandsche bevolking op Java en Madoera. Ia: Samentrekking van de afdeelingsverslagen over de uitkomsten der onderzoekingen naar de vischteelt en visscherij*. Batavia: Landsdrukkerij.

Palm, C.H.M.
1962 'Vaartuigen en vischvangst van Anjar Lor, Bantam, West-Java', *Bijdragen tot de Taal-, Land-, en Volkenkunde* 118:217-47.

Schippers, J.A.W.
1928 'De zeevisscherijen van Nederlandsch-Indië', *Koloniaal Tijdschrift* 17:15-37.

Soegiarto, A.
1977 'Introduction to the regional oceanography of the South-East Asian waters'. Paper presented at the Fifth FAO/SIDE Workshop on Aquatic Pollution in Relation to Protection of Living Resources, Manila, 17 January-27 February 1977.

Soegiarto, A. and S. Birowo
1975 *Atlas oseanologi perairan Indonesia dan sekitarnya*. Jakarta: Lembaga Oseanologi Nasional, LIPI. 2 vols.

Soeriaatmadja, R.E.
1956 'Seasonal fluctuations in the surface salinity off the north coast of Java', *Marine Research in Indonesia* 1:1-19.

Spall, P.W.A. van
1865 'Onderzoek omtrent het aanleggen van zout-zoetwater vischvijvers op Java', *Tijdschrift voor Nijverheid en Landbouw in Nederlandsch-Indië* 11:1-96.

Sunier, A.L.J.
1914 'De betekenis van het natuurwetenschappelijk visscherij-onderzoek voor Nederlandsch-Indië', *Mededeelingen van het Visscherij-Station te Batavia* 10:1-44.

Verloop, G.N.
1904 'Zeevisscherij naast de inlandsche landbouw', *Tijdschrift voor Nijverheid en Landbouw in Nederlandsch-Indië* 69:320-36.

Vink, W.C.A.
1912 'Opmerkingen en beschouwingen over de Inlandsche hoogzeevisscherij van Java en Madoera', *Mededeelingen van het Visscherij-Station te Batavia* 9:20-30.

Vischrijkdom
1933 'De "vischrijkdom" der Javazee', *Algemeen Weekblad* 18-16:225.

White, A.T., Purwito Martosubroto and M.S.M. Sadorra (eds)
1989 *The coastal environmental profile of Segara Anakan-Cilacap, South Java, Indonesia*. Manila: International Center for Living Aquatic Resources Management, on behalf of the Association of Southeast Asian Nations.

Weel, K.M. van
1923 'Meteorological and hydrographical observations made in the western part of the Netherlands East Indian archipelago', *Treubia* 4:1-559.

Zeeën
1922 *De zeeën van Nederlandsch Oost-Indië*. Leiden: Brill.

Zeevischvangst
1905 'De zeevischvangst in Oost-Indië', *Het Nederlandsche Zeewezen* 4:337-41.

Zeevisscherijen
1882 'Zeevisscherijen langs de kusten der eilanden van Nederlandsch-Indië', *Tijdschrift voor Nijverheid en Landbouw in Nederlandsch-Indië* 26:77-370.

BERNICE DE JONG BOERS

Sustainability and time perspective in natural resource management
The exploitation of sappan trees in the forests of Sumbawa, Indonesia, 1500-1875

In this article I investigate the history of sappan wood exploitation on the island of Sumbawa. I also try to assess the environmental effects of this exploitation by calculating the number of sappan trees that must have been felled on the island over two centuries. Finally I proceed to the question of whether the production of this wood was sustainable. My argument here is that the answer to this question is largely a matter of time perspective.

The island of Sumbawa

Sumbawa is a volcanic island located within the chain of islands referred to during the Dutch colonial era as the 'Lesser Sunda Islands'. Nowadays these islands form the provinces Bali and West and East Nusa Tenggara (Lombok, Sumbawa, Sumba, Flores, Roti, Sawu, and Timor). The island of Sumbawa has an extraordinary rugged and indented shape with numerous bays. Its total area is 15,600 km^2.

According to the classification of Köppen, Sumbawa has a tropical rainy climate (class A). This is not, however, a hot damp forest climate (class Af) as in many other parts of Indonesia, but a periodically dry savannah climate (class Aw), characterized by a mean rainfall of less than 60 mm in the driest month (Whitmore 1985:53). The average precipitation on Sumbawa is 1250 mm (Brewer 1979:18). The seasonal distribution of the rainfall is very uneven. During the east monsoon (lasting from April until November) Sumbawa is very dry and often looks arid and barren.

Sumbawa's physical landscape consists of mountains, terraces, plains, valleys, and rivers. At lower altitudes there are large grass plains punctu-

* I would like to thank Frances Gouda, David Henley, Rosemary Robson-McKillop, and Henk Schulte Nordholt for their useful comments on earlier versions of this article, and Youetta de Jager for helping with the production of the map. Research in the National Archives (ANRI) in Jakarta in 1996 was made possible by a grant from the Netherlands Foundation for the Advancement of Tropical Research (WOTRO).

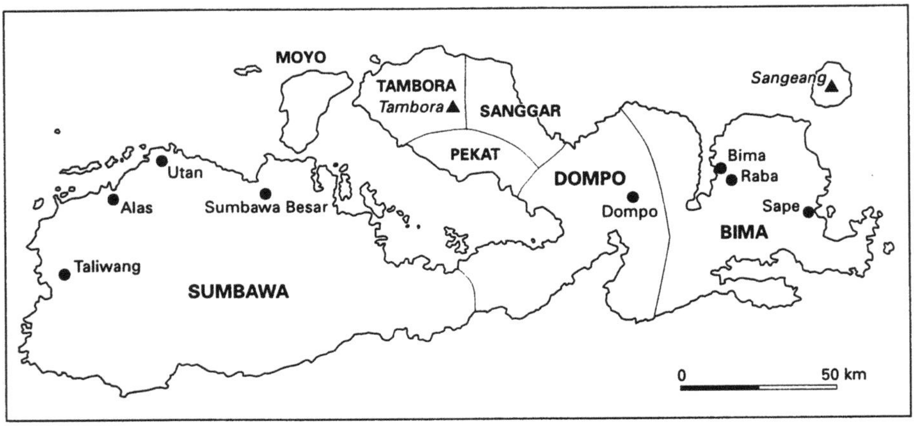
MOYO
TAMBORA
Tambora
SANGGAR
PEKAT
Sangeang
Bima
Raba
DOMPO
Dompo
Sape
BIMA
Utan
Alas
Sumbawa Besar
Taliwang
SUMBAWA
0
50 km

Sumbawa

ated by brushes and trees; residues of ancient forests also occur here. It is in these lowland areas, located primarily on the north coast and around the rivers, that most agricultural activities (including irrigated farming) take place and most of the population is concentrated. The hilly uplands consist of savannah and forests. Because the average annual rainfall is limited, most of the forest on Sumbawa is monsoon forest.[1] Higher in the mountains where the precipitation exceeds 1800 mm, however, rain forests can be found. In these regions the population is sparse and mostly practices *ladang* farming.

The population is currently almost one million, or about 58 people per km^2 on average; a relatively thin population. In the past the population density was even lower. Although no reliable population figures are available for Sumbawa before 1930, we can estimate the population density of Sumbawa at between 11 and 19 persons per km^2 at the beginning of the nineteenth century.[2] That is, before the volcanic Mount Tambora located on the Sanggar peninsula of the island erupted in April 1815. As a result of this eruption the population declined drastically due to famine and migration as well as the many immediate deaths. Zollinger, for example, estimated the number of inhabitants around 1830 at 45,000 (1850:176), while a general report of 1845 put it at 68,100.[3]

At the beginning of the twentieth century the population had grown to 311,287, raising the average population density to 20.38 per km^2 (*Volkstelling 1930*) – still a low density. In general this meant that on Sumbawa land was abundant and labour scarce. Under such circumstances it is not surprising that *ladang* cultivation was always the most important means of existence during the period 1500-1900. Supplementary economic activities included hunting, the collection of forests products such as timber, wax, and honey, and animal husbandry (especially horse breeding).

Even in 1984 *ladang* cultivation was still the most important form of agriculture on Sumbawa. About 36% of the total land surface of Sumbawa was then used for this activity (forest fallow included) against 10.5% for *sawah* cultivation. The main type of vegetation was forest (including *belukar* and *hutan* types).[4] Forests covered 65% of Sumbawa's surface, of which 32% represented fallowed swidden land (Dove 1984:112).

Around the year 1500, six tiny realms had formed on the island: Sumbawa, Bima, Dompo, Sanggar, Tambora, and Papekat (Pekat). At this early stage Sumbawa already had established trading contacts and exported

1 'Monsoon forest' refers to forests of the tropical Far East in which the water supply is periodically seriously limiting to plants (Whitmore 1985:197).

2 According to Francis (1856:131) the population of Sumbawa numbered around 300,000 people before Mount Tambora erupted, whereas Zollinger (1850:176) thought the number had only been around 170,000.

3 H.F.G. Trouwerbach, Algemeen verslag afdeling Bima 1845 (ANRI Makassar 3.8).

4 *Belukar* and *hutan* are Malay words for bush and forest respectively.

several of its products: rice, horses, wax, and the very important sappan wood.

What is sappan wood and why is it so special?[5]

Sappan wood, also known under the names Brazil, brazilwood and redwood, originates from a small tree or shrub which has the botanical name *Ceasalpinia sappan*. The word sappan comes from the Malay name for the tree: *sepang*. From this wood small boats were made, named after the wood: *sampan*.

The sappan tree is a fast-growing, shrubby tree, about five to ten metres high, with thorny branches and yellow flowers. It grows mainly in upland, stony, but not too cold areas. The tree is distributed over the entire Southeast Asian region and can also be found in India. It is often planted but also grows in the wild. In Indonesia it is found predominantly in the Moluccas and on Sumbawa.[6] Greshoff states that sappan is indigenous to Sumbawa (1894:121), and the sappan wood formerly exported from Sumbawa probably came from wild trees. The naturalist Heinrich Zollinger (1850:103) referred explicitly to 'plants growing in the wild' and the *Daghregisters* of VOC Batavia mention that Sumbawa sappan wood was cut 'in the forests' and 'on the mountains'.[7] It remains possible that at least some sappan trees were deliberately planted, but historical evidence for this is lacking.

As far as I know, Sumbawa was the only island in the archipelago that exported the wood on a large scale. Certainly the sappan wood of Sumbawa was often praised as the best of its kind in the archipelago. Other regions that produced and exported sappan wood were the Philippines and Siam (Thailand). Sappan wood from Siam was regarded as the best in Southeast Asia, being longer and thicker than that of Sumbawa. The Portuguese writer Tomé Pires noted this difference as early as 1513:

> 'The island of Bima [...] has a great deal of brazil, which they take to Malacca to sell, and they go there from Malacca for it because it sells well in China, and the Bima brazil is very thin. It is worth less in China than that from Siam, because that from Siam is thicker and better.' (Cortesão 1944:203.)

So the sappan trees in Siam probably achieved greater height and girth than those on Sumbawa. Most likely this must be attributed to differing climatic conditions. In Thailand the average rainfall is higher than on

5 This section of this paper is based on the following sources: Heyne (1927:753-4); Greshoff (1894:121-4); Burkill (1966, I:394-7); Lemmens and Wulijarni-Soetjipto (1992:60-2) and Rumphius (1741-50, IV:56-9).

6 In the western part of the island of Sumbawa the tree has the name *sepang* and in the eastern half *supa*.

7 *Daghregister* 15-10-1663, 10-7-1664.

Sumbawa, so that the trees can grow faster and the tree-rings are correspondingly thicker.

In Indonesia the sappan shrub is often used for fencing. It suits this purpose because of its many thorns and prickles. A hedge made of sappan trees is virtually impenetrable (*Encyclopaedie* 1917:434). The best features of the wood are its hardness and durability. Even when it comes into lengthy contact with (sea-)water it will not decay; hence its suitability for use in shipbuilding. Within this industry its primary use was for making wooden pins and nails. Masts were probably also made of sappan trunks. Outside shipbuilding the wood was used for building furniture and crafting adornments and frames. Because of its hardness it was also appropriate for making such items as walking sticks.

Finally, and most importantly, a red dye could be extracted from sappan wood by boiling it in water. This was used for colouring textiles and cakes, and according to Crawfurd (1820, I:462) it was the best red dye known in the archipelago. The paint industry in Europe also used sappan dye. In the Netherlands the wood was rasped to coarse powder by delinquents in the *tuchthuizen* ('houses of punishment'). For this reason it was also called 'raspwood'. As one of the special characteristics of the wood is its hardness, having to rasp it was a real punishment. The coarse powder was then sent to the dye factory, where the red pigment was extracted from it (Greshoff 1894:121-2; Noorduyn 1987:121). In China, sappan dye was used for colouring paper. By adding certain leaves, bark or chalk while boiling the wood, the colour could be made either lighter or darker. Because of its valuable dye, the sappan tree became a trading commodity at an early date.

In addition, Europeans also used the wood as dunnage – that is, for loading their ships properly to prevent the other cargo from shifting during the sea voyage (Crawfurd 1820, III:422; Bruijn et al. 1987, I:189-91). According to Boomgaard, its use as dunnage was almost the most important reason for bringing it to Europe:

> 'Compared to most other commodities shipped from the East to Europe, brazilwood had a surprisingly low value per unit of weight or volume, and one might very well ask why the Dutch (and the English) could be bothered to transport these enormous and bulky quantities all the way home. The answer is that it was used for stowage, and that as such its weight/value ratio was not unsatisfactory.' (Boomgaard 1995:4.)

I do not fully agree with this view, however, because a report from 1627 suggests that profits from sappan wood were quite considerable (Coen 1919-53, V:794). In Holland it sold at *f* 48 to *f* 58 per 100 Amsterdam pounds (49.5 kg), which was judged to be a high price. Pepper, by comparison, sold at *f* 24 for 100 Amsterdam pounds in 1622 (Coen 1919-53, IV:562). The high price of sappan wood was the main reason why the Dutch tried to collect at least 200,000 pounds (about 100 tonnes) of it each year (from Siam and

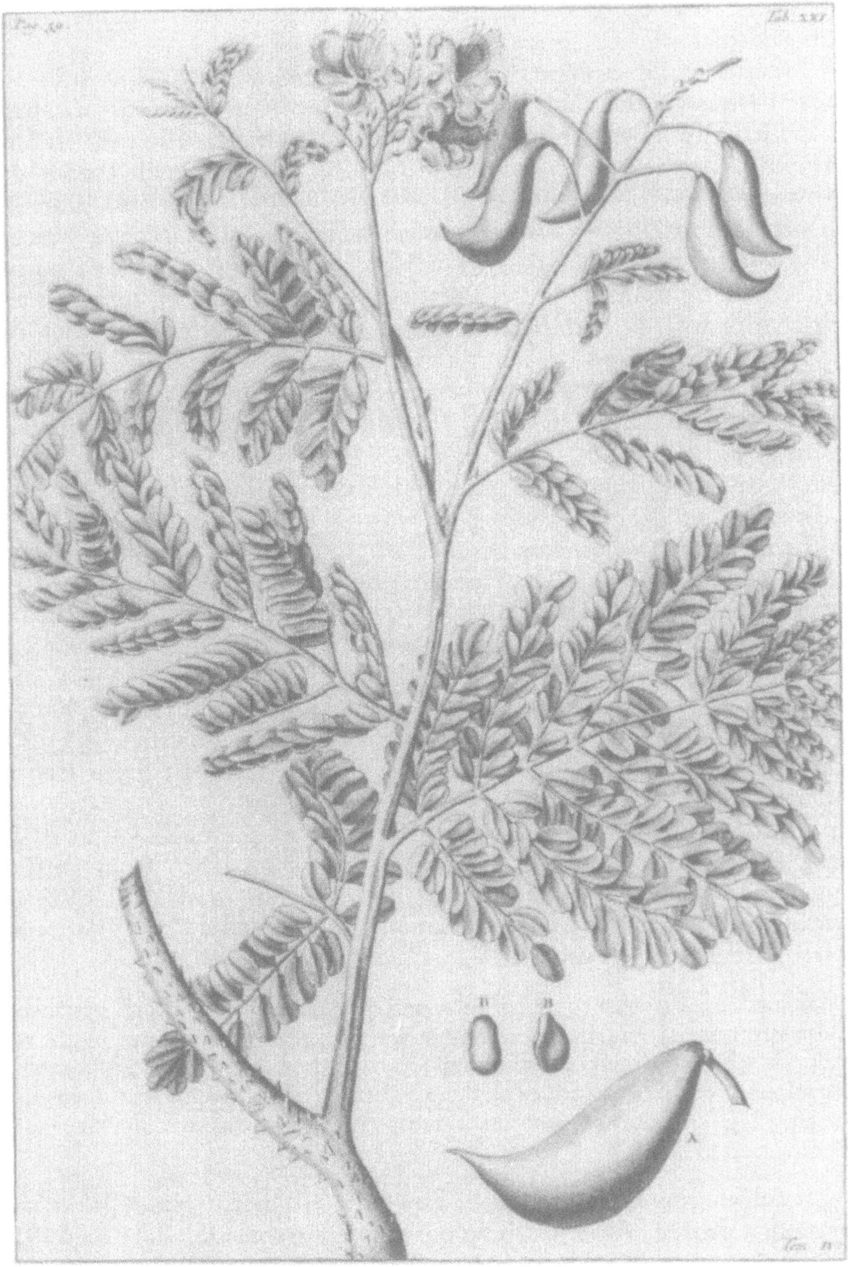

Sappan (Rumphius 1741-55, IV, opposite page 59.)

Sumbawa combined). The higher the price rose, of course, the more desirable it became as a dunnage material. In 1749, profits in the Netherlands were high enough to prompt the Dutch to collect the wood in East Java (especially Panarukan) as well as Sumbawa even though East Javanese sappan wood was worth about 25% less on the market than wood from Bima (Van der Chijs in Brascamp 1918:297).

A report from 1701 does mention 'a quantity of sappan wood for dunnage under the cargo of the ships bound for the home country', but I see dunnage as a supplementary rather than a central reason for transporting the wood to Europe.[8] The main reason was the extractable dye and the profits that it could generate on the European market. If this had not been the case, the Europeans would probably not have bothered to ship the wood home at all. From the moment artificial substitutes for sappan dye were invented in around 1870, production and shipping of the wood ceased completely. At that point other woods replaced it for dunnage purposes too.

The history of sappan felling on Sumbawa

The Javanese, in the thirteenth and fourteenth centuries, were probably the first to arrive on the island of Sumbawa to collect sappan wood. It is probably owing to this early wood trade that we can find two places on the north coast of Sumbawa called 'Utan' and 'Alas' (Goethals 1961:9). These names are a Malay and a Javanese word respectively, both meaning 'forest' and probably referring to the rich timber resources of the island. It is not known whether the Javanese were indeed collecting sappan wood or other special woods like teak (*Tectona grandis*). But this last possibility is not very likely as the island of Java was full of teak trees itself at the time. As cited above, according to Pires sappan wood was already exported from Sumbawa to China via Malacca at the beginning of the sixteenth century. Europeans, too, were interested in this wood, because they could either sell it to other countries at high profit margins, or send it home (as mentioned above) for use in the dye industry.

From the beginning of the seventeenth century, the Portuguese visited Bima annually to acquire a quantity of sappan wood.[9] This annoyed the Dutch who preferred to keep the wood for themselves because they could sell it to countries such as Japan, China, and Persia at high profits.[10] At the time, however, they could do nothing to prevent the Portuguese from sailing to Sumbawa because the island was under the rule of Makassar. The Sumbawanese even paid tribute to Makassar in the form of sappan wood. In Makassar the wood was mainly used in shipbuilding.

8 ARA VOC 1647:103.

9 *Daghregister* 9-11-1616.

10 Coen 1919-53, II:45, 64, 204; Paul 1984:57; *Daghregister* 7-1-1616, 15-2-1678, 27-6-1678, 25-8-1678, 4-2-1679, 30-6-1679, 21-9-1679, 24-11-1679.

From the 1620s onwards the Dutch came to Sumbawa, more or less regularly, to obtain a share of the sappan wood. What quantities were involved is not exactly known. The sultans of Sumbawa were able to deliver sappan wood by sending some of their male subjects to the forests in the mountains to cut sappan trees. The felled trunks were divided into smaller pieces and then carried to the shore, where local people assembled the smaller lengths while awaiting the ships that would come to pick it up. Before the ships were loaded the wood was first weighed and paid for.[11]

From 1660 onwards the Dutch aimed at obtaining two shiploads of sappan wood yearly (that is, roughly 5,000 *pikul* or 310 tonnes). In 1663 the VOC merchant Michiel van Heyst investigated how much wood would be available in Bima yearly, but the Sultan of Bima could not estimate the wood resources in his territory. Moreover, the Sultan informed Van Heyst that he was not able to guarantee a regular delivery due to insufficient labour. This shortage of labour was related, among other things, to the Sultan's obligation to pay a yearly visit to Makassar accompanied by many of his followers. Nevertheless, most Bimanese informed Van Heyst that sappan trees were abundant in the forests of Sumbawa, so that Van Heyst concluded that the Dutch could count on two shiploads annually.[12]

In 1667 Makassar fell to the Dutch. From that moment the VOC viewed Sumbawa as one of its own dependencies and immediately tried to make permanent contracts with the various realms of the island. In 1669 the first such contract was concluded with Bima and Dompo. Sumbawa followed in 1674. The realms of Sanggar, Papekat (or Pekat), and Tambora fell under Dompo. By the sixth article of these contracts the Dutch acquired the monopoly on sappan wood (together with two other products, wild cinnamon and tortoiseshell). This meant that the sultans were obliged to sell all sappan wood they had felled to the VOC. In these first contracts, however, nothing was decided yet about the amount of wood that had to be cut, nor about the price that the Dutch would pay for it.[13]

These contracts meant that the unimpeded trade in sappan wood ended. From then on the Dutch set all the conditions, the quantity, and the price. During this period the VOC paid one reaal or one rixdollar per *pikul* of sappan wood. These payments were sometimes rendered in cash, but more often in cloth (*lijnwaden*). On occasion the wood was also delivered in exchange for firearms and ammunition or lead and old iron. These items were all highly valued by the sultans because every now and then they had to fight a local war.[14]

The Dutch, of course, were eager not to have to pay too much. VOC representatives were ordered by Batavia to tell the Sultan of Bima that

11 ARA VOC 1663:32-9; Zollinger 1850:103.
12 *Daghregister* 15-10-1663.
13 *Corpus Diplomaticum* 1907-55, II:419-26, 492-8; ARA VOC 1269:695.
14 *Daghregister* 28-12-1673; ARA VOC 1287:1247-50, 1294:206, 1647:103, 113, 1663:152.

the Dutch did not really need the sappan wood from Bima because they were also able to get it elsewhere, where the wood was even better, thicker and more abundant.[15] In reality, however, the Dutch tried to collect as much wood as possible from Sumbawa and only invoked the 'better and thicker' argument in order to keep prices down.[16]

It is striking that from the moment that the VOC made contracts with the rulers of Sumbawa with regard to the production and sale of sappan wood, exports of this wood from Siam dropped sharply. From figures collected by G.V. Smith (1977:80, 82, 88) it appears that exports of sappan wood from Siam to Japan, China (Formosa), India, and Europe dropped precipitously after 1664. I assume that to a large extent this decrease in Siamese exports can be explained by the rising quantity of sappan wood exported from Sumbawa, especially after the first contracts with this island were concluded in 1669.

The lion's share of sappan wood from the island of Sumbawa came from the realm of Bima. The other realms delivered much smaller quantities (Table 1). The size of the deliveries varied greatly, although the Dutch did their utmost to obtain a regular supply. Time and again they summoned the sultans to meet their obligations, pressing them to sent their subjects to the forests to cut wood, or promising various rewards, such as pieces of fine red cloth or purple velvet, if the wood was cut in time.[17] On one occasion VOC representatives tried to encourage regular cutting by promising a double stiver as an extra incentive to the cutters.[18]

There were a number of reasons for the irregularity of the sappan deliveries. In the first place the shortage of labour on the island was responsible. Sumbawa was rather thinly populated, and of course the people who comprised the already limited labour force were often involved in subsistence activities such as planting and cultivating rice. On occasion, moreover, considerable numbers of them were dispatched to other places, for example to Makassar as members of the entourage of the sultan or to Flores to 'protect' the people there from Portuguese incursions.[19] The second reason was that it was sometimes too dangerous to cut wood in the forests due to war and banditry. A third factor seems to have been an occasional shortage of cutting knives (*parang*), although reports of this may have been a subterfuge of the sultans. And finally, interruptions in supply were sometimes caused by a shortage of sappan trees in the forests.[20] It is to this last factor that I now want to turn.

15 ARA VOC 1281:772, 785.
16 ARA VOC 1281:772.
17 ARA VOC 1311:224-39, 1647:113; Radermacher 1786:186.
18 A double stiver is equal to one tenth of a guilder.
19 ARA VOC 1246; *Daghregister* 10-7-1664. A part of Flores was under the rule of Bima.
20 ARA VOC 1647:113, 243-87; *Daghregister* 29-9-1665 and 29-9-1680.

Shortages of sappan wood

The international demand for sappan wood caused sappan trees in the forests of Sumbawa to be felled on a massive scale. This led to what might be called Sumbawa's first 'ecological crisis': a shortage of sappan trees. The first time that Batavia received news of such a shortage was in 1665, when the Sultan of Bima wrote to Governor General Maetsuyker that not much sappan wood was left in the forests.[21] In that year Batavia received only 242 *pikul* (15 tonnes) of wood from Bima. It should be noted that this was even before the Dutch had made permanent contracts with the realms of Sumbawa.

In 1672 came another report of a sappan wood shortage, this time from Dompo. Having delivered 1,000 *pikul* the Sultan of Dompo informed Batavia that he would not be able to supply any more wood and excused himself by emphasizing that his realm and its forests were very small. His subjects, he reported, had been looking for the wood for some days but had not been able to find any trees.[22] Sappan deliveries from Dompo after 1672 were indeed limited and sporadic (Table 1).

Table 1. Sappan wood exports to Batavia by region and year (in *pikul*)

	Bima	Dompo	Sumbawa	Tambora	Sanggar	Pekat
1634			++			
1636	++					
1663	2000	2000*		1500*	500*	500*
1664	1500					
1665	242					
1670	++					
1671	1554	1500				
1672	2440	1000				
1673	++		300			
1674	1400	474			649	
1675	4930					
1676	6173					
1677	5475					
1678	4820	700				
1679	2237					
1680	577		2063			
1681	540		2514	304		

All figures are based on data from the *Daghregisters*, except for those marked * which are from ARA VOC 1663. ++ indicates a large but unspecified quantity of sappan wood.

21 *Daghregister* 29-9-1665.
22 *Daghregister* 8-11-1672, 9-11-1672; ARA VOC 1287.

In the second half of the 1670s, however, Batavia once again imported large quantities of sappan from Bima. Boomgaard estimates that during this period Bima shipped about 350 tonnes to Batavia annually. Given this resurgence of exports, he claims, 'it cannot be doubted that Bima's problems [...] were of a temporary nature' (1995:4). The boom of the late 1670s, however, was itself only to be temporary. In my opinion Boomgaard's estimate of 350 tonnes of wood holds true for only two years, 1676 and 1677. In both 1675 and 1678 the figure was about 300 tonnes and after 1678, as before 1675, exports were much lower again (Table 1).

The years 1675 to 1678, then, were rather exceptional. Possibly the VOC applied extra pressure in this period to keep deliveries high. Certainly the realm of Sumbawa was forced to deliver an additional quantity of 15,000 *pikul* (926 tonnes) in 1677 (*Realia* 1882-5, III:161). This huge amount was still not completely paid off in 1681.[23] Very soon it became clear that given the ecology of Sumbawa's forests, the delivery of such large quantities of sappan wood on an annual basis was simply impossible.

The shortages of wood reported in 1665 and 1672 were not the last. In fact, they marked the beginning of an unremitting cycle of crises followed by upsurges in production which continued until far in the nineteenth century. Sappan shortages, then, turned out to be a recurrent problem. This provides a final argument in the discussion of whether the trees were wild or cultivated. Shortages, of course, suggest that the people of Sumbawa did not plant the trees – or at least if they did, they obviously did not plant enough of them.

The first reports of sappan shortages in Bima and Dompo should have been cause for alarm, but initially the Dutch did not heed the warning signals. They did complain, however, about the diminishing quality of the wood. The pieces delivered became smaller and smaller, a development to which the VOC responded simply by lowering the price paid per *pikul*.[24] It was, in fact, evidence of incipient overexploitation. The most mature trees had already been cut and the harvesting was proceeding at such a fast pace that young trees did not get the chance to mature. At that time the sole concern of the Company was short-term profit and no countermeasures were taken.

Only at the beginning of the following century did this attitude change. In 1702 the Dutch expected to need a quantity of 8,000 or 9,000 *pikul* of sappan wood. When they arrived on the island they found that the realm of Sumbawa had 3000 *pikul* ready for shipping, leaving 6,000 *pikul* still to be found.[25] Of this amount, the Dutch decided in April 1702 that Bima had to deliver 1,000, Sumbawa 1,500, Dompo 1,000, Tambora 1,500, Sanggar 500,

23 *Daghregister* 3-2-1681.
24 *Daghregister* 18-5-1680.
25 ARA VOC 1663:32-9.

and Papekat 500 *pikul*. In July the Dutch noted that the realms were slow in meeting their obligations and offered a bonus to the cutters, especially to those bringing in the more massive trunks. Their reasoning, in the words of a Company official, was that 'those few stivers do not matter as long as the goods are worth having'.[26]

In the meantime a certain Sandertz was ordered to find out what the forests of Sumbawa looked like, where they were located, and how much wood they could yield annually. It was discovered that the realms of Pekat, Tambora, and Sanggar had no forests of their own but had to make agreements with the Sultan of Dompo in order to obtain the wood. In return for wax and other products they were allowed to do their felling in the forests of Dompo, which were located near the coast and were very rich in wood. In October 1702 it was also discovered that the forests of Bima were far away and totally depleted.[27] This time the Dutch really could not turn a blind eye to the situation.

On 22 January 1703 it was therefore decided in Batavia that because the forests of Bima were 'too chopped out', it would be forbidden for anyone to cut sappan there again for a period of six years. In this way young trees would get the chance to mature and reach a useful size again (*Realia* 1882-5, III:162). In February 1704 it was also decided to impose a limit of 2500 *pikul* (155 tonnes) on the total annual sappan wood harvest.

So far I have found only scattered information about sappan production and export in the period 1710-1790. In 1717 and 1718 the Dutch again complained about the small, narrow pieces of wood delivered to them by Bima (*Generale Missiven* 1960-88, VIII:357). Still, VOC representatives continued to accept and pay for these pieces because they could still be employed in the dye industry (*Realia* 1882-85, III:162). We have already seen that around 1750 the Dutch started looking for sources of sappan wood in areas outside Sumbawa, such as East Java, because selling the wood in the Netherlands was still extremely lucrative and probably also because the deliveries from Sumbawa were too irregular. In December 1759, for example, the Sultan of Bima was (again) allowed to suspend his deliveries of sappan wood for a period of three years (*Realia* 1882-85, III:162).

From December 1760 onwards sappan wood was no longer dispatched to the Netherlands (*Realia* 1882-85 III:162). This related to the increased availability and import of other similar but even better species of dye wood. From then on the foremost importer of the sappan wood from Sumbawa (via Batavia) was Japan. According to the harbourmaster's records for Batavia (which can be found in the VOC archives) during the 1770s, about 7,000 *pikul* of sappan wood came from Nusa Tenggara annually (it is most likely of course that this wood came from Sumbawa) (Gerrit

26 ARA VOC 1663:32.
27 ARA VOC 1663:108, 152.

Knaap, personal communication). From the General Report for the year 1792 it is known that in that year 9,174 *pikul* of sappan wood arrived in Batavia from Sumbawa.[28] Of this amount, Bima supplied 7,140 *pikul* and the realm of Sumbawa 2,074. In 1794, a respectable 7,429 *pikul* of wood again arrived from the island, 5,358 from Bima and 2,071 *pikul* from Sumbawa.[29] So it seems that there were few problems with the production of sappan wood during these years.

It is almost certain that at the beginning of the nineteenth century part of the sappan wood production from Sumbawa was used on Java for colouring batik and other textiles. According to Crawfurd sappan wood was still exported to China and Europe in this period (Raffles 1817, I:204; Crawfurd 1820, I:462, III:422). Through the years the price paid for the wood continued to be the same, about 1 rixdollar per *pikul*. The English, who took over the sappan monopoly from the Dutch during the interregnum of 1811 to 1816, also paid 1 rixdollar or about *f* 2.50 per *pikul*.[30]

In April 1815 Mount Tambora on the island of Sumbawa exploded with a ferocity that hurled some 100 km^3 of debris into the atmosphere and caused the infamous 'year without summer' throughout the world in 1816 (De Jong Boers 1995). As a consequence the production of sappan wood came to a complete standstill. Some 10,000 Sumbawans died immediately, and many more in the aftermath as a result of famine and disease. Others fled to surrounding islands. There was therefore an enormous shortage of labour during the first decades after the eruption.

The sappan trees themselves were relatively unaffected by the eruption because they mostly grew in the uplands, where the choking layer of ash was more quickly removed by rain and wind than in the lowlands.[31] In 1821, in fact, sappan wood was one of the few products still available on Sumbawa (Reinwardt 1858:320-1). But because there were too few people to cut the wood, almost none was exported. The labour shortage was particularly acute in the region of Dompo, where the sappan trees were now able to mature undisturbed. As a result the forests were restocked with an abundance of large trees (Zollinger 1850:142).

During the late 1820s the export of sappan wood was resumed. The Dutch wanted Bima to supply 2,000 *pikul* per year. Very soon (1831), however, another shortage of sappan trees was reported in Bima.[32] The colonial government therefore requested the rulers of Bima to leave the trees alone for one year. The price paid per *pikul* was around *f* 4.00 in 1831 and *f* 4.25 in 1843 – in other words higher than ever, perhaps because there

28 ARA OIC 92 (Makassar 1793).
29 ARA OIC 62 (Makassar 1794).
30 Memorie P.T. Chassé 1816 (ARA Schneither 126).
31 Goethals 1961:19; Noorduyn 1987:125-30; Vetter in Kuperus 1937:22.
32 C.H. Steijns, Generaal verslag afdeling Bima en onderhorigheden 1833 (ANRI Makassar 362.14).

was not much wood available at the time.[33] All sappan wood from Sumbawa was sold to Japan for *f* 7.23 per *pikul.*

Around 1850, deliveries became highly irregular again even though Dutch merchants continued to value the wood as a profitable commodity in their trade with Japan. In 1850 the Governor of Celebes and Dependencies made new contracts with Sumbawa in which it was specified that Bima had to deliver 65 *koyang* (160 tonnes), Sumbawa 40 *koyang* (100 tonnes) and Dompo 20 *koyang* (50 tonnes) of sappan wood yearly (*Koloniaal Verslag* 1849:193). It quickly became apparent, however, that these targets were too optimistic. In new contracts signed in 1857 and 1858, the amount to be delivered by Bima was lowered to 50 *koyang*.[34]

Immediately after these new contracts were concluded, the value of sappan wood on the world market began to fall steadily until the Dutch were making losses rather than profits on the trade.[35] By 1872 a *pikul* of sappan wood was worth only *f* 1.51 on the world market. The invention of artificial red dyes such as alizarine in England was responsible for this. Natural red dyes were no longer in demand (Noorduyn 1987:121). In a new agreement concluded with the realm of Sumbawa in 1875, the clause stipulating forced deliveries of sappan wood was omitted. A similar revised agreement with Bima followed in 1886.[36]

The abolition of the sappan deliveries was probably welcomed by the Sumbawan population. It had always been the ordinary people who were forced to do the felling, whereas the sultans received the lion's share of the payments. The woodcutters themselves probably received little more than 0.75 rixdollar per *pikul*.[37] Locally the wood is still used today for the construction of houses and boats. An extract of the wood is also used for medicinal purposes (Hitchcock 1983:117, 1996:145).

The environmental impact of sappan exploitation

If we estimate the average production of sappan wood for the period 1675-1875 at 100 tonnes (about 1,600 *pikul*) yearly – a rather low estimate – then about 20,000 tonnes of wood were exported over the two centuries in question. If we take a higher estimate of 200 tonnes (3,200 *pikul*) per year, we arrive at 40,000 tonnes of wood over the same period. At first sight these figures seem impressive, and certainly they indicate a much higher rate of production than, for example, that of sandalwood in Timor and surrounding areas (Zwart 1937:927).

33 Francis 1856-59, II:134; ARA MvK Zakelijke Aantekeningen 872 (Gouvernement van Makassar):372-7.

34 ARA MvK Zakelijke Aantekeningen 877 (Celebes en onderhoorigheden).

35 Collectie A. Ligtvoet, *Historisch woordenboek van Zuidwest-Celebes* (KITLV Hisdoc H 606 C III):983.

36 ARA MvK Gouvernement Celebes en Onderhorigheden 1870-1905, index 14.

37 ARA VOC 1647:103.

What we need to know, however, is the actual number of sappan trees that were cut on the island of Sumbawa through the centuries. Unfortunately, precise data regarding the weight of wood supplied by the average tree are lacking. However, we do know that sappan wood is a fairly heavy wood weighing between 600 and 780 kg per m^3 (Lemmens and Wulijarni-Soetjipto 1992:61). Zwart estimated the specific gravity of this wood at 700 kg per m^3 (1937:926). In addition, we know that the trunks of sappan trees reach a maximum diameter of 14 cm (Lemmens and Wulijarni-Soetjipto 1992:61). In Sumbawa, however, most must still have been considerably smaller when they were felled. According to H.F.G. Trouwerbach in his General Report on Sumbawa for the year 1845, the diameter of trunks of sappan trees in the realm of Bima was seldom larger than 5 cm.[38]

If we take these figures of 700 kg per m^3 and a diameter of 5 cm as a starting point for further (admittedly rather speculative) calculation, and if we estimate the average height of a tree on Sumbawa at 5 metres, then the weight or yield of the average tree was:

pi x r^2 x usable height x specific gravity
or
x 0.025 x 0.025 x 5 x 700 = 6.87 kg

If the yearly average production of sappan wood was 100 tonnes, this would mean that the number of trees felled per year was 100,000/6.87 or 14,556. If we take the higher estimate of 200 tonnes per year, on the other hand, we arrive at a figure of 29,112 felled trees. The likely range, then, is between 15,000 and 30,000 trees yearly, or between 3 and 6 million trees for the whole of the period 1675-1875. Assuming that every sappan tree occupied 80 m^2 of the land surface (corresponding to a distance of about 5 metres between each tree), then each year an area of between 1.2 km^2 (15,000 x 80 m^2) and 2.4 km^2 (30,000 x 80 m^2) must have been deforested. This is between 0.0077 and 0.015% of the total land area of Sumbawa (15,600 km^2) – in other words, an insignificant proportion. The environmental impact, at least in terms of hydrological effects and erosion, must therefore have been minimal.[39]

Had the forests been homogeneous, then the felling might still have led to appreciable erosion in particular localities. The forests of Sumbawa, however, did not (and do not) consist only of sappan trees. They were strongly heterogeneous, containing a wide variety of trees and shrubs of which only the sappan trees were targeted for harvesting. Other species were left untouched, which minimized the potential even for local erosion.

38 H.F.G. Trouwerbach, Algemeen verslag afdeling Bima 1845 (ANRI Makassar 3.8).

39 It was definitely negligible compared to that of the commercial logging that has taken place since 1972 on the slopes of Mount Tambora. The homogeneous rain forest (of *Duabanga moluccana* trees) is being felled at great speed for commercial purposes, while at the same time the *ladang* cultivators are blamed for the deforestation (Liman 1983:6; Hitchcock 1984:32).

The question of sustainability

If a sustainable harvesting system is defined as one which has the ability to operate indefinitely, or at least for a very long period, at a constant level of production, then sappan exploitation on Sumbawa was clearly not sustainable. Production was not constant but fluctuated, and shortages of wood were a recurrent problem. In the short term, however, they were also a temporary problem. Every time Sumbawa experienced such a shortage, production always revived again fairly quickly thereafter (usually after a few years of respite from felling activities). This resilience is related to the natural vigour of the sappan, which is a fast-growing tree and can be harvested after only a few years' growth (ideally after six to eight years when its trunk diameter approaches 14 cm).[40]

The vigour of the tree is probably also the reason why the Dutch never evinced any great anxiety about running out of sappan wood. Initially they simply tried to obtain as much wood as they could lay their hands on in order to maximize immediate returns, and the attitude of the sultans of Sumbawa seems to have been similar. The predictable result was over-exploitation and shortages of the wood, but the Dutch were slow to take countermeasures. Later on they did develop some sort of policy, but this was reactive rather than anticipatory in nature. Every time a shortage occurred their response was simply to prohibit the felling of sappan trees for a few years, then resume cutting again until the next shortage loomed. Although systematic planting of sappan was technically feasible and would have enabled them to prevent such periodic crises, they presumably regarded it as uneconomic given the speed with which the sappan stock could replenish itself naturally when given the chance.

Looking back at the whole record from today's vantage point, paradoxically, the continuity of production over a very long period is more impressive than its short-run fluctuations. After all, reasonable amounts of wood continued to be produced for more than two centuries, albeit not at a constant level. Due to the high regenerative power of sappan trees, the harvesting system never needed to be changed in any radical way.

By way of comparison, it would have been a completely different story had sandalwood rather than sappan trees been involved. Sandalwood can be felled only after 50 years of growth. In the last century sandalwood trees with a trunk of more than 100 cm in girth still grew on Timor, but by about 1920 almost all of these larger trees had been cut down (De Voogd and De Grijp 1937:202) and they have never reappeared since. Nowadays only sandalwood trees of very small girth can be found on Timor. In this case the harvesting system was clearly unsustainable both in the short term and in the long term.

40 It takes about six to eight years for the heartwood to become fully developed (Lemmens and Wulijarni-Soetjipto 1992:61).

The conclusion, therefore, must be that whether a system of timber exploitation is sustainable or not depends first of all on the regenerative power of the tree itself. Secondly, it is also a matter of time-scale. Viewed from a short-term perspective, sappan harvesting as practised on Sumbawa could not continue indefinitely and was thus unsustainable. In the long run, however, the system survived more than two centuries and suddenly looks sustainable after all.

Today, I think, there is probably hardly a trace of the once so important trade in sappan wood to be seen on Sumbawa. Given that commercial interest in this wood is virtually nil nowadays, it might be assumed that sappan trees on Sumbawa are bigger and more beautiful than ever. Alas, as I discovered in 1996, the opposite is true. Sappan trees are now very few and far between. Although their wood is still occasionally used for specific house and boat construction jobs, in the eyes of the modern Sumbawans they are more or less worthless. Unaware of their glorious past, they simply chop them down and use them for firewood.

Abbreviations and glossary

ANRI	Arsip Nasional Republik Indonesia
ARA	Algemeen Rijksarchief, Den Haag
Hisdoc	Afdeling Historische Documentatie (KITLV)
KITLV	Koninklijk Instituut voor Taal-, Land- en Volkenkunde
koyang	a unit of weight of 40 *pikul* (thus about 2500 kg)
Makassar	Makassar residency archive (ANRI)
MvK	Archive of the Ministerie van Koloniën (ARA)
OIC	Archive of the Oost-Indisch Comité (ARA)
pikul	a unit of weight of 61.76 kg (literally: a shoulder load)
Schneither	G.J.C. Schneither collection (ARA)
VOC	Verenigde Oost-Indische Compagnie; (in ARA) Overgekomen brieven en papieren van de VOC

References

Boomgaard, P.
1995 'The VOC trade in Asian forest products in the 17th and 18th centuries'. [Unpublished paper.]

Brascamp, E.H.B.
1918 'De bosschen in de memorie van overgave van den gouverneur van Java's Noord-Oost-Kust Johannes Vos in 1771', *Tectona* 11:294-7.

Brewer, J.D.
1979 *Agricultural knowledge and cultural practice in two Indonesian villages*. [PhD thesis, University of California, Los Angeles.]

Bruijn, J.R., F.S. Gaastra and I. Schöffer
1987 *Dutch-Asiatic shipping in the 17th and 18th centuries*. Vol. 1. The Hague: Nijhoff. [Introductory volume.]

Burkill, I.H.
1966 *A dictionary of the economic products of the Malay Peninsula*. Kuala Lumpur: Ministry of Agriculture and Cooperatives. 2 vols. [First published 1935.]

Coen, J.P.
1919-53 *Jan Pietersz. Coen; Bescheiden omtrent zijn bedrijf in Indië*. Edited by H.T. Colenbrander and W.Ph. Coolhaas. 's-Gravenhage: Nijhoff. 7 vols.

Corpus Diplomaticum
1907-55 *Corpus Diplomaticum Neerlando-Indicum; Verzameling van politieke contracten en verdere verdragen door de Nederlanders in het Oosten gesloten, van privilegebrieven aan hen verleend, [...]*. Edited by J.E. Heeres and F.W. Stapel. 's-Gravenhage: Nijhoff. 6 vols.

Cortesão, A. (ed.)
1944 *The Suma Oriental of Tomé Pires; An account of the East, from the Red Sea to Japan, written in Malacca and India in 1512-1515 [...]*. London: Hakluyt Society. 2 vols. [Works Hakluyt Society, second series 89-90.]

Crawfurd, J.
1820 *History of the Indian archipelago; Containing an account of the manners, arts, languages, religions, institutions, and commerce of its inhabitants*. Edinburgh: Constable. 3 vols.

Daghregister
1888-1931 *Daghregister gehouden int Casteel Batavia vant passerende daer ter plaetse als over geheel Nederlandts-India, 1624-1682*. Edited by J.A. van der Chijs et al. Batavia: Landsdrukkerij, 's-Gravenhage: Nijhoff. 31 vols.

Dove, M.
1984 'Man, land and game in Sumbawa, Eastern Indonesia', *Singapore Journal of Tropical Agriculture* 5-2:112-24.

Encyclopaedie
1917 *Encyclopaedie van Nederlandsch-Indië*. Vol. 1. 's-Gravenhage: Nijhoff, Leiden: Brill.

Francis, E.
1856-59 *Herinneringen uit den levensloop van een 'Indisch' ambtenaar van 1815 tot 1851*. Batavia: Van Dorp. 3 vols.

Generale Missiven
1960-88 *Generale missiven van Gouverneurs-Generaal en Raden aan Heren XVII der Verenigde Oostindische Compagnie*. Edited by W.Ph. Coolhaas [vol. 1-8] and J. van Goor [vol. 9]. 's-Gravenhage: Nijhoff. [RGP, Grote Serie, 9 vols.]

Goethals, P.R.
1961 *Aspects of local government in a Sumbawan village (eastern Indonesia)*. Ithaca, New York: Modern Indonesia Project, Cornell University. [Monograph Series.]

Greshoff, M.
1894 *Nuttige Indische planten*. Amsterdam: De Bussy.
Heyne, K.
1927 *De nuttige planten van Nederlandsch-Indië*. Second edition. Batavia: Departement van Landbouw, Nijverheid en Handel in Nederlandsch-Indië. 3 vols.
Hitchcock, M.J.
1983 *Technology and society in Bima, Sumbawa, with special reference to house building and textile manufacture*. [PhD thesis, Oxford University.]
1984 'Is this evidence for the lost kingdoms of Tambora?', *Indonesia Circle* 33 (March):30-5.
1996 *Islam and identity in eastern Indonesia*. Hull: University of Hull Press.
Jong Boers, B. de
1995 'Mount Tambora in 1815; A volcanic eruption in Indonesia and its aftermath', *Indonesia* 60:36-60.
Koloniaal Verslag
1849 *Koloniaal Verslag 1849*. 's-Gravenhage: Algemeene Landsdrukkerij.
Kuperus, G.
1937 *Het cultuurlandschap van West-Soembawa*. Groningen/Batavia: Wolters. [PhD thesis, Rijksuniversiteit Utrecht.]
Lemmens, R.H.M.J. and N. Wulijarni-Soetjipto (eds)
1992 *Plant resources of South-East Asia. Vol. 3: Dye and tannin-producing plants*. Bogor: PROSEA.
Liman, Puspa Dewi
1983 'Tambora Utara sebagai suaka margasatwa', *Kompas* 25-9-1983:6.
Noorduyn, J.
1987 *Bima en Sumbawa; Bijdragen tot de geschiedenis van de sultanaten Bima en Sumbawa door A. Ligtvoet en G.P. Rouffaer*. Dordrecht: Foris. [KITLV, Verhandelingen 129.]
Paul, H.
1984 *Nederlanders in Japan, 1600-1854; De VOC op Desjima*. Weesp: Fibula-Van Dishoeck.
Radermacher, J.C.M.
1786 'Korte beschrijving van het eiland Celebes en de eilanden F[l]oris, Sumbauwa, Lombok en Bali', *Verhandelingen van het Bataviaasch Genootschap van Kunsten en Wetenschappen* 4:143-96.
Raffles, T.S.
1817 *The history of Java*. London: Black, Parbury and Allen. 2 vols.
Realia
1882-85 *Realia; Register op de generale resolutiën van het Kasteel Batavia* 1632-1805. Leiden: Kolff. 3 vols.
Reinwardt, C.G.C.
1858 *Reis naar het oostelijk gedeelte van den Indischen archipel, in het jaar 1821*. Edited by W.H. de Vriese. Amsterdam: Muller.

Rumphius, G.E.
1741-55 *Het Amboinsche kruid-boek [...].* Amsterdam/'s-Hage/Utrecht: Changuion. 7 vols.

Smith, G.V.
1977 *The Dutch in seventeenth-century Thailand.* Dekalb, Ill.: Northern Illinois University. [Center for Southeast Asian Studies Special Report 16.]

Volkstelling 1930
1936 *Volkstelling 1930. Vol. V: Inheemsche bevolking van Borneo, Celebes, de Kleine Soenda Eilanden en de Molukken.* Batavia: Departement van Economische Zaken/Landsdrukkerij.

Voogd, C.N.A. de, and A. de Grijp
1937 'Iets over het sandelhoutbedrijf op Timor', *Tectona* 30:202-16.

Whitmore, T.C.
1985 *Tropical rain forests of the Far East.* Second edition. Oxford: Oxford University Press.

Zollinger, H.
1850 'Verslag van eene reis naar Bima en Soembawa en naar eenige plaatsen op Celebes, Saleyer, en Floris gedurende de maanden mei tot dec 1847', *Verhandelingen van het Bataviaasch Genootschap van Kunsten en Wetenschappen* 23:1-224.

Zwart, W.
1937 'Uit de boschgeschiedenis van Java en Madoera, Deel II: handelshout-soorten en boschproducten', *Tectona* 30:917-41.

LESLEY M. POTTER

A forest product out of control
Gutta percha in Indonesia and the wider Malay world, 1845-1915

This chapter focuses on the latex gutta percha, the most valuable commodity harvested from the 'Malesian' dipterocarp forests in the second half of the nineteenth century.[1] It examines the collection and marketing of 'gutta', and the efforts of the colonial powers, especially the Netherlands and Britain, to control its supply and quality. The gutta percha experience had an important influence on the development of policy towards both the forests and their indigenous inhabitants, in whose hands its collection largely remained. Evidence is presented on the ecological impact of harvesting and the role of this commodity in local and wider economies and in cultural interaction. Various strategies - indigenous, trader and colonial - are interwoven in the political ecology of gutta percha, while parallels are drawn with aspects of modern theory concerning non-timber forest products.

Collecting from the forests of insular Southeast Asia: some background

Forest products have long been harvested for local use in Southeast Asia and traded both inside and outside the region. Dunn believed such trading to have taken place along the Malaysian coast from as early as 4000 years ago (Dunn 1975), although Wolters believed that the important Indonesia-China links were not established until about the fifth century AD, the trade with India two centuries earlier (Wolters 1967). China imported camphor (*Dryobalanops* Sp.), benzoin gum (*Styrax* Sp.), and pine resin from Indonesia, while aromatic woods such as sandalwood and *gaharu* (*Aquilaria* Sp.) were mentioned in Indian sources. Animal-derived commodities such as rhinoceros horn, bear's bile, and bezoar stones (from the kidneys

* I would like to acknowledge the assistance of Carin van Empel (Amsterdam) and Elisabeth Kat (Canberra) in locating additional materials, and of Johanna van Woerkom (Adelaide) in translating sections of some documents.

1 'Malesia' includes the biogeographical realm of present Malaysia, Indonesia and the Philippines, in which dipterocarp forests form the most characteristic element of the natural vegetation.

Kayan working gutta-percha (Hose and McDougall 1912:plate 86.)

of certain monkeys or porcupines) were important in Chinese medicine.

Most forest products were not perceived to be 'appropriate' to any European market (Wong 1960:21). The most sought-after items attracting European interest, notably pepper, cloves, and nutmegs, rapidly became domesticated (Pelzer 1978). Exceptions which could be marketed in Europe were dyestuffs such as sappanwood (*Ceasalpina sappan*) and a few specialty items including rattans, benzoin (sometimes also domesticated), sandalwood, and lac.

The monopoly control of much of the spice trade secured by the Dutch East Indies Company (VOC) during the 'age of commerce' from the mid-fifteenth to mid-seventeenth centuries (Reid 1993), was followed during the eighteenth century by its decline and ultimate demise. The temporary eclipse of the Dutch was accompanied by the rise to prominence of Britain, with extension of the interests of the East India Company eastwards to Tenasserim (Burma) and the Straits of Malacca. The founding of Penang in 1786 and Singapore in 1819, together with the British/Dutch exchange of Bengkulu in western Sumatra for Malacca in 1824, established Britain's strategic position in the Straits. By 1824 the most important products of the archipelago, exchanged in Singapore for cloth from Britain and India and tea from China, included a mixture of planted and collected items: pepper, coffee, spices, rattans, camphor, gold, benzoin, and beeswax (Wong 1960:39). It was particularly tea which drew China more firmly into European trading networks, as demand for the beverage increased through the eighteenth century and into the nineteenth (Warren 1981:3). A wider range of forest and marine products from Southeast Asia, desired in China in exchange for tea, were thus integrated indirectly into the new trade system dominated by England, with Singapore as the hub.

With changes in technology in Europe, new products became important and old ones were revalued. At times some products became the focus of frenzied 'rushes', following increases in European demand and prices. As collection often resulted in the destruction of the host plant, such rushes into the forests would be followed by attempts at official control. Colonial authorities feared the imminent disappearance of the resource, yet found making and implementing policy in the little-known forests, towards people they considered 'wild tribes', both difficult and costly. The most interesting new forest products were the various wild rubbers, for which Europe discovered an increasing need from the 1840s.

The gutta percha case-study

Gutta percha is a latex extracted from various species of the genera *Palaquium* and *Payena* of the family sapotaceae, tall trees scattered through the dipterocarp forests of equatorial Southeast Asia. The substance has been described as 'the best non-conductor known for both heat and electricity' (Foxworthy 1922:162). It will soften with heat, when it may be moulded, and hardens again when cooled. The latex occurs in a number of intercellular cavities in the tree rather than in connected tubes, which makes it more difficult to extract than, for example, Para rubber (*Hevea brasiliensis*). Methods of collection of gutta percha invariably began with the felling of the tree. Incisions were then made in the bark right around the trunk, allowing the latex to coagulate in the cuts. In 1840 its properties were known to some inhabitants of the Malesian forests, but many did not collect the latex; the tree, which was fairly common, had other uses. Malays made handles, buckets, and whips from gutta percha: after European interest was aroused new applications were found. Dr Montgomerie, Surgeon to the Straits Settlements, suggested its trial in surgical instruments in 1843; the first specimens to reach Europe arrived in London in the same year. Experiments followed in insulating telegraph wires, and by 1847 European demand was growing. This interest expanded rapidly after 1849, following the successful testing in the English Channel of a submarine cable coated with the latex (Foxworthy 1935:1625; Wong 1960:99). The market for gutta percha was thus tied to the growth of cable communications throughout the world. Demand fluctuated, becoming especially strong during periods of concentrated laying of undersea cables towards the end of the nineteenth century. At the time there were no substitutes, although several other latex varieties were tried. The area of natural production was largely confined to the Malaysian Peninsula, Sumatra and Borneo, with parts of Sulu, Southern Mindanao, and Sulawesi being of lesser importance. During the whole of this period, in spite of efforts at raising the tree on plantations, and in creating 'reserved forests' for its preservation, almost all the harvesting was in the hands of indigenous forest people. Their activities were financed by merchants who increasingly were Chinese with close links to the centre of the trade in Singapore.

The first twenty-five years, 1845-1869

The collecting of gutta percha for the international market began around 1845 in the forests of Singapore Island. By 1847 all the large 'Gutta Tuban' trees were said to have been felled and the centre of activity moved to Johore (Oxley 1847). The trees had various names in the places to which the trade extended, leading to considerable confusion, especially as there were many varieties in each location and inferior guttas abounded, lacking the special characteristics of gutta percha. All the 'guttas' contain a proportion of resin, the best as little as 10%, which determines how brittle they will eventually become. It was many years before it was definitively established that a total of 23 species of *Palaquium* and *Payena* were responsible for the commercial-grade latex (Foxworthy 1922:162, table).

In 1847 the indigenous Binua people were one of several groups in Johore working gutta percha for the Malays controlling the river mouths. Their industry was studied by Singapore scientists Logan and Oxley, who noted the sizes and distribution of the trees and yields of latex. The trees were both large 'from 60 to 80 feet in height, and from two to three feet in diameter' (Logan 1847:262) and widespread 'forming, in many spots, the principal portion of the jungle' (Oxley 1847:24). They grew luxuriantly, especially on alluvial fans, but were felled for harvesting, as it was argued that the latex would not run by tapping. Yields per tree averaged 3-5 catties (1.9-3.1 kg), up to a maximum of 18 catties (11.2 kg). While the people insisted that their felling would not 'extirpate' the trees as they took only the largest specimens, Oxley disagreed, commenting that 'despite its apparent abundance and widespread diffusion, the Gutta will soon become a very scarce article if some more provident means be not adopted in its collection' (Oxley 1847:24).

Concern was thus felt from the beginning that harvesting techniques were unsustainable, yet there was every encouragement to expand the trade. The Temunggung of Johore declared the latex a government monopoly and organized its collection energetically, bringing a group of hereditary serfs from Batam and sending out parties into the surrounding islands and north into Pahang. He provided incentives for local Malays and Chinese to make large profits in barter exchanges with forest-dwelling groups (Logan 1848:529). The Binua collected gutta percha as one of a number of commodities, including rattan, camphor, and damar, which they traded sporadically for cloth, articles of earthenware and iron, tobacco, betelnut, and lime. It appeared to have no specific place in their culture, although the seeds were eaten. Being a new product, it was not surrounded by ritual like the collecting of camphor, which required the use of a special language.

In 1848, the gutta percha supply area extended from Perak and Pahang south to Palembang and the Riau-Lingga archipelago. It was estimated that 270,000 trees were felled during the first three and a half years (Logan 1848:533). The collectors gradually spread the information to the other districts, Logan noting 'the slow indigenous process by which useful knowledge is diffused' (Logan 1848:531). During the 1850s the island of Borneo and the more remote parts of the Malay Peninsula became the main sources as the forests nearer to Singapore became exhausted. By 1853 the trade from the Riau-Lingga archipelago was already much diminished and exports had decreased. The trees were sprouting again, but it would be 30 years before they could be re-tapped (De Bruijn Kops 1853). Table 1, on the next page, indicates the rapidly changing supply situation to the mid-1860s.

Gutta percha collecting was thus introduced into parts of Indonesia very soon after the trade had begun. Unlike the Johore enthusiasm, early Dutch descriptions mention resistance to the new product and a careful appraisal of its profitability. In Palembang an army officer, commissioned to test its

Table 1. Imports of gutta percha into Singapore by value, 1846-1866.

Year	Sumatra	Malay Peninsula	Borneo
1845-1846	Nil	Nil	Nil
1847-1848	$7,177	$5,239	$30
1848-1849	$62,074	$42,735	$4,914
1852-1853	$67,127	$46,988	$40,440
1855-1856	$22,502	$73,050	$99,436
1865-1866	$48,231	$139,317	$237,871

Source: Wong 1960:99

usefulness in manufacturing water bottles, reported that the native inhabitants could not be persuaded to deliver the latex, even at *f* 25 per *pikul*.[2] They cited difficulty in finding the trees, the small yield (only 10 catties), and the fact that the tree was used by bees, giving honey and wax. It was actually valued more for beeswax than gutta percha, so would only be cut when very old (*Berigten* 1851; Teijsman and Binnendijk 1854).

A report from the Sukadana region of the Western Division of Borneo (*Nadere berigten* 1858) revealed an equal sophistication among the inhabitants in comparing the relative returns from various forest products. The Dayaks and Malays would not collect indiarubber (*Willughbia* Sp.), as its price was too low compared with what they could receive daily from rattan, damar, wax etc., which they could always find quickly. The higher price of gutta percha was the only reason for them to work with this new product, the trees of which were found singly or in small groups growing in uninhabited forest. The gutta seekers would leave for the bush in the dry season in groups of five or six and stay there for one or two months, by which time such a group would have collected one *pikul* (61.8 kg) of latex. For each *pikul* one hundred trees would have been felled (*De Javasche Courant*, 6 July 1853, in De Vriese 1855-56).

Collecting also spread throughout the Kapuas basin of West Borneo. From Sintang, far upriver, came an authoritative report by the explorer Von Gaffron, who added thoughtful comments on the general state of the forests and the future of the gutta percha tree. Because of Dayak shifting cultivation practices, in which a new piece of cleared forest was opened every two years, Von Gaffron believed that original forests could no longer be found in Borneo (Von Gaffron 1858:224). Despite Dayak clearing, gutta percha trees were still common, so obviously regenerated quite easily. They did not become productive until 35 years old, so only the oldest were worth cutting, while the others continued growing and would eventually take their place. This was a clear statement of the theory of selective harvesting, later to be applied in the timber industry. Von Gaffron

2 *f* 25 was equal to 8.26 Mexican dollars or Straits dollars in 1851 (Van der Eng 1993).

reiterated that the gutta percha tree must be felled to collect the latex, preferably during the period of the new moon and in the morning, when the sap was most liquid. An energetic workman, cutting two trees per day, could extract from each up to 12 catties of gutta percha. If the method was tried of leaving the tree standing and boring holes in the bark, only one catty would be collected and the tree was likely to die from fungal infections.

Von Gaffron listed eight types of latex-bearing tree, classifying their product according to quality. The three least marketable species were usually mixed with the better varieties, in what had become standard adulteration practices (Von Gaffron 1858). A further report from Banjarmasin in Southeast Borneo identified nine varieties of *ngiatoe*, the common local name. English botanist and mining engineer Motley described a process of indigenous experimentation in which many kinds of gum had originally been mixed together. Those products were rejected by the international market, forcing the natives 'who had already begun to look to the collection of gutta percha as a principal means of subsistence' to find and identify species with the required characteristics (Motley in De Vriese 1860:305). The original '*gitta taban*', of Singapore and peninsular Malaysia, although known, was quickly worked out. Surrogates such as '*getah merah*' and '*getah durian*', from different species of *Palaquium*, eventually formed the main export varieties (Ozinga 1940:139). Attempts to plant the approved species on low-lying land near Pontianak, Sambas, and Sukadana all failed. In 1856 a number of plantings were tried in Java (as in Singapore ten years before), but the tree was slow-growing and proved to need special care in its seedling stages so that light conditions in the forest were replicated (see note 7).

A survey undertaken in West Sumatra provided further details of harvesting practices. Dutch officials observed that it was possible to collect the latex without felling the tree, but the amounts obtained were small, and there were many difficulties involved. The tree trunks were very wide, straight, and tall, the trees virtually 'unclimbable' (In 't Veld in *Mededeelingen* 1866:60). In general no replanting was done and some reports gave a clear indication that the resource was declining. While there used to be many trees near villages, now people had to travel two or three days into the primary forest to find them (Visser in *Mededeelingen* 1866).

One novel attempt to alter collecting techniques involved sending Ngaju Dayaks (from the Lower Barito in Southeast Borneo) to Malacca to learn improved tapping methods. Following the convulsions of the Banjarmasin War (1859-1863), exports of gutta percha were low in Southeast Borneo, most of the harvest coming from the Upper Barito, unsafe for Europeans because of continuing rebel activities (Tromp 1874). However, the Lower Barito and its Dayak population was under Dutch direct rule. The experiment failed, according to official sources, because the people were too 'lazy and indolent' and too homesick. The reverse procedure, introducing Malays to do the instructing, was rejected as risky because of the Dayak reputation for head-hunting (Perelaer 1870:202).

The first twenty five years of gutta percha export saw considerable experimentation with the new product. Initial excitement was followed by more sober assessments as it was clear that there were good reasons why tapping methods were not adopted, and that the resource in its marketable form, that of the large trees, was disappearing from the more accessible forests. Although the level of tree destruction was lamented, little could be done about it by either the British or Dutch in this period, because they lacked administrative authority over most of the collecting areas. In 1865 the Dutch set up a forest service in Java, but their administration was nominal in most of the Outer Islands and they had only sketchy knowledge of many interior districts until later in the century.

One might compare the position of camphor, another popular forest product. The Nederlandsch-Indische Maatschappij van Nijverheid en Landbouw (Netherlands-Indies Company of Industry and Agriculture) was concerned that local Bataks in North Sumatra were annually destroying a number of camphor trees without any replanting. However, the idea that government should undertake their cultivation was not well received. The Resident of Tapanuli stated that the trees were reproducing satisfactorily and were not very suitable for plantations: not only did they grow slowly, but only one tree in every fifty contained the valuable product (*Kamferboom* 1872:101). Van Gorkom, in charge of the cinchona cultivation in Java, was negative: 'If one requires that the Government be responsible for the care of all naturally occurring vegetable products of the country which can be useful for trade and industry, then one demands perhaps a bit much. Besides camphor, would the rattan, wild cinnamon [...] and numerous other forest products equally deserve attention?' The answer was to bring some seedlings to Java and grow them there 'independent from the lawless Batak' (Van Gorkom in *Kamferboom* 1872:105, 110).

Gutta percha and other forest products, 1870-1915

In the late 1860s, prices began to rise for gutta percha. The first successful trans-Atlantic undersea cable was laid in 1866; that achievement, together with the opening of the Suez Canal in 1869, brought an impetus to further construction. The great expansion in demand meant that all types of gutta increased in value. The price rise of 1870 caused a rice shortage in Sarawak 'by tempting the population away from their usual occupations, such as farming, gardening etc.' (*Sarawak Gazette* 15-2-1872). Gutta percha became Sarawak's leading forest products trade item by value, with several agents of Singapore firms moving there to begin dealing in the product (Smythies 1961:174, table; Wong 1960:99). It was later estimated that three million trees were felled in Sarawak between 1854 and 1875, the rate of harvesting increasing towards the end of the period.[3] In Sintang, West Borneo, the high prices for gutta had led to a putting aside of all other work (*Koloniaal Verslag* 1870). Figures available from Banjarmasin also indicated rising production: the harvest of gutta in 1870 was worth almost four times that of rattan, reversing the position of a few years before, when rattan had been far more significant (Tromp 1874).

In 1877-1879, two years of severe drought followed by widespread flooding brought almost complete loss of crops in many districts. These disasters resulted in greater intensity of forest products collection. In 1879 Dayaks in the Western Division of Borneo bartered gutta percha for rice provided by Chinese traders. The figures for latex exported from Pontianak showed a jump from 237,556 kg in 1878 to 1,060,971 kg in 1879 (Van Gorkom 1884:603). Governor Senn van Basel, travelling up the Kapuas in 1879, encountered a group of Dayak gutta percha gatherers in the Tayan District. Although it was planting season, the group had been unable to resist the advances offered by Chinese merchants (Senn van Basel 1880:86). The Dayaks had assembled in a large body to defend themselves against warlike groups, for they were venturing inland 'along [...] unbeaten paths, far from their own territories from which the trees had already been stripped (Senn van Basel 1880:89). In Kutai (East Borneo) the story was similar. According to Bock (1881:118), the Malays (Banjarese) of Kota Bangun went into the forest to seek gutta percha in groups of twenty or thirty 'for mutual protection', afraid of being attacked, and often accompanied by friendly Dayaks. While noting the distress caused by drought, flood, and famine, Bock did not make any association between the collecting activities he was observing, and the shortage of food, although he was informed that many Dayaks had subsisted for over a year on roots and wild fruits.

The argument continued in all the producing territories as to whether some method of harvesting could not be found which would make it unnecessary to fell the tree. Differences could sometimes be distinguished between the behaviour of groups collecting from their own forests, and that of others, either working with advances from outside employers or ranging independently, seeking sources of dry-season income. Explorer Witty, travelling in 1881 near Marudu Bay in the early years of British North Borneo (Sabah), noted the destruction of 'gutta shrubs', which he blamed on 'loafers crossing into Labuan and Brunei'. The indigenous Dusun population south of Kinabalu had perfected their own tapping system, making cuts in the bark of their trees, which were very large, up to a hundred feet high and six feet in girth. A fine of a buffalo was imposed on any one of

3 *Sarawak Gazette*, 1-5-1875. There seems to be some doubt as to which period this figure refers to, however. In a Minute appended to letter 17938, 10-10-1894, Secretary Kew Gardens to Secretary of State for the Colonies, by 'E.F.(airfield)', the three million trees were said to have been felled in ten years 'in the early days of the trade', CO 273/200 Great Britain, Colonial Office, 1894.

their number who felled a tree.[4] The Dusun system is of interest as it included regulations aimed at conservation, but this applied only over a limited area. Burbidge, another commentator on the North Borneo scene, had also been impressed by the size of the trees, but he did not find any 'conservation ethic' in the forests he surveyed. Instead he noted that not only gutta percha, but the *Willughbia* vines yielding indiarubber, had been cut down. His description was graphic: 'It is most deplorable to see the fallen gutta trees lying about in all directions in the forest, and the rubber-yielding willughbeias are also gradually, but none the less surely, being exterminated by the collectors here in Borneo, as indeed, throughout the other islands and on the peninsula, where they also abound' (Burbidge 1880:75). Unless they could protect their trees, indigenous collectors would be justifiably fearful that others would invade their territories and help themselves.

There were unsuccessful attempts to promote tapping in parts of the Malay Peninsula. Following the establishment of British administration in Perak in 1874, Iban Dayaks were brought from Sarawak to look for gutta (Foxworthy 1935:1628). The Resident realized by 1880 that the larger trees were being destroyed and prohibited exports (Wray 1883:208). When trade was resumed in 1889, would-be cutters were supposed to obtain a tapping licence, but there was a rapid increase in extraction rates and more signs of forest decline. In more distant Pahang, prohibitions on working gutta percha were issued in 1895, but it was found that the productive forests lay near the borders of Kelantan and Terangganu (outside direct colonial intervention). The forests were simply invaded by cutters from those states and the latex smuggled out. In 1896 the Pahang forests were let out to Chinese traders in yearly leases, giving the holders an incentive to prevent outsider incursions. A prohibition was placed on 'foreign Malays and Dyaks', said to be the most destructive. However, the Forest Officer commented in 1899: 'It is, at present, just as impossible for the Government to stop the collection of gutta as it is to supervise its collection. It has already been tried and the result was that the gutta was still collected and exported via Kelantan and Terangganu'.[5] A report issued in 1901 by Mr Hill of the India Forest Service noted that the protection of gutta trees anywhere in the Federated Malay States had totally failed. He added: 'it is very difficult to get about the forests, for there are few roads, no reliable transport and little labour available. The indigenous tribes and parties of seasonal immigrants, Dyaks from Borneo have it, therefore, all their own way. Prohibitions are of no effect as they cannot be enforced' (Hill 1901, as quoted in *The Indian Forester* 27-2:101).

[4] F. Witty 1880-1. Diary, Journey to Marudu Bay (8 and 9 May). CO 874/74 1880-81, British North Borneo Company Papers, Singapore University Library.

[5] A. Butler, 1899. Report on Pahang getah, Federated Malay States, Forest Conservancy. Great Britain, Colonial Office, CO 273/263, no. 460.

The establishment of a central forest administration for the Federated Malay States and Straits Settlements (which took place in 1902, incorporating the limited services operated by individual states) was directly attributable to the perceived need to protect gutta percha. Almost immediately, a high export duty was placed on the product, the felling of trees for latex extraction was prohibited and reservations were declared over all the valuable *Palaquium* areas (Burn-Murdoch 1905). A modern writer on forest management in Malaysia has characterized the period from 1900 to 1922 as 'the gutta percha era' (Thang 1987:7).

To ensure compliance with teak forest regulations, a contingent of forest police had been formed in Java in 1880, to be incorporated in the Java Forest Service ten years later (Peluso 1992). For the Dutch territories outside Java, however, both a forest department and any kind of enforcement agency were still some way off. It was recognized as being a hopeless task to try to protect gutta percha in the Outer Islands without an organized forest service.[6]

In a report on gutta percha in August 1885, the *Java Bode* claimed that the collection of this product yielded very little advantage to the Netherlands, as almost all was forwarded to Singapore and the British market. The newspaper, evidently believing that gutta percha was easy to reproduce in plantations, strongly recommended the setting up of plantations 'in the chief towns of Netherlands India' (translated version in *Sarawak Gazette*, 1 September 1885). But this had already been tried without success.[7] Nevertheless, the Botanic Garden laid out a plantation in 1885 at Cipetir in West Java, using seedlings from Sumatra. The *Palaquium* varieties did not acclimatize well and many seedlings died. When the Forestry Department took over in 1890, they found the plantation in poor condition and filled the gaps with faster growing *Payena* species. This move was criticized ten years later by the director of the Botanic Garden, as *Payena* species were considered inferior (Van Romburgh 1900a:185, 200). In response, forester Ham replied that the longer dry season in Java, plus the elevation of the plantation (1700 m) were not very suitable for *Palaquium*, while *Payena* performed very well (Ham 1900). It was later suggested that trees taken out of their normal climatic zone would not produce good latex, so that the *Palaquium* at Cipetir yielded a resinous product, while the despised *Payena* (an upland variety) was superior in its natural conditions (Combanaire 1910:202-3).

Another possible solution to the question of gutta percha supplies lay in

6 W. Buurman, 'Nota voor den Directeur van Binnenlandsch Bestuur betreffende het in bescherming nemen van *Getah-pertja* produceerende boomen in de bosschen der buitenbezittingen (23-4-1890)', ARA MvK, verbaal 28-7-1890, no. 5.

7 In 1856, on the initiative of Teysmann, camphor, cinnamon, and gutta percha plantations were established in four residencies of Java. Twenty years later a search was made for these trees: in only one district had any gutta percha trees survived but they were hardly producing any latex (Van Gorkom 1884).

a different direction: stripping the bark and/or leaves from the tree and extracting the latex by chemical or mechanical means. After many atttempts, in 1890 two French scientists were successful using mechanical extraction from the leaves, which they brought from Sarawak to their small factory in Singapore. Two Dutch brothers associated with this venture petitioned the Minister of Colonies in 1892 for a concession near Palembang to plant gutta percha and harvest the leaves and boughs. The Director of the Botanic Garden rejected the request, pointing out that experiments in gutta percha planting were already being undertaken in Java.[8] The process of mechanical extraction from the leaves was eventually perfected in a much larger plant associated with the plantation at Cipetir (Van Gelder 1950:479; Foxworthy 1935:1632). Another factory operated by British capitalists was established in Pahang (Foxworthy 1935) and there was one in Sarawak. An observer commented that although the Sarawak factory processed leaves, it would not prevent tree destruction as the collectors would simply fell the trees and collect both leaves and latex at the same time. The only solution was to obtain the leaves from a supervised plantation (Sherman 1903:32). The Singapore plant and others in West Borneo soon suffered from shortages of leaves and were forced to close (Ozinga 1940; Van Gelder 1950). The Cipetir operation survived. The quality of the product was much improved and the tall gutta percha tree was eventually reduced to a low bush from which the leaves were collected (Department of Agriculture, Industry and Commerce 1926). A similar technique was tried by the forest department of the Federated Malay States in 1913-1914, in which the crowns of a group of trees were cut and the stems and leaves harvested. Five months later new leaves had grown, so that continuous harvesting was demonstrably possible (Z 1915).

Sustained rises in the price of gutta percha in the last decade of the nineteenth century, culminating in a 'boom' from 1899-1901, led to concerns from other nations about Britain's monopoly on the product, and prompted those with tropical colonies, such as the French in Indochina, to seek control of their own supplies. The French sent observers to study the industry in Malaya, while a 'Gutta Percha Commission' established in Saigon was instructed to seek out sites for plantations and suitable varieties (De la Pinardière 1894; Lourme 1899). Adventurer and gutta percha expert Adolphe Combanaire's secret mission to West Borneo in 1899 was to find gutta percha seedlings. He was interested not in *Palaquium*, but *Payena*, which grew in remote mountain locations. He avoided contact with Dutch officials and moved through the forests with small parties of mainly Dayak assistants. A keen biologist, Combanaire was able to throw light on the essential role of bats in the propagation and distribution of

8 'Verzoek van twee Heeren Ledeboer om medewerking van de Regeering voor de uitvoering van hun plan omtrent de aanleg van getah pertcha aanplantingen in Palembang', ARA MvK, verbaal 20-8-1892, no. 57, and 31-8-1892, no. 36.

gutta percha species (Combanaire 1910:197-8). Eventually he found the trees he was seeking in the almost untouched Schwaner Mountains, describing with joy 'the legions of gutta percha trees raised up in their majestic splendour [...] the smooth trunks reaching three metres in circumference' (1910:302). Gathering a hundred seedlings and wrapping them in bamboo leaves, the party after some difficulties managed to descend the Pembuang River to the coast of Southeast Borneo.

Combanaire's book provides a unique glimpse of the Dayaks, their culture and forest products, and of the Chinese traders who were buying their latex, at a time when prices in Singapore were reaching record levels. The author was astonished by the contemptuous and shameless way in which the Dayaks were cheated by the traders. He understood why the Dayaks preferred to travel long distances to trade across the Sarawak border, where, although cheating did occur, officials were more careful to check scales and would imprison merchants whose behaviour was too outrageous (1910:195, 235-6). Beccari, quoting from published accounts in the *Sarawak Gazette* of 1901, had noted similarly that a party of 500 Kayan had crossed from Bulungan (East Borneo) to the Upper Rejang, a journey which took five months, to trade their gutta percha and indiarubber 'valued at thousands of dollars'. They found the Rejang markets preferable to those in their own district, 'as long as they were on good terms with the hivan (Iban) Dayaks' (Beccari 1904:362). The Iban, who had been encouraged by the Brooke regime to travel widely in search of the 'jungle products', on which the revenue of the state largely depended, were known for behaving in a 'truculent' manner towards other groups and retained a reputation for head-hunting (Hose and McDougall 1912, I:150; Pringle 1970:267-79).

One reaction to the high prices of gutta percha at the end of the nineteenth century was the beginning of indigenous planting out of the trees, noted in both West Sumatra and West Borneo. In Padang a consortium of merchants, protesting government plans to tax forest products, argued that as in earlier times the population had domesticated nutmeg, benzoin, damar, copra, and gambier, which were then traded as agricultural products, they were now beginning to lay out regular plantings of gutta percha and indiarubber, which should be treated the same way (*Padangsche handel* 1899). Dayaks in the Sanggau District of the Kapuas basin planted gutta percha trees in their forest gardens or *tembawang*, old longhouse sites containing many domesticated forest fruits (Ozinga 1940:143; Van Gelder 1950:479).[9]

Ozinga argued that the subsequent decline in gutta percha prices restricted the further spread of this cultivation. Probably more important

9 Kantu' people near the Sarawak border in West Kalimantan also recall their ancestors planting *Palaquium*, though this was presumably the lower quality *jangkang* species (*Palaquium leiocarpum*), which grows on low ground liable to flood (Van Romburgh 1900b). Dove gives the Kantu' name for gutta percha as *jangkang* (Dove 1994b, 1995).

was the availability within a few years of Para rubber (*Hevea brasiliensis*). Despite negative attitudes by the local administration, by 1915 the coastal Malays and Chinese of Sambas were enthusiastically planting rubber, and this cultivation spread widely among the inland Dayak populations of the Kapuas basin during the 1920s, using seeds brought by Chinese merchants in return for a portion of the eventual crop (Ozinga 1940:329; Jackson 1970:78). Given the long time needed for its maturation and low yield on tapping, gutta percha was not very suited to domestication, unlike 'jungle rubber', which could be easily accommodated into small-scale agricultural systems (Dove 1993a, 1994b; Gouyon et al. 1993).

The second forty years of the gutta percha trade saw more attempts to understand the different species involved. Other rubber-producing plants were also studied because of an increased demand for indiarubber. In Indonesia the research of Burck (1883, 1884) was comprehensive, and he was later followed by Van Romburgh (1897, 1900a, 1900b). Burbidge (1879) and Wray (1883) were carrying out similar studies in the Malay Peninsula and North Borneo. Wray noted 'the mystery in which [...] botanical identification has been hitherto so completely enshrouded' (Wray 1883:209-10). The mystery was compounded by the inaccessibility of many of the forests. As the best varieties disappeared, even in the more distant areas, the quality of the product declined. Van Romburgh (1900b) drew attention to the large amounts of 'inferior guttas', such as *jelutong* and *jangkang* or *hangkang*, among the exports from Banjarmasin and Pontianak in 1897 and 1898. Eventually these substances were separated from the 'genuine' gutta percha in the trade figures.

Considerable organization went into this trade. While the basic collection continued to be carried out by indigenous people, the products were exchanged by barter with Malay or other traders, then sold to merchants, many of them Chinese from Singapore. By 1900 the direct role of Chinese had greatly increased (Dunn 1975:108). In some cases the Chinese actively sought to wrest control from the Malays by offering higher prices to local collectors, as occurred in the Baram River in 1872 (Denison 1882:176). In the 1890s numbers of Chinese from the former gold-mining districts of West Borneo had turned their attention to forest products and other trade along the Kapuas, with Sintang becoming an important collecting centre (Jackson 1970:76). In areas such as Southeast Borneo, the Banjarese were the active traders, with the Chinese visible only in the towns. Banjarese gradually replaced Buginese along the Mahakam (Magenda 1991) and were to be found along all the rivers of what is now Central Kalimantan. European trading houses, such as the Borsumij (Borneo-Sumatra-Maatschappij), also began to handle exported forest products such as gutta and rattan from the 1880s (Lindblad 1988:11-4).

To what extent were indigenous people able to profit from the system? There is no doubt that extensive cheating occurred, particularly over the weight of the product and the value of the goods given in return. It was claimed that people in the more remote districts received from Chinese traders only 7.5 cents worth of goods per pound of gutta percha (Z 1915). This example concerned trade with Punan or other hunter-gatherers, who did not seek face-to-face contact but left goods along river banks to be bartered for salt and other commodities. Some of the Iban migrating long distances from and within Sarawak were experienced collectors, quite conversant with the traders and their wiles.[10] To the Iban, wealth came in the form of ceramic jars; for Kenyah and Kayan aristocrats it meant items such as rare beads, for which they were willing to pay high prices. Lian argues that many moved into the Baram River from the Apo Kayan and other remote areas during the 1880s because prices were higher downriver for forest produce (Lian 1987:67). Within the stratified groups (which did not include the Iban), it was often the lowest class, captives from battles or hereditary slaves, who actually collected forest products. In the Upper Barito, a mini-sultanate set up by rebel princes after the Banjarmasin War was found to be still using Dayak slaves and *pandelingen* (debt-bondsmen) in forest-products collection in 1901 (Engelhard 1901:194). Many compensated for the low returns by using their knowledge of the diverse species to adulterate the product 'the natives guard the secret of the different kinds of Gutta-percha trees and their locality' (Sherman 1902). Such adulteration became more widespread as prices increased (Senn van Basel 1880:90), but so did the vigilance of the traders (Combanaire 1910:161). The practice of adulteration was perfected further by Chinese middlemen in Singapore, so that eventually the product became unreliable and the price declined.

Table 2 compares the relative importance of gutta percha and other leading commodities in the Straits Settlements' (largely Singapore's) east-west export trade over the period. It is notable that tin dominated throughout, though Para rubber came close to equalling it in 1915, having risen quickly from a very low base. Gutta occupied its highest position in terms of percentage in 1870 and was reasonably high again in 1900, but had dropped very considerably by 1915. Eighty per cent of the gutta percha trade passing through Singapore at the end of the nineteenth century came from the Dutch East Indies territories of Borneo and Sumatra (Chiang 1978:199, table).

Figure 1 shows the changes in prices over time for gutta percha and other major commodities. It is clear that while gutta percha was the highest-priced item, it also experienced the most extreme fluctuations.

10 The Iban travelled to 'North Borneo, Dutch Borneo, Sumatra, the Malay peninsula and even as far as Mindanau' (Baring-Gould and Bampfylde 1909:376).

Table 2. Major commodities in Straits Settlements east-west exports 1870-1915 (%) (Chiang 1978).

Year	Tin	Gutta	Rubber	Rattans	Pepper	Gambier	Total
1870	37.15	16.63	0.34	4.41	21.01	14.01	93.55
1875	30.68	2.34	0.86	3.91	29.12	23.92	90.83
1880	27.46	7.80	1.63	8.00	14.36	16.91	76.16
1885	31.88	6.55	1.18	5.47	20.43	11.57	77.08
1890	39.56	10.65	1.67	5.58	18.28	13.12	78.86
1895	48.43	4.83	2.91	5.37	8.08	12.71	82.33
1900	53.09	13.19	1.44	5.92	8.38	5.83	87.85
1905	62.52	3.41	4.38	3.54	7.72	4.74	86.31
1910	48.41	6.38	15.26	2.55	5.14	3.69	81.43
1915	43.13	1.30	36.91	2.08	4.35	0.61	88.43

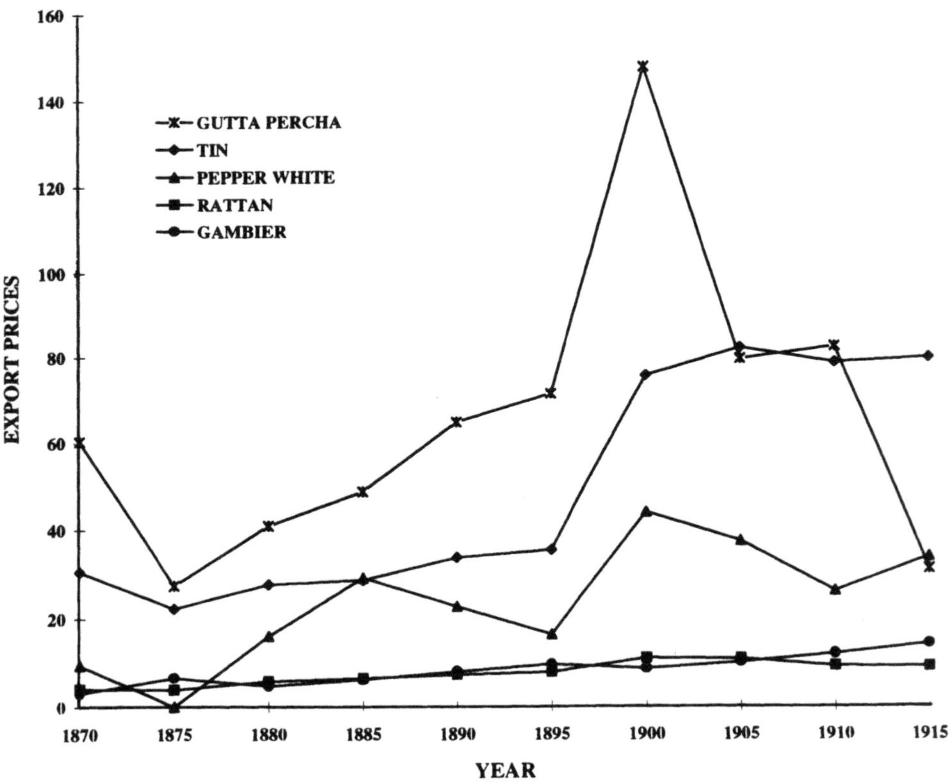

Figure 1. Average export prices (in Straits dollars) per *pikul* in Singapore for leading commodities, 1870-1915 (Chiang (1978).

A note on jelutong[11]

One forest product over which the Dutch Colonial administration took more direct action was *jelutong* (*Dyera* sp.). A tall swamp timber yielding a true rubber, it grew abundantly in the low-lying coastal districts of eastern Sumatra, Sarawak, West Borneo, and Southeast Borneo. During the rubber boom of the early 1900s, *jelutong* became valuable, leading to a rush of outsiders (mainly Banjarese) into the area around Sampit on Borneo's south coast, especially after 1908. Hearing reports that large numbers of trees were being destroyed, the local administration at first tried licencing tappers. When that failed, the area was opened to concession by European firms considered to have an interest in proper tapping methods and in tree protection. The chairman of the Borsumij, prominent in developing and financing the industry, conducted a vigorous campaign against such concessions, claiming that local livelihoods were being destroyed. The concession did not function for long, but it had the assistance of special '*jelutong* police', provided by the local government to keep poachers out of its lease. As the area was huge, the boundaries poorly marked, and local tappers ignorant of the new rules, many were imprisoned and their produce confiscated. This action caused uproar and the subject was raised in the Dutch Parliament. Wider questions were asked concerning peoples' free rights to gather forest products versus the rights of the state to dispose of forest land. Eventually the immediate 'problem' solved itself: prices fell for *jelutong*; the concessionaire failed and departed; the outside labourers returned home and a less damaging tapping method was introduced. A Forestry Department was established in Banjarmasin incorporating the *jelutong* police, but the ramifications of the questions of state rights versus peoples' rights did not go away. These issues were argued over fiercely for the rest of the colonial period and inherited by the Indonesian Government. Their echoes are still to be found in contemporary struggles in the forests.

The '*jelutong* question' is perhaps an obvious finale to the thwarted attempts by the Dutch colonial authorities to implement policy towards forest products in the Outer Islands. Here was a resource which appeared easily accessible (or at least coastal) and valuable, where a conservationist argument could justify the creation of concessions and the police could control indigenous activities. The tappers were largely Banjarese ('Malay') with no obvious traditional links to the area. They were seen as simply a labour force, capitalizing on a boom, with a little help from the Borsumij and the Chinese money-lenders financing them. Yet there was a pre-existing resident population in the area, including Dayaks who identified the first tapper of a tree as its traditional owner. If these forests were their main source of income, did they not have a right to be there?

11 See Potter 1988:130-4 for greater detail.

Relating the case-study to the modern theories

Discussion will follow under four headings: 1. ecology and environmental impacts of harvesting; 2. markets, technological change, and domestication; 3. culture, economy, and indigenous strategies; and 4. colonial policy and its implementation. In each case the evidence from the gutta percha experience will be compared with one or more aspects of modern theory and conclusions drawn.

Ecology and environmental impact of harvesting

It has been claimed that the collection of non-timber forest products (NTFPs) can have serious ecological impact, particularly high-intensity harvesting which 'can gradually eliminate a species from the forest' (Peters 1996:6). If the product is in the form of large adult trees, it will disappear much faster. Peters' conclusion is that the ecological impact 'can be as devastating as logging in causing the disruption of local populations and species extinction' (1996:9).

The gutta percha case-study provides qualitative information about the impact of intensive harvesting, mainly of large adult trees, on particular species. Given the enormous numbers of trees claimed as being felled and the long period over which gutta percha was in demand, it is perhaps surprising that the two important genera, *Palaquium* and *Payena*, were not eliminated completely from the forest. With the diversity of species, it is impossible to know whether there was actual extinction of some of the more marketable types, as was claimed: Van Romburgh (1900b:578) certainly believed that the proportions of inferior guttas were increasing. It also appears that natural hybridization could occur, making it more difficult to assess changes.

The few studies of growth and regeneration carried out on any kind of scientific basis were largely confined to the gutta percha reserves of the Malay Peninsula, in which the trees were allowed to re-grow from about 1902 under more or less natural conditions (Burn-Murdoch 1905:36-7; Foxworthy 1922:164). Before the establishment of such reserves, it was stated that regeneration had been impossible: 'The natives cut every available tree, and repeat the process as fast as they spring up again; they have thus suppressed for the last 40 years their reproduction and multiplication' (Serullas 1891). By this time tree sizes had become progressively smaller in heavily worked-over areas: 'Poles thirty to forty feet high are fairly common [...] but large trees are rarities' (Burn-Murdoch 1905:311). That there was very considerable re-growth after 1902 is unquestionable, but there was also less pressure on the species, allowing a tapping regime to be perfected by the 1920s. Mead (1925) wrote that the tapping method was being slowly introduced into Sarawak, while Mjöberg (1930) has a plate depicting a tapped gutta percha tree (plate 61,

p. 149).[12] There were no comparable studies in the Outer Islands of Indonesia. The plantation at Cipetir in Java was more of a commercial operation in a controlled environment, although limited experimentation with different species was attempted.

It is also important to note the thorough way in which the Malesian forests were worked over by human populations during the nineteenth century. Von Gaffron claimed in 1858 that there were no longer any 'natural' forests remaining, the oldest trees he saw being only 150 years old. If one adopts that kind of time frame, then the equatorial forests of Malesia may well be viewed as a human artefact (see Brookfield this volume).

Markets, technological change and domestication

Ruiz-Pérez (1995) introduces the idea of a 'backbone' or 'attractor' product, 'articulating the extractive economies and supporting the market around them'. The disappearance of such a product would have a serious impact, affecting all trading and market relations. Gutta percha was undoubtedly an 'attractor' during the study period, largely because of its high price, inspiring great feats of collection. As the price declined after 1902, in a number of areas it came to be replaced quite quickly by *jelutong*, which partly explains the enthusiasm with which that commodity was sought, even though the growing and collecting conditions were different.

Homma's (1992) paper is concerned with changes in extraction through time, suggesting three distinct developmental phases: expansion, stagnation, and decline. While 'boom-and-bust cycles' have often typified NTFP, Homma goes further in claiming that all products will either be domesticated or replaced by substitutes in the final phase. Ruiz-Pérez (1995) suggests that product depletion and market demand lead to domestication, while technological change is largely industry-driven. Both issues were of great importance in the political ecology of gutta percha. While tapping techniques were eventually workable, so was collecting the sap from the leaves, which necessitated a quite different kind of product, low-growing and domesticated, backed up by the necessary machinery and far removed from 'wild forest people'. The desire of the market to secure both continuity and reliability of the product became a driving force behind the experiments in alternative sources.

Acclimatization of every useful commodity in Java, where it could be controlled and manipulated like a teak plantation, was the unvarying answer of the Dutch to all questions on the handling of forest products. Local domestication *in situ*, while attempted, was eventually overtaken by smallholder rubber, a more appropriate product for a sedentarized

12 Yet exports from Sarawak in 1920 included only 24 tons of gutta percha as against *gutta jangkar*, 71 tons and jelutong, 5,227 tons (Smythies 1961).

society.[13] Homma's domestication thesis is thus partially appropriate, but some products are obviously better candidates for domestication than others. In addition to *Hevea brasiliensis*, there are local forest products, such as rattan and benzoin, which are well suited to planting in fallows and ready to harvest in a few years. More interesting are the slow-growing trees, such as *tengkawang*, ironwood and gutta percha which give no immediate reward but constitute an investment in the future. The fact that Kapuas Dayaks were planting *Palaquium* in their *tembawang* reveals its importance in their economy. The fact that they were doing this in 1908, a boom year for *Hevea* and well after gutta percha prices had begun a terminal slide, reveals that they were probably still unaware of the possibilities of smallholder rubber, or else had a touching faith in the future of *Palaquium* (*Koloniaal Verslag* 1908).

Culture, economy, and strategies of forest people

A controversial thesis expounded by Dove states that 'whenever a forest product becomes valuable in international markets, élites are likely to appropriate it and leave only products of little value to the forest dwellers' (Dove 1994a:1). In the gutta percha situation it was not possible for colonial interests to control the product, given the isolation and difficulties of the forest environment. On that ground the people might be considered to have 'won', although the way the trade was organized, not by the state but by individual commercial interests, ensured that their gains were as low as possible. To modify Dove's thesis, the valuable commodity (gutta percha) was largely appropriated by outsiders: the Chinese traders with their Singapore connections. The Dutch colonial state was weak and ineffective in securing a larger share of the profits, but had neither the resources nor the inclination to intervene. The Brooke regime in Sarawak made some attempts to protect the collectors, but was more reliant on the Chinese merchants for its revenue.

The forest people of Malesia, while always involved in collecting activities, assessed the possibilities of this new product rather carefully and did not react in a uniform manner. There is more material on the Iban than on other groups, because of their special relationship with the Brooke regime, but they should not be seen as necessarily typical of other forest people such as Penan hunter-gatherers, of hierarchical societies like the Kenyah, nor of non-Dayak and Islamic people elsewhere, about whom there is less information. Universal strategies involved techniques to minimize personal danger and to compensate for the low rewards provided by traders. People do not appear passive victims in the face of the machinations of the more powerful, although slaves and those lowest in the hierarchy undoubtedly had fewer options. While fitting in well with

13 Such transitions were slow and uneven: smallholder rubber did not reach the Baram until the 1930s (Lian 1987), although its initial plantings in Sarawak began twenty years earlier.

the swidden farming cycle, forest-product collecting in times of drought or flood became vitally important in obtaining an exchange entitlement to essential food.

It was also something which young men enjoyed doing. Of the Iban collectors, one correspondent wrote: 'These young fellows, for those who go are nearly all under eighteen years of age, are the working pioneers of jungle produce wherever it exists, from the mouth of Banjir Masin round northward to the Kina Batangan river [...] about halfway round the whole island of Borneo' (*Sarawak Gazette* 1-12-1885). It was high adventure, testing one's strength against the perils of the wilderness, including dangerous enemies. The red-rubber rhinoceros and its human rider constructed by the successful Murung collectors of the Upper Barito (Lumholtz 1920, I:124-6) was a celebration of precisely that. As Ruiz-Pérez suggests, collection was important to social and cultural identity. Its modern equivalent has been the scouring of the forests for *gaharu* (*Aquilaria* Sp.), also an occupation indulged in by young men.

Colonial policy and its implementation

There is no doubt that forest products, as exemplified by gutta percha, were affected by the economic, trade, and environmental policies of the states concerned. This product is a good example of the direct involvement of the British in the Federated Malay States and the Brookes in Sarawak in forming environmental and social policy on the ground, as compared with the 'hands-off', more indirect attitude of the Dutch (at least until the *jelutong* affair of 1910). Yet the British policy, relying on prohibitions and export taxes, resulted in failure to protect gutta percha trees. The Dutch, not even attempting to introduce similar regulations, were left with the indirect response: acclimatize the plant on Java, then utilize the new technology of lopping a continually growing supply of leaves. Eventually the same strategy was copied in the reserved forests of the FMS. There was thus a concerted effort to remove the supply of gutta percha from the hands of indigenous people and place it under government control. Such activities appear to support Dove's thesis (1993b, 1994a) that 'little' people always lose control of valuable products. However, the controls exercised by the colonial authorities were very partial and only extended to a portion of the producing area. Despite the existence of government plantations, most gutta percha continued to come from forest collection, carried out by indigenous people in the most remote districts.

Conclusion

In this article I have taken the example of one non-timber forest product, gutta percha, which was of paramount importance in the Malesian forest economies during the second half of the nineteenth century. I have subjected the history of the harvesting of this commodity to detailed scrutiny, examining the activities and strategies of all stakeholders in the forest arena. I have compared the case-study evidence with hypotheses developed to analyse modern collecting activities, in the hope that the past will illuminate the present and current theory will assist in organizing the diverse materials of the past. The forests have shrunk and been converted to other land uses, but the questions raised during the era of gutta percha and *jelutong* remain relevant, while the ancient search for the '*hasil hutan*' continues to be engaged in by local populations.

Abbreviations

ARA	Algemeen Rijksarchief, Den Haag
CO	Colonial Office Records, Public Records Office, London
MvK	Archive of the Ministerie van Koloniën (ARA)

References

Baring Gould, S. and C.A. Bampfylde
1909 *A history of Sarawak under its two white rajahs, 1839-1908.* London: Sotheran.

Beccari, O.
1904 *Wanderings in the great forests of Borneo; Travels and researches of a naturalist in Sarawak.* London: Constable.

Berigten
1851 'Berigten van verschillenden aard; Getah pertjah in Palembang, als artikel van industrie', *Natuurkundig Tijdschrift voor Nederlandsch-Indië* 2:184-8.

Bock, C.
1881 *The head-hunters of Borneo; A narrative of travel up the Mahakkam and down the Barito [...].* London: Sampsom Low, Marston, Searle and Rivington.

Bruijn Kops, G.F. de
1853 'Schets van den Riouw-Lingga Archipel', *Natuurkundig Tijdschrift voor Nederlandsch-Indië* 4:59-60.

Burbidge, F.W.
1879 'Notes on gutta percha and caoutchouc-yielding trees', *Journal of the Straits Branch of the Royal Asiatic Society* 3:52-9.
1880 *Gardens of the sun; or, A naturalist's journal on the mountains and in the forests and swamps of Borneo and the Sulu archipelago.* London: Murray.

Burck, W.
1883 'Eenige opmerkingen omtrent getah-pertja', *Tijdschrift voor Nijverheid en Landbouw in Nederlandsch-Indië* 28:330-59.

1884 *Rapport omtrent een onderzoek naar de getah-pertja produceerende boomsoorten in de Padangsche Bovenlanden.* Batavia: Landsdrukkerij. [Mededeelingen uit 's Lands Plantentuin 1.]

Burn-Murdoch, A.M.
1905 'Some facts about gutta percha', *Agricultural Bulletin of the Straits Settlements and Federated Malay States* 4-2:33-40.

Chiang Hai Ding
1978 *A history of Straits Settlements foreign trade 1870-1915.* Singapore: Memoirs of the National Museum.

Combanaire, A.
1910 *Au pays des coupeurs de têtes; A travers Bornéo.* Paris: Plon-Nourrit.

Denison, N.
1882 'Journal of a trip from Sarawak to Meri', *Journal of the Straits Branch of the Royal Asiatic Society* 10:173-88.

Department of Agriculture, Industry and Commerce
[ca 1926] *Products of the Dutch East Indies; Gutta percha.* Weltevreden: Albrecht.

Dove, M. R.
1993a 'Smallholder rubber and swidden agriculture in Borneo; A sustainable adaptation to the ecology and economy of the tropical forest', *Economic Botany* 47:136-47.
1993b 'A revisionist view of tropical deforestation and development', *Environmental Conservation* 20:17-24, 56.
1994a 'Marketing the rain forest; Panacea or red herring?', *Asia Pacific Issues; Analysis from the East-West Center* 13:1-8.
1994b 'Transition from native forest rubbers to Hevea brasiliensis (Euphorbiaceae) among tribal smallholders in Borneo', *Economic Botany* 48-4:382-96.
1995 'Political versus techno-economic factors in the development of non-timber forest products; Lessons from a comparison of natural and cultivated rubbers in Southeast Asia (and South America)', *Society and Natural Resources* 8:193-208.

Dunn, F.L.
1975 *Rain-forest collectors and traders; A study of resource utilization in modern and ancient Malaya.* Kuala Lumpur: Malaysian Branch of the Royal Asiatic Society. [Monograph 5.]

Eng, P.J. van der
1993 *The silver standard and Asia's integration into the world economy, 1850-1914.* Canberra: Australian National University. [Working Papers in Economic History 175.]

Engelhard, H.E.D.
1901 'De Afdeeling Doessonlanden', *Bijdragen tot de Taal-, Land- en Volkenkunde* 52:179-222.

Foxworthy, F.W.
1922 *Minor forest products of the Malay peninsula.* Singapore: Methodist Publishing House. [Malayan Forest Records 2.]

1935 'Palaquium' in: I.H. Burkill (ed.), *A dictionary of the economic products of the Malay peninsula*, vol. II, pp. 1623-43. London: Crown Agents for the Colonies.

Gaffron, W.G. von
1858 'Nota omtrent den getah-pertja-boom', *Natuurkundig Tijdschrift voor Nederlandsch-Indië* 16:224-8.

Gelder, A. van
1950 'Guttapercha', in: C.J.J. van Hall and C. van de Koppel (eds), *De landbouw in den Indischen archipel*, Vol 3, pp. 476-519. 's-Gravenhage: Van Hoeve.

Gorkom, K.W. van
1872 'Missive van den ambtenaar, belast met de kinakultuur, van 4 Mei 1871, no. 58', in: 'De kamferboom op Sumatra: stukken, gewisseld nopens de kwestie, of voorziening noodig is tegen de uitroeiing van dien boom', *Tijdschrift voor Nijverheid en Landbouw in Nederlandsch-Indië* 17:104-17.
1884 *De Oost-Indische cultures, in betrekking tot handel en nijverheid*. Amsterdam: De Bussy. 2 vols.

Gouyon, A., H. de Foresta and P. Levang
1993 'Does "jungle rubber" deserve its name? An analysis of rubber agroforestry systems in southeast Sumatra', *Agroforestry Systems* 22:181-206.

Ham, S.P.
1900 'De proefaanplantingen van caoutchouc- en getah-pertja-leverende houtsoorten te Tjipetir onder de leiding van het Boschwezen', *Tijdschrift voor Nijverheid en Landbouw in Nederlandsch-Indië* 61:64-77.

Hill, H.C.
1901 *Report on the present system of forest administration in the Federated Malay States, with suggestions for future management of the forests of those states*. Selangor: Government Printing Office.

Homma, A.K.O.
1992 'The dynamics of extraction in Amazonia; A historical perspective', *Advances in Economic Botany* 9:23-32.

Hose, C. and W. McDougall
1912 *The pagan tribes of Borneo [...]*. London: Macmillan. 2 vols.

In 't Veld, A.F.
1866 'Behoort bij missive van den Assistent Resident van Aijerbangies en Rau dd. 16 December 1864 no 1072', in: 'Mededeelingen omtrent de wijze van inzameling en bereiding van getah-pertja en gom-elastiek op Sumatra's Westkust', *Tijdschrift voor Nijverheid en Landbouw in Nederlandsch-Indië* 12:58-60.

Jackson, J.C.
1970 *Chinese in the West Borneo goldfields; A study in cultural geography*. Hull: University of Hull. [Occasional Papers in Geography 15.]

Kamferboom
1872 'De kamferboom op Sumatra; Stukken, gewisseld nopens de kwestie, of voorziening noodig is tegen de uitroeiing van dien boom', *Tijdschrift voor Nijverheid en Landbouw in Nederlandsch-Indië* 17:91-108.

Koloniaal Verslag
1870-1915 *Koloniaal Verslag*. Various years.

Lian, F.J.
1987 *Farmers' perceptions and economic change; The case of Kenyah farmers of the Fourth Division, Sarawak.* [PhD thesis, Australian National University, Canberra.]

Lindblad, J.T.
1988 *Between Dayak and Dutch; The economic history of Southeast Kalimantan 1880-1942*. Dordrecht: Foris. [KITLV, Verhandelingen 134.]

Logan, J.R.
1847 'The Binua of Johore', *Journal of the Indian Archipelago and Eastern Asia* 1:242-93.
1848 'Range of the Gutta Taban collectors, and present amount of imports into Singapore', *Journal of the Indian Archipelago and Eastern Asia* 2:529-33.

Lourme
1899 'Rapport sur les travaux de la commission de la gutta-percha', *Bulletin Economique de l'Indochine* 2-10:320-8.

Lumholtz, C.
1920 *Through Central Borneo; An account of two years' travel in the land of the head-hunters between the years 1913 and 1917*. New York: Scribner's. 2 vols.

Magenda, B.
1991 *East Kalimantan; The decline of a commercial aristocracy*. Ithaca: Cornell Modern Indonesia Project. [Monograph 70.]

Mead, J. P.
1925 'Forestry in Sarawak', *The Empire Forestry Journal* 4-1:92-9.

Mededeelingen
1866 'Mededeelingen omtrent de wijze van inzameling en bereiding van getah-pertja en gom-elastiek op Sumatra's Westkust', *Tijdschrift voor Nijverheid en Landbouw in Nederlandsch-Indië* 12:53-75.

Mjöberg, E.
1930 *Forest life and adventures in the Malay archipelago*. London: Allen and Unwin.

Nadere berigten
1858 'Nadere berigten omtrent het voorkomen en den aard der getah pertja van den Nederlandsch-Indischen [...]', *Tijdschrift voor Nijverheid in Nederlandsch-Indië* 4:396-410.

Oxley, T.
1847 'Gutta Percha', *Journal of the Indian Archipelago and Eastern Asia* 1:22-9.

Ozinga, J.
1940 *De economische ontwikkeling der Westerafdeeling van Borneo en de bevolkingsrubbercultuur.* Wageningen: Zomer en Keuning. [PhD thesis, Rijksuniversiteit Utrecht.]

Padangsche handel
1899 'De Padangsche handel over het heffen van een uitvoerrecht van boschproducten in de Buitenbezittingen', *De Indische Gids* 21, II:1510-3.

Peluso, N.
1992 *Rich forests, poor people; Resource control and resistance in Java.* Berkeley: University of California Press.

Pelzer, K.J.
1978 'Swidden cultivation in Southeast Asia; Historical, ecological, and economic alternatives', in: P. Kunstadter, E.C. Chapman and S. Sabhasri (eds), *Farmers in the forest; Economic development and marginal agriculture in northern Thailand,* pp. 271-86. Honolulu: University Press of Hawaii.

Perelaer, M.T.H.
1870 *Ethnographische beschrijving der Dajaks.* Zalt-Bommel: Noman.

Peters, C.M.
1996 'Observations on the sustainable exploitation of non-timber tropical forest products; An ecologist's perspective', in: M. Arnold and M. Ruiz-Pérez (eds), *Emerging issues in non-timber forest products Research.* Bogor: CIFOR/ODA.

Pinardière, L. de la
1894 'L'industrie de la Gutta-Percha', *Revue des Eaux et Forêts* 33 (deuxième série 8):71-2.

Potter, L.M.
1988 'Indigenes and colonisers; Dutch forest policy in South and East Borneo (Kalimantan), 1900 to 1950', in: J. Dargavel, K. Dixon and N. Semple (eds), *Changing tropical forests; Historical perspectives on today's challenges in Asia, Australasia, and Oceania,* pp. 127-53. Canberra: Centre for Resource and Environmental Studies, Australian National University .

Pringle, R.
1970 *Rajahs and rebels; The Ibans of Sarawak under Brooke rule, 1841-1941.* Ithaca: Cornell University Press.

Reid, A.
1993 *Southeast Asia in the age of commerce, 1450-1680. Vol. 2: Expansion and crisis.* New Haven/London: Yale University Press.

Romburgh, P. van
1897 'Getah pertja; Hare eigenschappen, haar voorkomen en de wijze waarop zij gewonnen wordt', *Teysmannia* 7:37-44, 134-42.
1900a *Caoutchouc en getah-pertja in Nederlandsch-Indië.* Batavia: Kolff. [Mededeelingen uit 's Lands Plantentuin 39.]

1900b 'Iets over djoeloetoeng en djongkang, twee getah-soorten, welke tegenwoordig in groote hoeveelheden gewonnen worden', *Korte Berichten uit 's Lands Plantentuin* 10:577-82.

Ruiz-Pérez, M.
1995 *A conceptual framework for CIFOR's research on non-wood forest products.* Bogor: Centre for International Forestry Research. [CIFOR Working Paper 6.]

The Sarawak Gazette
1870-1901 *The Sarawak Gazette.* Various issues.

Senn van Basel, W.H.
1880 'De onbebouwde gronden op Borneo's Westkust', *Tijdschrift voor Nederlandsch-Indië* 9, I(nieuwe serie):81-101.

Serullas, M.
1891 'Rediscovery of Gutta Percha Tree at Singapore', *Bulletin of Miscellaneous Information, Kew Gardens* 163:230-1.

Sherman Jr, P.L.
1902 'An American report on gutta percha', *Agricultural Bulletin of the Straits and the Federated Malay States* 1:263-6.
1903 *The gutta percha and rubber of the Philippine Islands.* Manila: Bureau of Public Printing.

Smythies, B.E.
1961 'History of forests in Sarawak,' *Sarawak Gazette* (30 September):167-74.

Teijsmann, J.E. and S. Binnendijk
1854 'Over eenige planten, uit de Indischen Archipel, welke gezegd worden getah op te leveren', *Natuurkundig Tijdschrift voor Nederlandsch-Indië* 6:115-8.

Thang, H.C.
1987 'Forest management systems for tropical high forest, with special reference to peninsular Malaysia', *Forest Ecology and Management* 21:3-20.

Tromp, S.W.
1874 'Missive betreffende de rotan en getah-pertjah cultuur in de Zuider- en Ooster-afdeling van Borneo', *Tijdschrift voor Nijverheid en Landbouw in Nederlandsch-Indië* 19:56-62.

Visser, W.G.
1866 [Letter Plangei, 7 November 1864], in: 'Mededeelingen omtrent de wijze van inzameling en bereiding van getah-pertja en gom-elastiek op Sumatra's Westkust', *Tijdschrift voor Nijverheid en Landbouw in Nederlandsch-Indië* 12:70-5.

Vriese, W.H. de
1855-56 *Tuinbouw-flora van Nederland en zijne overzeesche bezittingen [...].* Leiden: n.n. 3 vols.
1860 'Aanteekeningen betreffende getah-pertja-boomen (sapoteen) en getah pertja van Zuid-Oostlijk Borneo, naar aanleiding van ontdekkingen

van James Motley', *Natuurkundig Tijdschrift voor Nederlandsch-Indië* 21:299-315.

Warren, J.F.
1981 *The Sulu Zone 1768-1898; The dynamics of external trade, slavery, and ethnicity in the transformation of a Southeast Asian maritime state.* Singapore: Singapore University Press.

Wolters, O.W.
1967 *Early Indonesian commerce; A study of the origins of Srivijaya.* Ithaca: Cornell University Press.

Wong, L.K.
1960 'The trade of Singapore, 1819-69', *Journal of the Malayan Branch of the Royal Asiatic Society* 33-4:1-315.

Wray Jr, L.
1883 'Gutta-producing trees', *Journal of the Straits Branch of the Royal Asiatic Society* 12:207-21.

Z
1915 'Uit de geschiedenis van de getah-pertja', *De Indische Mercuur* 1915:739-40, 838, 977, 1022. [Translated from *The India Rubber World,* August-November 1915.]

FREEK COLOMBIJN

The ecological sustainability of frontier societies in eastern Sumatra

The swampy east coast of Sumatra is not an environment conducive to large-scale human settlement. The tidal mud with mangrove forest is fertile, but easily develops toxic acid when it is drained. The spongy peat soils more inland are not fertile and subside as soon as they are reclaimed; the impressive rain forest that grows on this soil lives on a balanced nutrient cycle (Furukawa 1994:25-33). Until the nineteenth century food production consisted mainly of fishing and some gardening and shifting cultivation. All main settlements were transit ports, pockets in large stretches of uninhabited land, that for their food supplies and trade goods relied on faraway places, either in the Sumatran highlands or overseas (Andaya 1993:14-8, 191-2; Furukawa 1994:101-11, 139-40). It is therefore not surprising that the first large-scale exploitation of natural resources in this area, which started in the second half of the nineteenth century, was not to yield food, but to exploit one specialized commodity: wood logged in so-called *panglong* (lumber camps). The *panglong* workers gradually opened up the forest, moving on when an area had been clear-felled. For their subsistence they depended wholly on external provisions.

A similar process had taken place in the Riau archipelago, off the Sumatran coast, a century earlier. Although the local soil offered better opportunities for agriculture, the main settlement, Tanjung Pinang on the island of Bintan, functioned as an international port of transit trade and a fishing place (Map 1). The settlement naturally required a site, but did not have its resource base on the land, and Tanjung Pinang needed substantial imports of rice. In Riau too, the first large-scale exploitation of natural resources occurred to produce a commodity, gambier. Once again the production was not sustainable, spurring the gambier planters on to clear more and more forests. The same story, with some modifications, can be told about tin-mining on Bangka, where miners gradually opened up the land.

* I am grateful to Clare Guenther for correcting my English and to Leonard Blussé, Michel Hockx, Tony Reid, and Mary Heidhues for other forms of intellectual input. The Netherlands Foundation for the Advancement of Tropical Research (WOTRO) provided financial support for archival research in Jakarta.

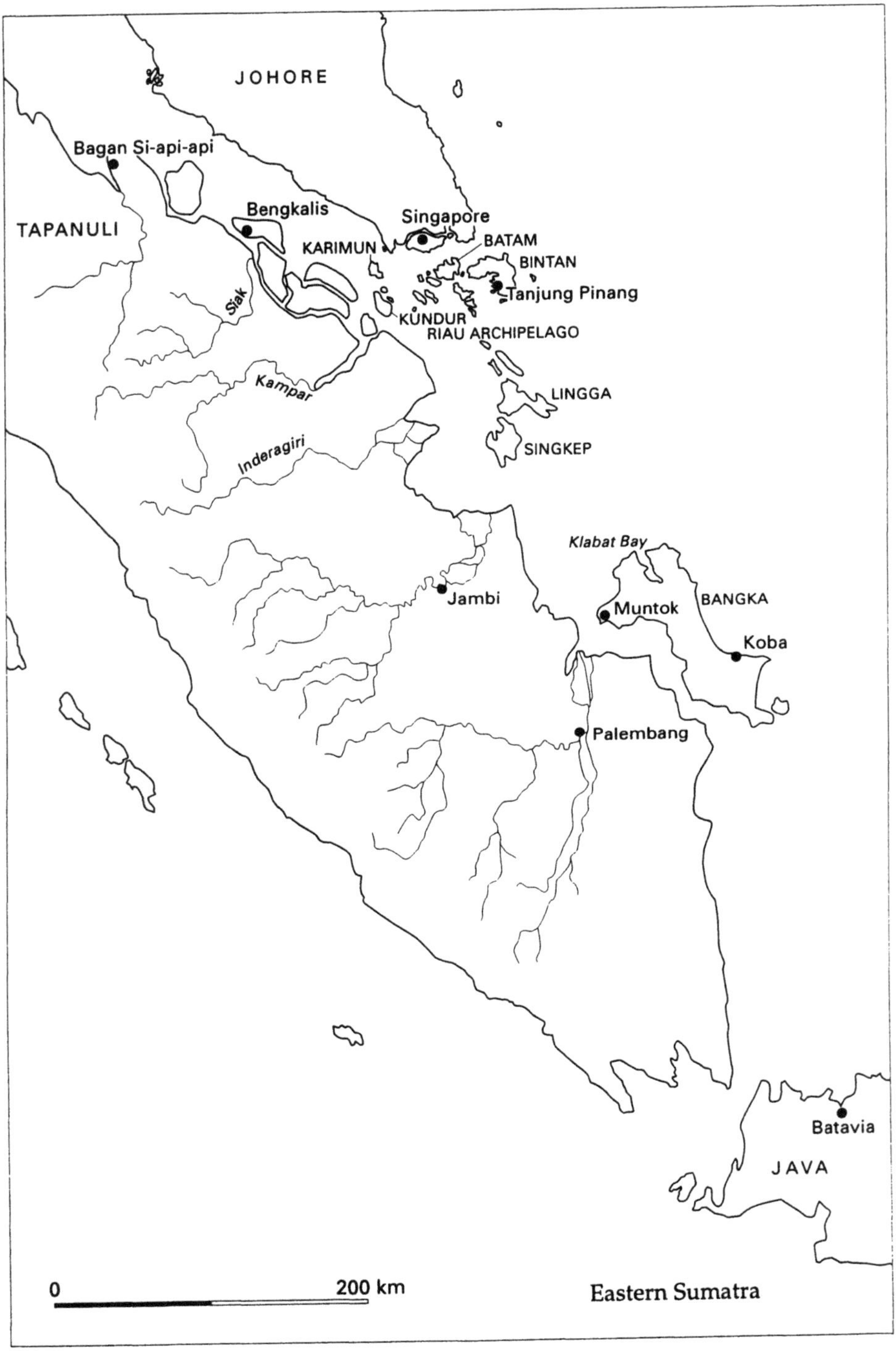

Eastern Sumatra

In these three cases, the continuing search for new resources when older ones were exhausted has inspired several authors to employ the word 'frontier'.[1] With the exception of James C. Jackson, they do so casually, without going explicitly into the question of what constitutes a frontier. Sometimes the term seems to have no particular purpose. The aim of this article is first to explore the applicability of the concept 'frontier society' to eastern Sumatra, and then to reflect upon the consequences of the frontier character of tin-mining, gambier cultivation and *panglong* for the ecological sustainability of these economic activities. Before I commence this exercise, I want to sketch the background of the labourers.

Chinese migration

Since the three geographical areas under discussion did not support a big population, a large labour force had to be brought in from elsewhere for the unexpected new economic opportunities. These labourers were Chinese, and the development of Bangka, Riau, and eastern Sumatra must be seen as part of a centuries-long process of Chinese migration into the South China Sea, or Nanyang. Their entrepreneurial skills, and willingness to take considerable risk, contributed considerably to the development of Southeast Asian economies (Reid 1997).

Chinese traders visited Southeast Asia as private merchants or sometimes as envoys on big imperial fleets. Many of these traders settled in this region either for just one season, for several years or a lifetime. The volume of the migration changed over the years, fluctuating together with the switching attitude of the imperial court towards private overseas trade. The bloody Taiping Rebellion in South China (1851-1864) drove thousands of refugees overseas. Most migrants came from the southern, coastal provinces of Fujian and Guangdong. These people, generally referred to as 'Chinese', actually came from different ethnic groups with various languages and ways of life. In the second half of the eighteenth century coolie migration became more important than merchant migration.[2]

In the nineteenth century Singapore became the hub for Nanyang-bound migrants; from there the migrants spread in all directions.[3] When Thomas Stamford Raffles founded Singapore in 1819, the island was practically uninhabited, but within a year Chinese began to arrive in great numbers (FitzGerald 1993:149; Freedman 1965:7). Migration expanded rapidly

1 Jackson 1965:101-4, 1969:34; Ng 1976; Somers Heidhues 1996:175. See also Breman (1987: 143) for tobacco plantations in Deli (Sumatra), Chew (1990) for Chinese in Sarawak (Borneo), Furukawa (1994:141, 163, 185) for various ethnic groups in eastern Sumatra, and Reid (1993:33-6) for pepper in Sumatra, and sugar cane, benzoinn and tobacco in Southeast Asia.

2 FitzGerald 1993:94-112, 146-7; Freedman 1965; Reid 1996; Somers Heidhues 1992:55.

3 Restrictions on immigration in the Netherlands Indies did not apply to Riau since the area could not be controlled (Cator 1936:34-7, 223).

Tin mine near Belinyu, Bangka, ca 1920 (KITLV photo collection 34570).

when poor peasants sought a living as coolie. They dreamed to return with a fortune in three or four years. The short-term profits sought in the Nanyang were meant to serve the long-term goal of survival of the family at home. The short-term perspective of the migrants contributed to the transient character of the frontier settlements. Few men learnt a language other than their own Chinese dialects, and the extreme imbalance of the sex ratio made the settlements ephemeral from the outset.[4] In reality, most migrants stayed longer than they had thought, often for the rest of their lives. By the time they arrived at their first work-place they had accrued a considerable debt for the fare of their passage from China. Their wages in the first year were usually spent paying off this debt. Payday was once every year, before Chinese New Year. Any savings could be lost to prostitutes and gambling. Both these attractions were lavishly provided at Chinese New Year with the aim of emptying the workers' wallets and compelling them to re-enlist in whatever work they were doing. The forced purchase of provisions at inflated prices from the financier and addiction to opium[5] kept the migrants in a state of permanent indebtedness. In the mid-nineteenth century about 10,000 Chinese arrived annually in Singapore, with 3,000 people returning home. There were also Chinese who stayed voluntarily in Singapore, Tanjung Pinang, and other towns. The urban Chinese population formed a more permanent community and had a more even sex ratio.[6]

Following Lynn White (1968:79-91), some people have argued that Asian traditions of thought foster a more harmonious exploitation of nature than the Judeo-Christian tradition. A few words about this argument seem necessary. The assumption is that each tradition of thought has its own perception of the human-versus-nature relationship, and that this perception strongly influences ecological behaviour of people. In the Judeo-Christian tradition there is a dualism between man and nature; God created plants and animals for the benefit of humans. Put simply: 'Nature, Mr. Allnut, is what we are put into this world to rise above' (Katherine Hepburn to Humphrey Bogart in *The African Queen*, 1951). In contrast, Daoist philosophers from the fourth century BC put forward explicitly anti-anthropocentric aphorisms: 'fish and game had certainly not been created for his [man's] benefit any more than he [has] been created for the benefit of mosquitoes and tigers' (Needham 1956:55). Daoism, the 'Order of Nature', stressed above all the unity of nature, or the continuity of being. Nature is viewed as an enfolding harmony of impersonal cosmic functions. The cosmos – that is, human beings, animals, plants, trees, rivers, rocks, etcetera – is all made of *qi*, 'vital force' or 'matter-energy'. Slightly more

4 For instance, in 1860 the sex ratio of the Chinese in the Riau archipelago was 1700:100. Arsip Nasional Republik Indonesia, Jakarta (ANRI), Riau 58-2, Politiek verslag 1860.

5 In the case of 45% of the *panglong* workers in the 1930s (Cator 1936:224).

6 Jackson 1969:48-9; Pastor 1927:41; Siah 1848; Heidhues 1992:42-3, 55, 152.

recent is the view that man and nature depend on each other as *yin* and *yang*, two complementary forces that presuppose and contain each other.[7]

The argument that Daoism might be a check on ruthless environmental behaviour has no qualms about the question whether nineteenth-century peasants shared views with literati living more than two millennia before. The little and great tradition influenced each other. According to De Groot (1901:272, 325), the idea of animated plants, animals, and objects is much older than the philosophical theory of the general cosmic origin of souls (*qi*). One of the sources for Daoism were the shamans working for the ordinary people. Conversely, the popular practice of *fengshui*, the geomancy used to select good construction sites without disturbing the flow of wind and water, has its roots in Daoism; it was popular both in Fujian communities and throughout those of the Southeast Asian Chinese (Bruun 1992; Needham 1956:33, 68-70, 132-9, 304). However, the argument is not sustainable, because alongside Daoist views there were many other ideas not expressing the human-nature relationship in harmonious terms. Confucianism, for instance, places people at a higher level than non-human beings. Although these contradictory views may be irreconcilable at a philosophical level, they were mutually acceptable in the minds of ordinary people (Bruun 1992). This is shown by the *Qing Bell of Poesy*, a collection of works from 911 authors, published in 1869, thus contemporary with our period of study. An analysis of this work demonstrates that there was no single Chinese perception of nature at that time, and that almost any perception could be proven by referring to some of these poems (Elvin forthcoming). If no unequivocal picture of the migrants' perception of nature can be constructed, it becomes futile to consider the impact of their perception on their environmental behaviour. The human ecological impact in eastern Sumatra can be more fruitfully studied from the angle of the typical social structure of the frontier societies.

Turner in the tropics

Sometimes the jargon from the social sciences, such as 'role' and 'plural society', becomes part of common language and in due course the meaning becomes blurred. In other cases colloquial words acquire new, narrowly circumscribed meanings and are catapulted into the academic vocabulary. The latter process happened with 'frontier', when the word was used by Frederick Jackson Turner in his seminal paper 'The significance of the frontier in American history', delivered at the meeting of the American Historical Association in 1893 (Turner 1920:1-38). In the first half of the paper he describes the advance of American settlement into an area of 'free' land. At the frontier the European colonist had to strip off his civil-

7 Bruun 1992; Hjort and Svedin 1992; Needham 1956:33-9, 45–54, 472–5; Tu 1989.

ization and adapt to the harsh conditions of the wilderness and primitive society. Gradually he transformed the wilderness into a 'civilized' area, but in the process the colonist himself had become American. This process repeated itself again and again when the frontier moved westward. In fact there was not one single frontier, but a number of successive waves across the continent: first of the trader and trapper; then of the rancher and, where mineral deposits were found, of the miner; followed by the farmer; and finally the townsman. When one frontier moved west, it gave way to the next frontier. Each area was a 'palimpsest' on which the American history was written layer over layer. In the second half of his paper Turner developed the view that the expansion, with the recurrent return to primitive conditions and the subsequent evolution, explains much of American development. Above all, the frontier shaped the American national character with high values placed on individualism and democracy.

Turner's tremendous impact has been no doubt partly due to his visual style of writing. 'Stand at Cumberland Gap and watch the procession of civilisation marching single file – the buffalo following the trail to the salt springs, the Indian, the fur-trader and hunter, the cattle-raiser, the pioneer farmer – and the frontier has passed by' (Turner 1920:12). The second part of his paper, which discussed how frontier society shaped the American character, has been widely discussed and has become known as the 'Turner thesis'. For our purpose, however, it is the first part that is relevant.[8]

Turner himself has remarked that the allure of the frontier was the availability of natural resources: beasts for the hunter and trader, immense grasses for the rancher, and virgin land for the farmer. The pioneer could pick any one of these attractions and, when returns diminished because the soil was exhausted, there was other vacant land to move to at the frontier for nominal prices. Many regions showed an actual decrease of population after the first settlers had moved on. The pioneers did not care if they owned the land or not; they occupied it for a while and felt independent. Later settlers purchased the land (Turner 1920:18-22). Turner, who no doubt considered the expansion of the frontier as a great period in American history, shows here misgivings about the treatment of nature.

An essential characteristic of a frontier is its 'potential'; it excites the imagination and everything seems possible (Cooper 1993:157-8). Put in economic terms, the attraction of a frontier is the financial windfall to be made. The windfall of the frontier consists of one or more of the ordinary production costs partially or completely removed. The greatest of all

8 The Turner thesis cannot be applied to Indonesia, where only a minority of the population were overseas migrants. Nevertheless, it is noteworthy that the Chinese *kongsi*, associations of migrants who pooled labour and capital, were in the initial phase remarkably democratic (Heidhues 1992:75-6).

windfalls was the cheap land for those who occupied it first; other windfalls were, for example, the furs, ready to be trapped, and free prairies with abundant feed for cattle (Webb 1953:180-202). Most natural resources of the frontier ranked highly on the global commodity markets, so that the wider geographical context should be taken into account in the study of frontiers (Bankoff 1995:32; Reid 1993:34; Wood 1985:3-9). Physical access to frontiers is difficult and improvement and expansion of transportation facilities are a fundamental determinant of the development of frontier societies (Pichón 1996:364).

To rephrase, I would say that the biggest windfall was the fact that only production costs were taken into account, but not replacement costs of the diminished natural resource. No wonder then that natural resources were depleted rapidly at the frontier. In America, the state sold land for next to nothing, which resulted in quick and massive deforestation. Little was done to protect the forests against plunder and theft, and even if there were government patrollers, they were shunned by the public (Bartlett 1974:237-52). In the present Amazon basin, the vastness of the frontier area makes it difficult to enforce forest administration, and the natural resources are essentially a free good (Pichón 1996:350).

From this brief excursion into American history, we can gather the characteristics of a frontier society that help explain the pioneering activities in eastern Sumatra, namely: the allure and potential of natural resources, the availability of free land or land without a deed, the absence of an established government, the neglect of replacement costs and other windfall profits, a resource consumerism with an exploit-and-move-on mentality, the succession of waves of pioneers each exploiting a different resource, and the exploitation of the natural resources for the international market. The frontier concept bears an ambivalent connotation: the heroic pioneers opening land for others, but at the same time diminishing natural resources.

Tin mining in Bangka

Tin on Bangka is found in alluvial deposits along the beds of rivers, close to the surface. Tin was used for tin foil on joss paper burnt at religious ceremonies in China; it was almost an essential item of consumption and demand was assured. Tin was also used to pack China tea and in the nineteenth century an industrial demand for tin developed in Europe and the United States, so that the market expanded considerably. Tin was discovered around 1710 when the Sultan of Palembang was overlord of Bangka. In 1722 he signed an exclusive contract with the Vereenigde Oostindische Compagnie (VOC) to deliver all tin to Batavia. In order to populate Bangka with loyal subjects and at the same time increase tin production, Sultan Mahmud Badaruddin invited over a thousand Chinese to the island in 1734. The Chinese population gradually grew to over 25,000 people in the third quarter of the eighteenth century. During the last quarter of the eighteenth century, exhaustion of tin deposits and attacks by Ilanun pirates spurred many Chinese to leave the island. A severe smallpox epidemic in 1798 caused a further loss of people, and at the turn of the century there were less than 2,000 Chinese. Tin mining almost ground to a halt. In 1812 the British asserted direct administration over Bangka, handing over the island to the Dutch in 1816. The colonial governments restored safety and reinvested in the mines, so that Chinese returned to the island. The Chinese continued to do the actual work in the mines (Andaya 1993:185-90, 218-26; Heidhues 1992:xv, 3, 8, 20-36). Europeans introduced technical innovations in the late nineteenth century. These are beyond the scope of the present article.

There were three methods of excavating ore in Bangka. The so-called *lebang Palembang* or *Palembangse putjes* were perpendicular shafts dug by one person. The ore was lifted up in baskets and washed in the nearest stream. Once the miner had found a productive stratum, he followed it underground horizontally (Horsfield 1848:818; Jackson 1969:42). Since iron was already being exploited on Bangka, the local residents knew how to smelt ore in crude furnaces (Heidhues 1992:8).

More effective were *kulit* mines, worked by teams of up to three men. They did not carry the waste soil away, but deposited it in a conduit that carried it off. When pay dirt was reached, the conduit could be used for washing the ore. A *kulit* mine was about five metres wide, and, due to dependence on rain water, only worked during the rainy season (Horsfield 1848:799, 812; Van Diest 1865:31-2).

A technical breakthrough was made with the *kolong* technique, the introduction of which is ascribed to a Chinese man called Oen Ah Sing (or Assing), the same man who invented the standard size for ingots; crucial for the success of Oen Ah Sing was that he brought experienced miners from China or the gold-mines of West Kalimantan (Andaya 1993:189). A stream was dammed and the water channelled to propel a water-wheel that powered a chain-bucket pump draining the mine of seepage and rain-water. The pumps, sometimes placed in tandem, allowed the miners to dig at least six metres deep. A mine was 30 metres long and almost just as wide. Via sluices, water from the main canal was led to conduits and used for washing ore. The miners started to work at the lower end of a valley. When the first site became exhausted or became too deep, the miners shifted to the adjacent site upstream. The topsoil of the second site was loosened up and then through a network of canals washed into the crater left over from the original site. Deeper layers of waste soil were thrown into wooden gutters to be carried off. The deepest layers were carried up in *pikul* along ladders. Layers with ore were brought to a separate corner and sifted in a strong current; the lighter clay and sand was carried away, while the ore sank to the bottom. When the second site became exhausted or too deep, a third was opened upstream, and in this way the whole valley was progressively turned up. Van Diest counted one hundred men working a *kolong*.[9]

The three techniques operated side by side. The technique of *kolong* mines, which dominated the production, was only mastered by Chinese and found on valley floors. A *kolong* mine required the provision and organization of a large work-force, which at that time could only be provided by a *kongsi*, an association of Chinese workers. *Kulit* mines, run by Malay, Buginese or Chinese miners, were found on hillsides, where they had less trouble with seepage of groundwater. They were also economical at the edges of abandoned *kolong* mines because of the smaller investment. For a long time autochthonous Bangkanese people, called *orang gunung*, continued to dig *Palembangse putjes* in remote spots, where transport of heavy equipment was unfeasible (Horsfield 1848:312, 818; Heidhues 1992:11-5). Van Diest (1865:67) encountered only remnants of *putjes*, so apparently the technique had been given up in the course of the nineteenth century.

Apart from the obvious depletion of a non-renewable resource, which is inherent to mining, the environmental impact of tin mining was twofold. In the first place, reaching the ore implied the removal of large quantities of soil and the disturbance of natural drainage patterns; what was left behind was a land stripped of vegetation and honeycombed with water-logged diggings (Jackson 1969:38). Travelling through the country in 1813, at the nadir of mining, Thomas Horsfield (1848:808) passed 'extensive surfaces which have been turned up and drained of their contents, exhibiting the remains of former mines and aqueducts'. Half a century later Van Diest (1865:30) still easily recognized abandoned mines. The land was overgrown with creepers (*Nepenthes*), the first plants to invade soil without a layer of humus, as he explained, and with shrubs that grew very slowly. *Kulit* mines removed the fertile topsoil equally well, but their surface area was far more modest. *Palembangse putjes* turned the landscape locally into a kind of sieve (Van Diest 1865:67).

In the second place, the production of charcoal, burnt in the furnaces to smelt the ore, caused deforestation.[10] In the early eighteenth century Bangka was covered with dense dipterocarp forest, but in the second half of the nineteenth century lack of wood was felt increasingly, especially near the mines. The colonial government responded by discouraging the slash-and-burn agriculture of peasants and by reserving 'several thousand'

9 Van Diest 1865:12-6; Horsfield 1848:811-3; Jackson 1969:42-5. The mine as such was called *parit*. The driving wheel of the water pump could also be moved by a man leaning on his arms and treading with his feet.

10 A lot of the charcoal burning was done by *orang gunung*, who also took care of transport for the mines by carrying goods on their back (Van Diest 1865:19; Heidhues 1992:88-9).

hectares of forest, most of it *belukar* (secondary regrowth), for charcoal production. The stress was relieved when in the 1920s the smelting was partly transferred to Singapore and western countries (Jackson 1969:29; Heidhues 1992:69, 96-7). At the outbreak of the Pacific war, 30% of the island remained tree covered, most of it with secondary forest. *Belukar* was repeatedly cut for fuel and rarely grew to more than an arm's width. But until then, after two centuries of tin mining, Bangka had managed to fulfil its own need for charcoal (Helbig 1940:196-7; Malmrös 1939:362). This indicates that the demand for wood by the mines had never been excessive.

Bangka showed most characteristics of a frontier with both the positive side of opening land and the negative side of wasting resources. 'The discovery of the tin now attracted numerous foreigners [...] and a commencement was made in cleaning the ancient forests, which had till lately not been disturbed.' Until then 'Bangka was covered with impenetrable forests' (Horsfield 1848:302). Eager for short-term profits, the *kongsi* creamed the most easily accessible ores, abandoned mines before they were completely depleted, and moved on to new sites. The frontier did not expand inland, like the big American frontier, but followed the coast in order to avoid transport problems over land. Mining started around Muntok, in the northwest, and then extended via the Klabat Bay in the north, down to Koba on the south end of the east coast. When miners left Bangka at the end of the eighteenth century, they discovered tin on Singkep and started to operate there; the frontier had just moved to another island.[11] Meanwhile, on Bangka, *kolong* mines concentrated on the most favourable sites. An interesting phenomenon is that the *kolong* mines were not succeeded by new frontiers of completely different activities, but by small-scale mining techniques of indigenous residents, who could profitably work the deserted *kolong* (Horsfield 1848:314, 797-807, 818; Jackson 1969:34-9). At some places, the frontier of mining was followed by permanent Chinese settlements. The miners who stayed in one place were invariably married and had taken up other occupations (Horsfield 1848:799-800). At the end of the nineteenth century pepper gardens were laid out on the mining concessions and later were also made on small holdings (Heidhues 1992:100). In the last quarter of the nineteenth century, Europeans introduced new techniques for surveying, extracting, transporting, and smelting ore. With these techniques they increased production in otherwise deserted mines (Heidhues 1992:65-71, 80, 127-31). They formed another wave of successive exploiters of resources.

The quick moves of the *kolong* mines, typical of a frontier society, exacerbated the ecological damage of mining. The government was fairly

11 ANRI, Palembang 36, Memorie van Overgave J. van den Bogaart, Resident van Palembang, to Aart Quirijn Palm, 30-11-1802.

strong, except during the end of the sultanate (Andaya 1993:186, 218-9). When Sultan Mahmud complained that he was unable to keep control of the tin trade on Bangka, the Dutch officials probably rightly interpreted this as a disguise for the Sultan's private trade outside the VOC-monopoly contract.[12] The *kulit* mines were easily supervised, because of their heavy and expensive capital goods that fixed the location of the mine, and due to the fact that they smelted ore once a year only, shortly before Chinese New Year.[13] It is likely that the firm hold on the mines has mitigated their environmental impact. There was less control over the woods. Forester A.H. Berkhout observed that Bangka had exceptionally rich forests and calculated that reserves should be more than enough for the provision of charcoal. He could only explain the loss of forest in the old mining districts due to careless logging habits (Berkhout 1895:14, 19, 45-8).

Gambier cultivation in Riau and Johore

Gambier is the boiled-down juice of leaves of a bush, called *Uncaria gambir*. It is used as an ingredient of the betel quid. It formed an export product from Palembang in the seventeenth century (Andaya 1993:120). In 1743 the *raja muda* (Viceroy) of Riau ordered two of his headmen to get seeds from the Sumatran mainland. The Buginese and Malay residents then laid out hundreds of gambier gardens, tended by Chinese coolies (Ali 1982:80). In 1787, after having attacked the VOC lodge, the *raja muda* and almost all the Buginese and Malay residents left Riau, trying to escape from Dutch retaliation. The *raja muda* returned in 1800. The Dutch gained suzerainty over the archipelago in 1818; direct administration was limited to the capital Tanjung Pinang, a peninsula of the main island of Bintan. In the meantime the Chinese had completely taken control of every aspect of gambier cultivation. From Riau the gambier spread to the island of Singapore, and from there into Johore. Gambier attracted thousands of Chinese cultivators to Riau and Singapore, so that gambier gardens dominated the agricultural area almost as a monocrop, like the Chinese dominated the population numerically.

Gambier plantations were straddled along streams (*sungai* or *kreken*) that provided transport and water for boiling gambier. A gambier plantation consisted of a central hut (*bangsal*), the actual gambier garden, a forest reserve for firewood, and invariably a small pepper garden that provided an additional income. Twigs of the gambier were cut at dawn and

12 ARA VOC 2481, Generale Missive Valckenier and Raden van Indië, 10-1-1741, f. 912v.

13 About 1800 the Sultan forbade work in *Palembangse putjes* as long as other mines were productive (ANRI, Palembang 36, Memorie van Overgave J. van den Bogaart, Resident van Palembang, to Aart Quirijn Palm, 30-11-1802). His prohibition may have been dictated by the fear not to control these isolated production units.

the late afternoon when the leaves were richer in juice than during midday. The twigs were stripped of leaves in the hut. The morning yield was boiled during the afternoon, and the afternoon yield was boiled the next morning. While some coolies collected twigs, others gathered dead wood or felled trees for fuel. The head of a *bangsal* took on the boiling. The complex boiling process of one portion took about five to six hours; one hour for extracting the gambier and the rest for evaporating the excess water. The resulting sludge was put in reservoirs to leak out. Then it was cut in pieces and dried in the sun or above fires. The residue of leaves was used to fertilize the pepper gardens. In Riau the plantations had an optimum size in area and number of coolies, so that they almost became assembly-line enterprises: one person boiled gambier, two cut leaves and two collected wood. They worked non-stop, keeping each other busy through the day. The gambier gardens had a standard area of approximately 30 hectares and an equal size of forest reserve. With this size, two cutters could harvest the whole garden in about four months, which was precisely the time the bushes needed to recover from the previous yield. By the time the cutters had gone through the whole garden they could start again at the beginning.[14] It seemed a perfect arrangement.

The whole system had one major disadvantage: boiling gambier consumed huge amounts of fuel wood. Although the bushes could produce 20-25 years, the plantations were deserted after 10-12 years, because the whole forest reserve was burnt up.[15] Conflicts about the standing stock of firewood around the boundaries between the plantations frequently flared up. In 1846 fights were on such a massive scale that 1,500 people fled from Riau to Singapore, only to return next year to destroy the plantations of their adversaries (Graafland 1888:518; Trocki 1976:145). The overall result was rapid deforestation of the Riau archipelago.

Figures from colonial reports permit us to calculate the loss of forests in Riau; the average annual loss of forests in the period 1821-1870 can be set at 6,900 hectare (Table 1).[16] This equals 3% of the total land area of those

14 ANRI, Riau 60-1, 60-2, Algemeen verslag Riau 1824, 1825, 1831; ANRI, Riau 63-1, Cultuurverslag Riau 1863, 1865; ANRI, Riau 71-3, Rapport Angelbeek 14-8-1825; Couperus 1780; Teijsmann 1874.

15 See for instance ANRI, Riau 60-2, Algemeen verslag Riau 1831.

16 Since in Riau the plantations had an optimum size in area and coolies and a standard productive time of 10-12 years (during which years the reserve forest was burnt up), the number of workers is an indicator of the rate of deforestation. The number of workers is known for only a couple of years, but the Chinese population was better registered and we can assume that half of the Chinese people were engaged in gambier production. From the Chinese population we can make a crude assessment of the deforestation in Riau. A less reliable assessment of the deforestation rate is possible with the export figure as indicator. The annual deforestation calculated with this method would be 40% lower than with the Chinese population as indicator. Since the export figures exclude unregistered trade ('smuggling') and the applied conversion rate from export figures to deforestation is conservative, this second method supports the idea that the outcome of 6,900 hectare is a fairly accurate estimate (Colombijn 1997).

Table 1. Estimated annual deforestation in the Riau Archipelago due to gambier cultivation

period	deforestation (hectare)
1821-1830	4800
1831-1840	4400
1841-1850	5100
1851-1860	8100
1861-1870	9800
1821-1870	6900

Source: ANRI, Riau 60-1, 60-2, 61-1, 61-2, 62-1, 62-2, 63-1, 63-2, 232-1 to 232-5, Algemene (administratieve) verslagen Riau 1824-1870; ANRI, Riau 58-1, 58-2, 59, 233-1 to 233-3, Politieke verslagen Riau 1855-1870; ANRI, Riau 63-1, 115, Cultuurverslagen Riau 1863-1864; ANRI, Riau 71-3, Rapport Ch. van Angelbeek 14-8-1825. Summary raw data can be found in Colombijn (1997).

islands where gambier spread prior to 1870.[17] If the figure is broken down for individual islands and shorter time spans, the pace of deforestation becomes astonishing. For instance, in 1859, 5% of Batam's total surface was cleared of forest. If we subtract cultivated and inaccessible land, and land naturally denuded of forest from the total surface area, the deforestation rate of the remaining stock would even be considerably higher. Of course, cutting firewood once did not preclude the regrowth of secondary forest, but this produced less wood than the primary forest. The decreased value of the woods is illustrated by the fact that in the 1860s the proportion of gambier garden to forest reserve changed rather abruptly from 1:1 to 1:4.[18] By the early twentieth century all forests had been replaced by *alang-alang* and *belukar*. Another ecological effect was loss of fertility, probably due to erosion. In the early twentieth century a gambier shrub produced good leaves only for 3-4 years, compared to 20-25 years a century earlier (Koster 1922).[19]

The ecological impact of pepper in Riau was limited, due to the small area under cultivation. Nevertheless it had a detrimental effect because at the start of a pepper garden the humus of a field was totally removed, mixed with wood, and burnt. This provided extremely rich fertilizer, but the dug-up field remained barren.

Gambier cultivation shows all the characteristics of a frontier society. Starting from Tanjung Pinang, gambier spread over Bintan and subsequently

17 Basically the Riau archipelago, excluding Karimun and Kundur (and the Lingga archipelago): 2307 km^2 (Riau 1985).

18 ANRI, Riau 63-1, Cultuurverslag 1863. This implies that Table 1 underestimates the acceleration of deforestation in the 1860s.

19 Several authors have stated that gambier exhausts the soil (for instance Heidhues 1996:178), but I have found no supporting evidence for this in agronomic handbooks (Ridsdale 1991; Zeijlstra 1949). Since, as far as the gambier was concerned, a shrub could produce 25 years, exhaustion of the soil apparently did not play a major role.

expanded into the archipelago, attracted by the available stocks of fuel wood. Around 1830 three quarters of Bintan was still covered with heavy, almost impenetrable forest, but already five years earlier Dutch civil servants had noted for the first time that settlements with old gambier gardens were deserted with the planters having moved on to other places to lay out new gambier gardens.[20] By 1839 Batam began to produce gambier as well and two decades later it had almost surpassed Bintan; by the next decade gambier production in North Bintan collapsed and more islands southwest of Riau had been opened (Colombijn 1997). In 1858, in an attempt to develop his realm, that lagged behind Riau, the Sultan of Lingga invited Chinese from Riau and Singapore to open gambier gardens in his territory. This was unsuccessful initially and serious expansion in this direction had to wait until after 1870.[21] In the 1860s the Dutch colonial civil servants almost panicked about the pace of deforestation and exclaimed that Chinese cut and burnt the forest with 'unparalleled recklessness'.[22] The gambier gardens opened up the forest for others, but it is not quite clear which second and subsequent frontiers followed. On Karimun the cultivators shifted to pepper when forests for gambier were burnt up (Graafland 1888:519). In Riau as a whole there seemed to be a diversification of primary production, with fishing and collection of *agar-agar* (seaweed), extraction of *sago*, cultivation of coconuts, shipping, trade, and also a continued but marginalized gambier cultivation (*Encyclopaedie* 1919:612).

A similar expansion took place in Malaya. The Treaty of London (1824) demarcated a British and a Dutch sphere in the Malacca Straits, but the Chinese did not adhere to boundaries drawn by diplomats. For them the region was one, with Singapore as the centre and Tanjung Pinang as a subcentre. The official worries about deforestation were fairly synchronized in Singapore and Tanjung Pinang. On both sides of the border complaints about the waste of forests and reports of abandoned gambier gardens started around 1830. From the town of Singapore the gambier cultivation expanded across Singapore Island, and from there spread into Johore, first along the Straits of Johore and then up along the west coast. There was also expansion into the Riau archipelago, especially in the 1860s when financiers from Singapore were taking over part of the business from the Chinese in Tanjung Pinang (Jackson 1965; Trocki 1976).[23] Where gambier planters had abandoned gardens, they were partially succeeded by a second frontier of European nutmeg, coconut, and sugar plantations,

[20] ANRI, Boschwezen 44a, Houtwerken op het eiland Bintang [1830]; ANRI, Riau 60-1, Algemeen verslag Riau 1825.
[21] ANRI, Riau 233-3, Politiek verslag 1858.
[22] 'weergalooze roekeloosheid', ANRI, Riau 63-1, Algemeen verslag 1870. See further the extensive report by Resident E. Netscher, who, however, no doubt gave a gloomy picture in an attempt to get more *controleurs* assigned to him (ANRI, Riau 58-2, Politiek verslag 1865).
[23] ANRI, Riau 58-2, Politiek verslag 1865.

which had great problems in extirpating the *alang-alang* (Balestier 1848:146). Jackson (1965:101-4) mentions as typical frontier conditions in Singapore and Johore: an expanding pioneer fringe, vast areas of almost unoccupied land, an absence of communication facilities, and a concomitant lack of administration.

Two frontier conditions exacerbated the ecological impact of gambier. Firstly, gambier was produced for the market, and at periods of rising prices forest was opened in a craze, whereas during a slump gardens were prematurely (from an agronomic perspective) abandoned. Consequently, the cultivators moved like migratory birds, coming or going with the rise or fall of the price in the main market, Java. After about 1836, when England developed a demand for gambier in the tanning industry, the fluctuation stabilized somewhat. Secondly, there was no strong government to stop the assault on the forests. Colonial civil servants quickly realized that a conservation policy for the forests was necessary. Their concern should not be mistaken for a 'green' awareness, but stems from the fear of eroding a tax base. According to the Dutch, the appropriate means of control was a restricted granting of new concessions.[24] Since the Dutch did not manage to protect the forests on the small peninsula of Tanjung Pinang, it is obvious that the *raja muda* in his far bigger territory was not able to enforce effective control either. The Chinese had shaped their production organization during the absence of power in 1787-1800 and had not been brought up to respect a Malay sovereign, like the indigenous population had been. In the 1860s, when gambier experienced another boom, the *raja muda* counted 112 gambier gardens that had been opened on Bintan and Batam without his permission.[25] With disregard of the rights of local people, the Chinese even opened gambier plantations in land already cultivated by Malays and cut the latters' fruit trees.[26]

To conclude this section I want to make an illuminating comparison with indigenous cultivation. There were no Malay gambier growers in Riau, but Minangkabau produced gambier in the Padang highlands (mainland of Sumatra). At least four differences can be noticed. First, gardens were laid out in primary forest of villages (*nagari*) and local heads granted the right to open up a plot to local people. Consequently, there was stricter supervision and the planters were more subject to social control. Secondly, once released, the land became communal property of the matrilineage of the cultivator, who thus got a personal interest in the conservation of the resource. In the Minangkabau perception of family land, each generation enjoys the usufruct, but must hand over the land intact to future generations, including those not yet born. This perception could have been taken verbatim from the definition of sustainable development of the

24 ANRI, Riau 60-1, Algemeen verslag 1825.
25 ANRI, Riau 58-2, Politiek verslag 1863.
26 ANRI, Riau 58-2, Politiek verslag 1865.

Brundtland Committee (Van Reenen 1996:97-102; World Commission on Environment and Development 1987:43). Thirdly, and this is partly explained by the second point, the extraction of the gambier was not done just by boiling. Although leaves in a rattan net were boiled and steamed for an hour, the gambier was mostly extracted by squeezing the leaves. The leaves were constricted in the net, and pressed between two halves of a tree trunk driven together by wedges. Fourth, the gardens were laid out in isolated places with much room between. There was no monoculture of gambier. Not all trees were felled for a new garden and weeds were permitted in the garden from the beginning. All these measures, whether deliberate or not, promoted the regrowth of secondary forest and prevented *alang-alang* in empty, sun-lit fields. These four points made the productivity per field in virgin territory less, but upheld sustainability (Encyclopaedisch Bureau 1915:59-71; *Schets* 1856; Zeijlstra 1949:591).

Panglong in East Sumatra

In the 1860s gambier was joined by *panglong* as a cause of deforestation. *Panglong* is the Chinese word for a shed were wood was sawn into planks. It came to designate all lumber camps working in East Sumatra in the late colonial period. The government appointed a ten-kilometre-wide strip from Bagan Siapiapi to Indragiri (with the banks of the Kampar considered 'coastal' up to 50 km from the mouth), the big islands off the coast, like Bengkalis, and the Riau-Lingga archipelago as *panglong* area. This area lay in an arc around Singapore and indeed all wood was shipped to that port, and partly shipped on to other destinations from India to Hong Kong. *Panglong* were notorious for appalling labour conditions. Logging was not an exclusively Chinese business. There were a few *orang laut* (sea gipsies), who, unlike the Chinese coolies, worked on short contracts. Local Malays pointed out the richest forests to the Chinese loggers (*Panglongs* 1915:1020-7; Pastor 1927:2-5, 20, 50-63, Appendix D). There were also a couple of *panglong* run by Malays, who proved stiff competitors during the years of the Depression (Erman 1994:24; *Panglongs* 1915:1027).

Logging in Riau for the international market was centuries old, but the operation of *panglong* started in the mid-nineteenth century. The first mention of a *panglong*, to my knowledge, stems from 1854, when the *raja muda* of Riau issued twelve licences to Chinese entrepreneurs to cut wood for the Singapore market.[27] *Panglong* became prominent in the 1860s, when Singapore constructed big harbour works, and reached its peak at the eve of the Depression of the 1930s (Cator 1936:220; *Panglongs* 1915:1013). The

[27] Notably, he withdrew the licences and confiscated two small ships with wood, because he felt cheated. ANRI, Riau 73-2, Memorie van Overgave T.J. Willer, Resident van Riau, to J.H. Walbeehm, 16-12-1854.

first *panglong* produced mainly planks, but between 1898 and 1913 the sawmills moved to Singapore. Then the production of beams took over, and was later superseded by *panglong* for firewood and charcoal. In 1930 there were 139 beam-cutting works in Bengkalis and Riau, 104 firewood works and 231 charcoal burning plants (Cator 1936:220; Erman 1994:24-5). Planks came mostly from mangrove forest. The *panglong* for beams used forest giants in the swamp forests of Sumatra's mainland and some of the big islands, including those of the Riau-Lingga archipelago. The *panglong* for fuel wood and charcoal operated in mangrove forests on the islands and used trees with a maximum diameter of one foot (*Panglongs* 1915:1014, 1019; Pastor 1927:3-12).

T. Walter has left us a unique and vivid description of a *panglong* for beams in Lingga (paraphrased in *Panglongs* 1915:1021-7; see also Pastor 1927:6-10). From the coast the author followed a wooden trail (*knuppelbaan*) inland. The top layer of the trail was made of greased sleepers, fixed 60 cm apart, at right angles to the direction of the trail. The trail was used to haul the four metre long sledges (*kuda kayu*) loaded with beams to the coast, where they were taken aboard ships for shipment to Singapore. Following this trail in opposite direction, he reached the simple coolie shed with a small garden and pigsty. Further land-inward the trail split up into branches leading into the fields were the actual logging was done. A once forest-covered valley had been completely cleared of trees. One man had been hewing for two days to fell a big tree. His axe (*biliung*) had a flexible, narrow helve, so that the blade came swaying down in the trunk. To get to one good tree, four other trees that prevented the good one from falling free were sometimes cut first. A tumbling giant dragged down smaller trees.[28] Other coolies cut and sawed the trunk into beams on site, with a maximum diameter of 125 cm. The biggest branches were also used for beams. The heaviest work was towing the beams over the trail. Since the hauling teams of eight men sung to the rhythm of their footsteps, one heard them approaching from afar. Walter observed that the towing was done by families of *orang laut*, including women and children clinging to their mother's loincloth.

Charcoal burning formed an exception to the older style of logging for planks and beams in several ways. Out of necessity the oven stood inland on solid, dry soil. Since such plots were rare, the oven was never far from a Malay village. The village provided a check against abuses, so that labour conditions were much better. The coolies worked less hours and spent leisure time in the villages. These *panglong* took on a more permanent

28 Hauling out the big trunks necessitated cutting down other trees that stood in the way (ANRI, Riau 60-1, Algemeen verslag 1827). Dutch officials estimated vaguely that the Chinese destroyed 'surely ten times', or 'more than double' the number of trees considered necessary if adequate (meaning Dutch) supervision were present (ANRI, Riau 63-1, Algemeen verslag 1861 and 1862).

nature. Contrary to the almost 100% male composition of the other *panglong*, there were more women here and families lived in separate houses. The families laid out gardens and planted nipah trees, the leaves of which can be used for roof cover. The wood necessary for charcoal came from mangrove forests, and it was collected from a wide area during week-long trips. The biggest part of the wood was bought from Malays (Pastor 1927:12-3). The fact that a charcoal oven was run by only 3-5 men, compared to up to 40 in a beam-cutting mill (Cator 1936:219) may have contributed to better conditions and better integration with local society.

The obvious ecological impact of the *panglong* was deforestation, accompanied by disturbance of a wider ecosystem and loss of biodiversity. In 1916 Dutch civil servants noted that the state of the forests in Bengkalis became alarming and in 1933 the situation of the tidewater forests was deemed 'far from rosy' (Erman 1994:23; Jonker 1933:717). Production figures give a rough indication of the area logged per year. Production in 1928, when it was at its peak, totalled 475,000 m^3 of beams, 100,000 ton firewood and 37,000 ton charcoal (Endert 1932). I estimate that this would amount to 22,000 hectare of timberland, different types of forest taken together.[29] This is little compared to the total land area of eastern Sumatra, but had a considerable impact on the coast and islands, where the *panglong* were concentrated.

The expansion in the Riau-Lingga archipelago clearly had a frontier element. The loggers displayed resource consumerism: they logged and moved on. 'When the nearer forests [closest to Singapore] were exhausted, one started to work further away; one island after the other was destroyed, until one finally reached Lingga and Singkep.'[30] In the early 1870s the estuary of the Siak was opened by loggers and in the twentieth century a few *panglong* were opened in Jambi and in Aceh, the far north of

29 For firewood and charcoal in theory all wood collected could be used, but beams and planks needed far more wood than the final produce for two reasons. Firstly, as said, felling one big tree destroyed several smaller ones. Secondly, to get good beams and planks a considerable part of the tree was chopped off and the chips thrown away. If we assume that both causes generated independently a 100% loss of wood (which seemed a liberal estimate for the first cause and a conservative estimate for the second), the beams had cost 1,900,000 m^3 of standing wood. Assuming the same average specific weight of fuel wood as on Java, 550 kg/m^3 (Smiet 1990), the fuel wood had an original volume of 181,818 m^3. When we generalize from the efficiency of charcoal ovens in Karimun, 3.2 m^3 of wood for each ton of charcoal (Jonker 1933:721), the coal had consumed 118,400 m^3 of wood. The total volume of logged wood must have been in the order of 2,200,000 m^3. Endert (1933:405) measured that the volume of wood in forests on Sumatra and Kalimantan varied from 50 to 500 m^3 per hectare, with the modal value in the range 100-250 m^3. Because the *panglong* area was 'not particularly rich' forest (Endert 1932:751), the lower limit of the range of modal values seems sound, 100 m^3. The total area needed would have been 22,000 hectare. This outcome does not take account of the serious under-reporting of output.

30 'toen de nabijgelegen bosschen waren uitgeput toog men verderop aan het werk. Het eene eiland voor het andere ging er aan totdat men tenslotte op Lingga en Singkep belandde' (*Panglongs* 1915:1014).

Sumatra.[31] Notably, *panglong* never became important on islands previously used for gambier, because those were already bereft of forests. The expansion was into virtually uninhabited land (Pastor 1927:4). The pioneers were attracted by a natural resource that suddenly proved high in demand on the international market and therefore became a commodity ready to be reaped. Whoever could lay their hands on the stocks could easily make windfall profits. Replacement costs were ignored; the colonial government took no measures to reforest, promote general regeneration, or protect valuable species, although local officials repeatedly expressed concern about the wasteful ways of exploitation.[32] The official worries about the operations of the *panglong* were ambivalent, because as long as forest land was not yet exploited, the civil servants praised the potential that was, regrettably, left unutilized.[33] It is unknown whether others used the inroads into the jungle for a second frontier of, for instance, agriculture.[34]

The frontier conditions aggravated the ecological havoc. The apparent abundance of forests and the haste to supply the market with wood stimulated careless logging and wasteful processing of the felled trees. *Panglong* operated in a virtual power vacuum. The operations along small creeks were hidden behind coastal vegetation, so that the government patrols, rare anyway, could pass by at short range without noticing them (*Panglongs* 1915:1019; Pastor 1927:17). The loggers in the mangrove forest swarmed out over an area far wider than the boundaries of their concession (Pastor 1927:11). Local leaders on faraway islands issued logging concessions themselves, by-passing their sovereign, the Sultan of Lingga. The more the lumbermen moved into islands far away from the centres of power, the less control could be exercised.[35] Occasionally Chinese loggers were apprehended and their proceeds confiscated when they violated the conditions of their concession.[36] The Dutch government did take measures to improve labour conditions (mainly to maintain a good relationship with the British in Singapore) and prevent tax evasion in the *panglong*, but not in order to prevent total clearing of the forests. The Depression and the devaluation of the Singapore dollar, rather than protective measures, eventually took the ecological pressure off (Cator 1936:220-1; Pastor 1927: 89-147).

31 ANRI, Riau 1873, Politiek verslag 1873; Cator 1936:218.
32 See, for instance, ANRI, Riau 63-1, Algemeen verslag 1861.
33 ANRI, Riau 63-1, Algemeen verslag 1861.
34 One indication is that Chinese laid out sugar-cane gardens in Bengkalis for the local market. ANRI, Riau 59, Politiek verslag 1873.
35 ANRI, Riau 58-2, Politiek verslag 1860 and 1863.
36 ANRI, Riau 233-18, Testimonies of Seng Toon and Sie Seok, 17-9-1869.

The role of intermediate cities

In the three above cases, the haste to open up the forest was due to new opportunities on the global market. Until now this market has been dealt with as an abstract, outside thing. But the relation between the pioneers on the frontier and the market goes via concrete exchanges of goods between people of flesh and blood. A frontier cannot function without an intermediate town to facilitate the linkage with the international market (Wood 1985:3). Contrary to William Wood's use of the word, I do not mean by 'intermediate town' or 'intermediate city', a medium-sized city but a town's broker position in the global urban network. An intermediate town is a place where the relationship between a frontier and the world market is being materialized. Despite the differences in production of the above three cases, tin mining, gambier cultivation, and *panglong*, the resemblances in the way they were linked to the market are striking. Therefore it is fruitful to discuss the linkages of the three means of exploitation together in one section. The intermediate town for tin mining was Palembang, for gambier Tanjung Pinang and Singapore, and for the *panglong* Singapore alone.

The three frontier societies were by definition located in a sparsely inhabited area and the human resources had to be brought in from elsewhere. Since these people began to work in an empty land, there was no subsistence base to rely on and no, or too few, neighbours to fall back on for their daily needs. Lacking savings that could help them through the start, they needed credit for their pioneering work. The general lending pattern was a chain of relationships. This chain consisted of: the financier (*tauke* or *teko*)[37] in the intermediate town, who was the original source of capital; an agent in the field who oversaw several pioneer enterprises or *kongsi*; the head of an individual enterprise; and coolies of an enterprise. The *tauke* took the initiative to get a concession for a mine, gambier garden or *panglong* from the ruler and then looked for people who could do the work on site.[38] He gave, via the intermediate persons, credit in the form of provisions, charged at inflated prices, to the pioneers who repaid by selling their produce, via the same go-betweens, to the *tauke* at a price below market price (Graafland 1888:520). The manipulation of prices implied that interest was paid both for the loan and for the repayment. Credit and debt was the substance that forged the link between frontier and intermediate town.

All three frontier societies showed variations on this pattern, which

[37] *Tauke* and *teko* are both equivalents for 'boss', but different words. *Tauke* is derived from the Chinese term for 'head of the guest workers', and *teko* is from 'big brother' (Mary Heidhues, personal communication).

[38] Graafland (1888:512-7) gives the text of the regulation for granting a concession issued by the *raja muda* of Riau in 1883.

were however of secondary relevance. In tin mining, there was an extra societal level above the *teko*, namely the Sultan. The *teko* were all court officials. When the European powers took direct control of Bangka, the top positions of Sultan, *teko* and agent were replaced by British and Dutch civil servants, but the structure of credit remained intact (Jackson 1969:39; Heidhues 1992:15, 28-45). The gambier cultivation on Singapore and Johore fitted the pattern precisely, with much power, economic as well as judicial, in the hands of the agents. These agents, called *kangchu* (literally 'lord of the river'), established themselves at the mouth of a stream or creek and controlled all gambier gardens upstream (Jackson 1965: 84-7; Trocki 1976). In Riau these riverlords, or *soengie-hoofden*, played a less dominant role. Police functions were assigned to the *majoor der Chinezen*, a functionary in Tanjung Pinang, and in the credit relationship the head of a *bangsal* could take the initiative to establish a direct link with a *tauke* (Graafland 1888:520; Teijsmann 1874:170-1). Typical of the financiers of *panglong* was that these *tauke* also had a forward linkage and owned wood-processing industries like sawmills, shipyards and furniture factories (*Panglongs* 1915:1014; Pastor 1927:3).

Hisao Furukawa (1994:185) remarked that gambling on a new life permeates life at the frontier and he even seems to consider this the most idiosyncratic feature of a frontier. I have not mentioned gambling as a characteristic so far, firstly because the people Furukawa writes about do not form frontier societies in the sense used here, secondly, because gambling is also common in non-frontier societies in Indonesia, and thirdly, because gambling implies running a risk voluntarily, whereas it is questionable that the coolies had much choice.[39] Nevertheless, this is the point to introduce 'risk' as a significant feature of frontiers. Especially when they were indebted, the coolies had to take risks as they came, and did not have the freedom to make a balanced assessment of risks like independent entrepreneurs in an established settlement could do (Knapen this volume).

Several authors have noted that credit was provided in the form of daily necessities, rice, salt, oil, and clothes,[40] but nobody has pointed out the implications of this practice for the pioneer settlers. For their subsistence they were directly dependent on provisions sent by the *tauke* in the intermediate town. This situation applied most to Bangka which was fully dependent on Palembang for its staple food, rice. At the end of the eighteenth century, the risk was made plain to the tin miners when pirates

39 The *tauke* who invested in gambier did gamble. They decided to invest in a sharply fluctuating market, and investing in a gambier garden, which took about eighteen months to produce its first yield, was a speculative business. Tin mining and logging were less of a risk. Tin had a guaranteed market with fixed prices, and *panglong* required less upfront expenditure, so that one could withdraw any time.

40 Cator 1936:219; Reid 1993:34; Heidhues 1992:35; Trocki 1976:144.

took their old stocks, ruined the *tauke* who could no longer deliver supplies, and completely cut off the supply of food from Palembang. Famine was the result. Also, the crop failure in Palembang of 1796 caused acute deprivation in the mines.[41] When the Chinese left Bangka in great numbers it was not in search of better business, but because they faced immediate starvation. Many miners preferred voluntarily slavery and exile above that imminent threat. The shortage of food was a central theme in the reminiscences of survivors of those dire times at the mines (Horsfield 1848:323-5). The situation was better in the Riau archipelago where more local rice was cultivated, and where sago stands provided a buffer. When gambier prices were low, the cultivators shifted to sago, which they considered second-choice food.[42] Nevertheless, when some *tauke* in Tanjung Pinang went bankrupt due to competition from Singapore entrepreneurs, their debtors on the plantations suffered from hunger as well.[43] In 1787 the first Chinese planters in the gambier gardens saw all their Bugis and Malay protectors abandon them. Over the years 1857-1859 two-thirds of the value of imports in Riau concerned rice.[44] On top of the problem of uncertain supplies, the lumbermen in the *panglong* were confronted with an even more perilous livelihood. They operated in an extremely hostile situation and when their *tauke* went bankrupt, they might be bereft of food and the means of transport to get away. They were also at the mercy of the head of a *panglong*, which entailed additional dangers. Loggers with a protracted disease could eventually be thrown out of the shed and chased into the jungle to look after themselves; this meant almost certain death (Pastor 1927:51-63, 75).

It remains unclear as yet whether the dependence on the intermediate town for vital supplies affected the exploitation of natural resources on the frontier in any sense, and if that was the case, how.

Conclusion

Tin mining, gambier cultivation, and *panglong* have all been detrimental to their immediate natural surroundings. Almost any economic activity affects the environment, but these three production systems caused severe degradation. Tin miners excavated mineral deposits, and churned the soil, so that the humus layer disappeared and watercourses changed; charcoal production slowly depleted Bangka's forests. Gambier cultivators clear-

41 ANRI, Palembang 36, Memorie van Overgave P. Walbeeck, Resident van Palembang, to J. van den Bogaart, 12-11-1794; Andaya 1993:228; Jackson 1969:37-8.

42 ANRI, Riau 60-2, Algemeen verslag 1831.

43 ANRI, Riau 58-2, Politiek verslag 1865.

44 ANRI, Residentie archieven Pasar Ikan 1211, Beknopt overzicht van de handel en scheepvaart in Riau, 1857-1859.

felled whole islands at a high pace, with the concomitant problem of erosion. *Panglong* deforested as well.

The three production systems experienced their boom years in three different centuries and used a different kind of resource: tin mining, a non-renewable resource; gambier, a resource that required investment in the ecosystem first; and *panglong,* a resource that is in principle renewable. Nevertheless, the social structure of the production and the process of expansion shows striking similarities.

The pioneers in gambier gardens and *panglong* were archetypical frontier societies, with typical features, such as: expansion, an inherently migratory production system, consumption of natural resources, windfalls made from ignoring replacement costs, empty land, a weak government, and a booming but fluctuating international demand. Most characteristics applied to tin mining as well, but two of them were less prominent. Firstly, the use of tin in joss paper, which came close to being a daily necessity in China, assured a steady demand for tin. Secondly, there was no lack of political power, except at the end of the eighteenth century, when royal power was waning. The easy surveillance of tin mines contrasts with the small-scale, year-round producing gambier gardens and *panglong* that escaped scrutiny. The geographical extension of the frontier was limited by logistics; in the three cases this meant that frontiers did not stretch far beyond the navigable end of the creeks and rivers. The frontier societies could only subsist with an intermediate town in the environs that supported them with everyday necessities. The dependence of the pioneers on this intermediate town formed a major risk for them. The supporting 'intermediate town' and 'risk' are two characteristics that perhaps can be added to the list of features of a frontier society.

The frontier conditions go a long way in explaining wasteful production methods. Abundance of free bounties when the exploitation of older resources became more laborious, the craze of a rising international market, lack of government control, and the short-term outlook of the pioneers and ensuing indifference to what would happen with the land in the future contributed to a reap-and-run exploitation. Reliance on a town for provisions enabled the frontier societies to maximize commodity production and reduced the need for a varied use of the environment. Considering the prolonged period of tin mining, over two centuries, this must have been more restrained than the gambier gardens and *panglong,* despite its gradual spatial expansion and eventual damage to nature. Notably, tin mining was not a frontier activity in all aspects, and especially the stricter grip of the government seems relevant. A last condition to be mentioned is that scarcity of labour impedes conservation measures that often require an investment in labour (Pichón 1996:360).

Against the background of today's environmental worries an implicit condemnation of the environmental record of frontier societies is almost inevitable, but at the same time anachronistic. The contemporaries of the pioneers considered them beneficial for they cleared otherwise inaccessible and hostile areas. For instance, the Chinese poet Na Tian Piet, who lived in Sumatra, Singapore, and Johore, saw the gambier cultivators as an important force in the development of Johore (Salmon 1992:23).[45] Also the Deli Planters' Association saw their tobacco plantations as a positive contribution, considering the title of a public-relations book of theirs (Volker 1928): *Van oerbosch tot cultuurgebied; Een schets van de beteekenis van de tabak, de andere cultures en de industrie ter Oostkust van Sumatra* (From jungle to cultivated land; A sketch of the meaning of the tobacco [...]).

The general impression is that the first pioneers cleared the path for successive frontiers, but at the same time reduced the potential of succeeding frontiers. *Kolong* mines gave way to less intensive mining, and where gambier cultivation was continued, the yield was declining. Around some old pioneer settlements, where only one commodity had been produced, a more diversified economy including food production developed. These new forms of agriculture had to cope with reduced fertility of the soil. Nevertheless, after these areas had been opened up, the population size generally increased.

Frontier societies with environmental consequences were certainly not limited to Sumatra. Sugar cultivation around Batavia, which started its expansion in the late-seventeenth century, was a phenomenon similar to gambier cultivation in the sense that it consumed large amounts of firewood (Nagtegaal 1995:17-8; Raben 1996:58-62).[46] Tin mining on Bangka had its equivalent, and perhaps instructive model, in gold mining in West Kalimantan (Cator 1936:145-56; Chew 1990). With the partial exception of *panglong*, in all these cases environmental degradation was the result of the production process and not so much of the removal of the resource itself: the need for trees for firewood and charcoal, rerouting water flows for hydraulic power, and upsetting soils to reach layers of pay dirt.

The mention of sugar mills and gold-mines might give the impression that frontier societies and Chinese primary production are a one-to-one match. Such an equation would be erroneous on both sides. Firstly, there are numerous cases where people other than Chinese constituted a frontier society, often with the same negative results for the environment. I already indicated the European plantations for tobacco and other crops in

45 'Bertambah banjak orang Tjina / Datang dari negri mana mana / Pinda di Muar poenja tana / Bertanam gambir banyak disana. Muar bernama Bandar Maharani / Soeda ditoekar nama nja kini / Sekarang ramei dahoeloe soeni / Banjak orang bertanam tani.' (Salmon 1992:23). (The Chinese people become more numerous / Coming from everywhere / They occupy land in Muar / And cultivate much gambier over there. Muar is called Bandar Maharani / It has been renamed like that / Now it is lively, it used to be dead / Many people till the soil.)

46 Blussé (1986:26) does not speak of a frontier society, but compares the expansion of sugar mills from Batavia to the town's surroundings with the 'Wild West'.

northern Sumatra (Pelzer 1978:32-65; Thee 1969:1-64). Acehnese growing pepper in Tapanuli (northwestern Sumatra) (Lee 1995:67-71, 98-9), Buginese growing pepper in Kalimantan (Vayda and Sahur 1985), and to a certain degree the gutta percha collectors (Potter this volume) are other examples. A recent, still current phenomenon are the spontaneous transmigrants, usually coming from Java, who are entering southern Sumatra (Charras and Pain 1993). Secondly, quite often Chinese pioneers settled permanently, married and started to invest in a sustainable environment. The charcoal burners in the *panglong* are a good example of this process of settling down. Another example were the Chinese who laid out irrigated rice fields in former gold-mine areas in West Kalimantan (Heidhues 1996:171). The example of Chinese pioneers who settled down permanently and the Minangkabau gambier cultivators suggest that people invest in their environment when they see their own direct interest in sustainability. This insight is not new, but nevertheless worth repeating, as long as tragedies of commons[47] keep taking place.

47 See Hardin 1968.

Abbreviations

ANRI	Arsip Nasional Republik Indonesia, Jakarta
ARA	Algemeen Rijksarchief, Den Haag
VOC	Archive of the Dutch East Indies Company (VOC), ARA

References

Ali Haji Ibn Ahmad, Raja
1982 *The precious gift (Tuhfat al-Nafis)*. Annotated by V. Matheson and B. Watson Andaya. Kuala Lumpur: Oxford University Press.

Andaya, B.W.
1993 *To live as brothers; Southeast Sumatra in the seventeenth and eighteenth centuries*. Honolulu: University of Hawaii Press.

Balestier, J.
1848 'View of the state of agriculture in the British possessions in the Straits of Malacca', *Journal of the Indian Archipelago and Eastern Asia* 2:139-50.

Bankoff, G.
1995 'Coming to terms with nature; State and environment in maritime Southeast Asia', *Environmental History Review* 19-3:17-37.

Bartlett, R.A.
1974 *The new country; A social history of the American frontier, 1776-1890*. London: Oxford University Press.

Berkhout, A.H.
1895 'Boschbouwkundige beschrijving van het eiland Banka; Een dienstreis van den houtvester', *Tijdschrift voor Nijverheid en Landbouw in Nederlandsch-Indië* 50:11-66.

Blussé, L.
1986 *Strange company; Chinese settlers, mestizo women and the Dutch in VOC Batavia*. Dordrecht: Foris. [KITLV, Verhandelingen 122.]

Breman, J.
1987 *Koelies, planters en koloniale politiek; Het arbeidsregime op de grootlandbouwondernemingen aan Sumatra's Oostkust in het begin van de twintigste eeuw*. Second edition. Dordrecht: Foris. [KITLV, Verhandelingen 123.]

Bruun, O.
1992 'Fengshui and the Chinese perception of nature', in: O. Bruun and A. Kalland (eds), *Asian perceptions of nature; A critical approach*, pp. 234-49. Copenhagen: NIAS.

Cator, W.J.
1936 *The economic position of the Chinese in the Netherlands Indies*. Oxford: Blackwell. [PhD thesis, Rijksuniversiteit Leiden.]

Charras, M. and M. Pain (eds)
1993 *Spontaneous settlements in Indonesia; Agricultural pioneers in southern Sumatra/Migrations spontanées en Indonésie; La colonisation agricole du sud de Sumatra*. Jakarta: Departemen Transmigrasi, Paris: ORSTOM/CNRS.

Chew, D.
1990 *Chinese pioneers on the Sarawak frontier, 1841-1941*. Singapore: Oxford University Press.

Colombijn, F.
1997 'Een milieu-effect rapportage van de gambircultuur in de Riau-archipel in de negentiende eeuw', *Tijdschrift voor Geschiedenis* 110:290-312.

Cooper, M.
1993 'Access to the waterfront; Transformations of meaning on the Toronto lakeshore', in: R. Rotenberg and G. McDonogh (eds), *The cultural meaning of urban space*, pp. 157-71. Westport/London: Bergin and Garvey.

Couperus, A.
1780 'Berigt, aangaande de gamber, derzelver planting en bewerking op Malacca', *Verhandelingen van het Bataviaasch Genootschap der Kunsten en Wetenschappen* 2:356-82.

Diest, P. van
1865 *Bangka beschreven in reistogten*. Amsterdam: Stemler.

Elvin, M.
forthcoming 'The bell of poesy; Thoughts on poems as information on late-imperial Chinese environmental history', in: *Festschrift for Professor L. Lanciotti*. Napoli.

Encyclopaedie
1919 *Encyclopaedie van Nederlandsch-Indië*. Vol. 3. 's-Gravenhage: Nijhoff, Leiden: Brill.

Encyclopaedisch Bureau
1915 *De gambircultuur in de Buitenbezittingen*. Weltevreden: Albrecht. [Mededeelingen van het Bureau voor de Bestuurszaken der Buitengewesten 7.]

Endert, F.H.
1932 'De proefbaanmetingen in de panglong-gebieden van Bengkalis (Sumatra's Oostkust) en Riouw', *Tectona* 25:731-85.
1933 'Eenige resultaten van de boschverkenning in de Buitengewesten', *Tectona* 26:391-422.

Erman, E.
1994 'Tauke, kuli dan penguasaha; Eksploitasi hutan panglong di Riau', *Sejarah: pemikiran, rekonstruksi, persepsi* 5:20-33.

FitzGerald, C.P.
1993 *The southern expansion of the Chinese people*. Bangkok: White Lotus.

Freedman, M.
1965 *The Chinese in South-East Asia; A longer view*. London: China Society.

Furukawa, Hisao
1994 *Coastal wetlands of Indonesia: Environment, subsistence and exploitation*. Kyoto: Kyoto University Press.

Graafland, A.F.P.
1888 'Schets der Chineesche vestigingen in de afdeeling Karimon', *Bijdragen tot de Taal-, Land- en Volkenkunde* 37:505-44.

Groot, J.J.M. de
1901 *The religious system of China [...]*. Vol. 4. Leiden: Brill.

Hardin, G.
1968 'The tragedy of the commons', *Science* 162:1243-8.

Heidhues, M. Somers
1992 *Bangka tin and Mentok pepper; Chinese settlement on an Indonesian island*. Singapore: Institute of Southeast Asian Studies.
1996 'Chinese settlements in rural Southeast Asia; Unwritten histories', in: A. Reid (ed.), *Sojourners and settlers; Histories of Southeast Asia and the Chinese; In honour of Jennifer Cushman*, pp. 164-82. St. Leonards: Allen and Unwin.

Helbig, K.
1940 'Die Insel Bangka; Beispiel des Landschafts- und Bedeutungswandels auf Grund einer geographischen "Zufallsform"', *Deutsche Geographische Blätter* 43:137-207.

Hjort af Ornäs, A. and U. Svedin
1992 'Earth-man-heaven; Cultural variations in concepts of nature', in: O. Bruun and A. Kalland (eds), *Asian perceptions of nature; A critical approach*, pp. 159-75. Copenhagen: NIAS.

Horsfield, T.
1848 'Report on the island of Bangka', *Journal of the Indian Archipelago and Eastern Asia* 2:299-336, 373-427, 705-25, 779-824.

Jackson, J.C.
1965 'Chinese agricultural pioneering in Singapore and Johore, 1800-1917', *Journal of the Malaysian Branch of the Royal Asiatic Society* 38:-177-105.
1969 'Mining in 18th-century Bangka; The pre-European exploitation of a "tin island" ', *Pacific Viewpoint* 10:28-54.

Jonker, H.A.
1933 'De vloedbosschen van den Riouw-Lingga archipel', *Tectona* 26:717-41.

Koster, H.J.
1922 'Gambircultuur (*Uncaria gambir* Roxb.) en boschtoestand in den Riouw-archipel', *Tectona* 15:883-7.

Lee Kam Hing
1995 *The sultanate of Aceh; Relations with the British, 1760-1824*. Kuala Lumpur: Oxford University Press.

Malmrös, F.
1939 'De houtskoolbereiding bij de Banka tinwinning', *Tectona* 32:361-7, 687-715.

Nagtegaal, L.
1995 'Urban pollution in Java, 1600-1850', in: P.J.M. Nas (ed.), *Issues in urban development; Case-studies from Indonesia*, pp. 9-30. Leiden: Research School CNWS.

Needham, J.
1956 *Science and civilisation in China. Vol. 2: History of scientific thought*. Cambridge: Cambridge University Press.

Ng Chin-keong
1976 *The Chinese in Riau; A community on an unstable and restrictive frontier*. Singapore: Institute of Humanities and Social Sciences, Nanyang University. [Research Project Series 2.]

Panglongs
1915 'De panglongs in de afdeeling Lingga der residentie Riouw en Onderhoorigheden', *Koloniaal tijdschrift* 4:1009-33.

Pastor, G.
1927 *De panglongs*. Weltevreden: Landsdrukkerij.

Pelzer, K.J.
1978 *Planter and peasant; Colonial policy and the agrarian struggle in East Sumatra, 1863-1947*. 's-Gravenhage: Nijhoff. [KITLV, Verhandelingen 84.]

Pichón, F.
1996 'Settler agriculture and the dynamics of resource allocation in frontier environments', *Human Ecology* 24:341-73.

Raben, R.
1996 *Batavia and Colombo; The ethnic and spatial order of two colonial cities, 1600-1800*. [PhD thesis, Rijksuniversiteit Leiden.]

Reenen, J. van
1996 *Central pillars of the house; Sisters, wives, and mothers in a rural community in Minangkabau, West Sumatra*. Leiden: Research School CNWS. [PhD thesis, Rijksuniversiteit Leiden.]

Reid, A.
1993 *Southeast Asia in the age of commerce 1450-1680. Vol. 2: Expansion and crisis*. New Haven/London: Yale University Press.
1996 'Flows and seepages in the long-term Chinese interaction with Southeast Asia', in: A. Reid (ed.), *Sojourners and settlers; Histories of Southeast Asia and the Chinese; In honour of Jennifer Cushman*, pp. 15-49. St. Leonards: Allen and Unwin.
1997 'Entrepreneurial minorities, nationalism, and the state', in: D. Chirot and A. Reid (eds), *Essential outsiders; Chinese and Jews in the modern transformation of Southeast Asia and Central Europe*. Seattle: University of Washington Press.

Riau
1985 *Riau dalam angka 1984*. Pekanbaru: Pemda Tk. I Riau.

Ridsdale, C.E.
1991 '*Uncaria gambir* (Hunter) Roxb.', in: R.H.M.J. Lemmens and N. Wulijarni-Soetjipto (eds), *Plant resources of South-East Asia. Vol. 3: Dye and tannin-producing plants*, pp. 125-8. Wageningen: Pudoc.

Salmon, C.
1992 'Na Tian Piet et sa vision du monde malais dans les années 1890', in: C. Salmon (ed.), *Le moment 'sino-malais' de la litterature indonésienne*, pp. 9-42. Paris: Archipel. [Cahier d'Archipel 19.]

Schets
1856 'Schets van de bereiding der gambier te Paija Komba (Padangsche Bovenlanden)', *Tijdschrift voor Nijverheid in Nederlandsch-Indië* 3:21-7.

Siah U Chen
1848 'The Chinese in Singapore; General sketch of the numbers, tribes, and avocations of the Chinese in Singapore', *Journal of the Indian Archipelago* 2:283-9.

Smiet, A.C.
1990 'Agro-forestry and fuel wood in Java', *Environment Conservation* 17-3:235-8.

Teijsmann, J.E.
1874 'De gambier kultuur', *Tijdschrift voor Nijverheid en Landbouw in Nederlandsch-Indië* 19:169-77, 295-311, 357-68.

Thee Kian Wie
1969 *Plantation agriculture and export growth; An economic history of East Sumatra, 1863-1942*. [PhD thesis, University of Wisconsin.]

Trocki, C,.A.
1976 'The origins of the kangchu system 1740-1860', *Journal of the Malaysian Branch of the Royal Asiatic Society* 49-2:132-55.

Tu Wei-ming
1989 'The continuity of being; Chinese visions of nature', in: J. Baird Callicott and R.T. Ames (eds), *Nature in Asian traditions of thought; Essays in environmental philosophy*, pp. 67-78. Albany: State University of New York Press.

Turner, F.J.
1920 *The frontier in American history*. New York: Holt.

Vayda, A.P. and A. Sahur
1985 'Forest clearing and pepper farming by Bugis migrants in East Kalimantan; Antecedents and impact', *Indonesia* 39:93-110.

Volker, T.
1928 *Van oerbosch tot cultuurgebied; Een schets van de beteekenis van de tabak, de andere cultures en de industrie ter Oostkust van Sumatra*. Medan: Deli Planters Vereeniging.

Webb, W.P.
1953 *The great frontier*. London: Secker and Warburg.

White, L.
1968 *Machina ex deo; Essays in the dynamism of Western culture*. Cambridge, Mass.: MIT Press.

Wood, W.B.
1985 *Intermediate cities in the resource frontier: A case-study of Samarinda and Balikpapan, East Kalimantan, Indonesia*. [PhD thesis, University of Hawaii.]

World Commission on Environment and Development
1987 *Our common future*. Oxford: Oxford University Press.

Zeijlstra, H.H.
1949 'Gambir', in: C.J.J. van Hall and C. van de Koppel (eds), *De landbouw in de Indische archipel, Vol. IIB: Genotmiddelen en specerijen*, pp. 578-619. 's-Gravenhage: Van Hoeve.

MICHAEL R. DOVE

The political ecology of pepper in the Hikayat Banjar
The historiography of commodity production in a Bornean kingdom

Introduction

Whereas the records of the participation of indigenous kingdoms in commodity production for colonial markets are very good, the records of how such kingdoms perceived this participation are scanty by comparison. This analysis is an attempt to interpret one such record, from seventeenth-century Indonesia. The commodity involved is black pepper (*Piper nigrum* L. [Piperaceae]), which merits attention not only because of the enormous role it played in Indonesia's trade in the early modern era, but also because it was perhaps the first introduced crop to undergo the transition from small-scale household cultivation to larger-scale production with state involvement (Andaya 1995:185-6). The dynamics of this transition – what motivated versus retarded it, and who reaped the benefits and who paid the costs – still pose challenging theoretical and policy-related questions, some insight into which may be offered by this ethno-historical analysis.

The pepper injunction

There is a remarkable passage in the *Hikayat Banjar,*[1] a seventeenth-century chronicle of the kingdom of Banjar in Southeast Borneo, in which its founder and ruler, King Ampu Jatmaka, issues an injunction against the cultivation of *sahang* or black pepper:

> 'And let not our country plant pepper as an export-crop, for the sake of making money, like Palembang and Jambi [two kingdoms in Sumatra]. Whenever a country cultivates pepper all food-stuffs will become expensive and anything planted will not grow well, because the vapours of pepper are hot. That will cause malice all over the country and even the government will fall into disorder. The rural people will become pretentious towards the townsfolk if pepper is grown for commercial

1 This is spelled *Hikajat Bandjar* in the original: I have modernized the spelling of all Indonesian (also Malay, Banjarese, Javanese, and Kantu' (Iban)) terms.

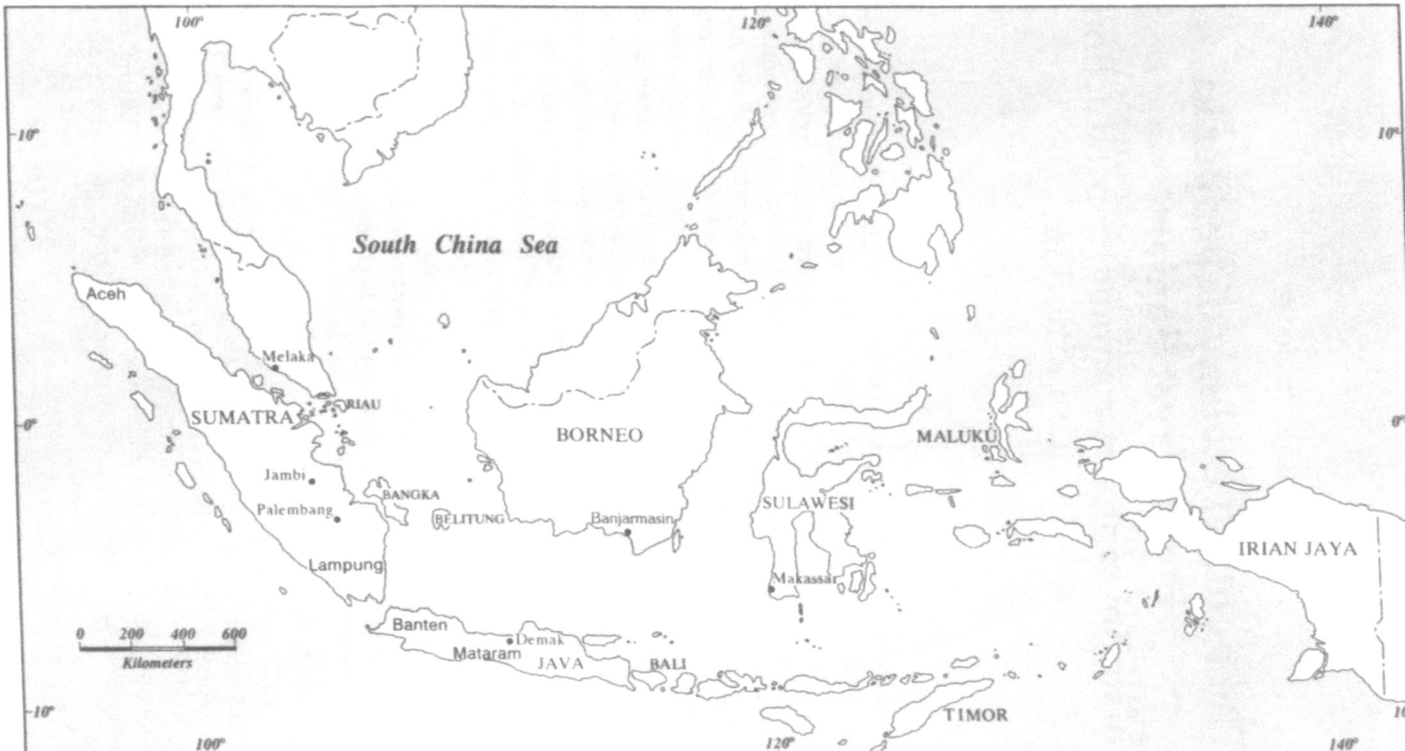

Figure 1. Indonesia (early modern era)

> purposes, for the sake of money. If people grow pepper it should be about four or five clumps per head, just enough for private consumption. Even four or five clumps per head will cause much vapour, owing to the great number of people involved, let alone if it is grown extensively as a crop; then the country inevitably would be destroyed.' (Ras 1968:265-7.)

This injunction is repeated further on in the *Hikayat*, in almost identical terms, by three subsequent rulers (Ras 1968:331, 375, 443).[2] A modern transcriber of the chronicle notes that these multiple references 'are completely stereotyped, yet the author [of the *Hikayat*] seems to attach great importance to them' (Ras 1968:58) – an importance that is attested to by their repetition.

These passages in the *Hikayat Banjar* are not unique. Reid (1993:298) reports a similar passage from the *Hikayat Pocut Mohamat*, an eighteenth-century Acehnese epic. This passage runs as follows:

> 'Marketing does not yield much profit, even if you grow pepper, my friends. If there is no rice in the country, nothing else will be of use. What is the use of a purple kerchief or a dagger with a hilt of pinch-beck? If there is no rice in the country, the standing of royalty will be lost. If there is nothing to eat, your children will starve, and you will have to sell all you possess.' (Drewes 1979:167.)[3]

Some of the region's rulers took concrete action against pepper. Reid (1993: 299-300) reports that early in the seventeenth century, the Sultan of Aceh ordered the destruction of pepper vines in the vicinity of the capital, because their cultivation was leading to the neglect of food crops and to consequent annual food shortages. He also reports that Banten cut down its pepper vines around 1620 in the hope that this would encourage the Dutch and English to leave the sultanate in peace, though self-sufficiency must have been an additional reason; and the Sultan of Magindanao told the Dutch in 1699 that he had forbidden the continued planting of pepper so that he could avoid conflict with foreign powers. Noorlander (1935:4-5, 124-5; cited in Hudson 1967:70) says that in the Banjar court itself, an isolationist faction sought to end the sultanate's foreign contacts by destroying the kingdom's pepper groves. The proscription or destruction of valuable resources by those without sufficient power to resist their

2 This injunction is given added weight by the fact that in two of these cases, it is the ruler's last act before he 'disappears' or departs the mortal world (Ras 1968:331-2, 373-5).

3 I also have found the opposite vision, as in Marsden's (1975:139) paean to the pepper gardens of Sumatra:
'A pepper garden cultivated in England would not, in point of external appearance, be considered as an object of extraordinary beauty [...]; yet, in Sumatra, I never entered one, [...] that I did not find myself affected with a strong sensation of pleasure. Perhaps the simple view of human industry, so scantily presented in that island, might contribute to this pleasure, by awakening those social feelings that nature has inspired us with, and which makes our breasts glow on the perception of whatever indicates the prosperity and happiness of our fellow creatures.'

Figure 2. The *kraton* of the Sultan of Banjarmasin (Schwaner 1853, I:opposite page 56).

exploitation by those with sufficient power to insist upon it, is not uncommon (Dove 1993a) nor is it, thus, all that noteworthy. What is noteworthy about the relevant passages in the *Hikayat Banjar* is the historic detail on the way that one such response was conceived and articulated.

The Hikayat Banjar

'The Story of Lambu Mangkurat and the Dynasty of the Kings of Banjar and Kota Waringin', more commonly known as the *Hikayat Banjar*, is the native, court-based chronicle of a coastal Malayic kingdom that existed in Southeast Borneo until 1860 (Figure 1),[4] although the chronicle itself only covers up until 1661 (Ras 1968:1, 3). According to its foremost contemporary scholar, J. J. Ras, the *Hikayat Banjar* was written, and rewritten, over the period of about a century between the mid-sixteenth and mid-seventeenth centuries (Ras 1968:177-81, 196), by three or four separate chroniclers, residing at different courts at different times. Ras (1968:2) says that the text was recopied several times in the first half of the nineteenth century, but there is no evidence that this involved substantive rewriting; thus, the chronicle can be taken to reflect the views of its sixteenth- and seventeenth-century authors.[5]

Following Ras, I regard the *Hikayat* as a chronicle produced by, and thus reflective of the sixteenth- and seventeenth-century Banjarese kingdom. I regard the references to pepper cultivation in particular as deliberate, meaningful, and meriting analysis. As Snouck Hurgronje (1888) writes (cited in Ras 1968:15), 'If one takes these myths and legends for what they are, one has in them precious sources of knowledge concerning the character of the people'. I will consider these references to pepper cultivation to be a unique source of knowledge on commodity production in old Borneo in the early modern period. I believe that this production represented a critical element in the historic development of the Banjar Kingdom and I suggest that the chronicle represents an indigenous articulation of this element.

Background: pepper cultivation, trade, and the Banjar Kingdom

Analysis of the *Hikayat Banjar* will be facilitated by reviewing, first, pepper cultivation today in Borneo, then the history of the pepper trade, and finally the history and economic background of the Banjar Kingdom.

4 I use the term 'Malayic' rather than Malay, because the Banjarese distinguish themselves from Malays, since they list Malays among the 'foreign traders' (Ras 1968:431; see also Saleh 1976:206). Ras (1968:8) describes the Banjarese language as 'the independent continuation of a rather archaic type of Malay, superimposed on a substratum of Dajak dialects, with an admixture of Javanese.'

5 Ras (1968:173) says that one version of the chronicle includes a sketchy continuation of dynastic history up to the beginning of the nineteenth century.

Figure 3. Borneo (contemporary)

Pepper cultivation

An example of pepper cultivation by a contemporary group of swidden cultivators is given by the Kantu', a tribal people of West Kalimantan (Figure 1). The Kantu' meet subsistence needs through the cultivation of upland rice and a wide variety of non-rice cultigens in swiddens; they meet market and trade needs through the cultivation of Para rubber (*Hevea brasiliensis*), called *getah*, and, to a lesser extent, pepper, called *lada*.[6] The Kantu' cultivate pepper in tiny gardens cleared from the forest, which average perhaps 500 m^2 in area and contain perhaps 225 plants (Figures 4, 5, 6). Cultivation is intensive, at least by comparison with the Kantu' swidden agriculture. De Waard (1989:227) sums up this system by saying that pepper cultivation in Borneo is characteristically associated with chemically poor soils, high inputs, and high productivity.

The Kantu' experience with pepper has turned most of them against it – in particular when compared with the cultivation of rubber, their other major trade or cash crop. There are several reasons for this. The cost of the proper chemical inputs is one reason; another is the impact of pepper cultivation on the land. Whereas land that is put under rubber cultivation can be subsequently used for swiddens (namely, after the use-life of the rubber is over), land put under pepper cannot.[7] Pepper is said to take all of the *lang-lemak* (fertility) from the soil, to eliminate its *bau* (aroma), to make it *kusi* (barren). It is traditionally the only land use in the Kantu' territory that precipitates *belayang madang*, a grassland succession of *Imperata cylindrica*.[8] These impacts are reflected in the fact that while the Kantu' will ordinarily lend land to one another for the purpose of swidden making, they will not lend land for pepper gardening.

Even more important than the consequences of pepper cultivation for land use are its consequences for labour use. Pepper requires comparatively much attention.[9] The Kantu' say of it: 'Kebun lada nadai tau' lekat

6 The Kantu' say that they have had pepper for 'generations' and that they first obtained it from Dayak living across the border in Sarawak.

7 Former pepper garden sites can be, and are, successfully planted in rubber, however. This may help to explain why pepper and rubber are often found within the same agricultural system, as among the Kantu'. Heidhues (1992:215) writes that degraded former pepper lands on Bangka are used not only for rubber but also for durian, coffee, and cloves.

8 A number of observers have noted a linkage between pepper cultivation and grassland succession (Blacklock 1954:42, 47; Brookfield et al. 1990:497; Burkill 1966, II:1779; Potter 1988:129; Reid 1995:101-4); and some have even suggested that the mark of historic pepper cultivation can still be seen in contemporary grasslands. However, whereas pepper cultivation may precipitate grassland succession, other factors are necessary to perpetuate these successions, which are inherently unstable on their own. In addition, succession to grassland – even to the much-maligned *Imperata cylindrica* – does not necessarily reflect environmental degradation (Dove 1986a, 1986b). *Imperata* grasslands may serve a wide variety of economic and ecological functions, including that of being opened for further cultivation (which is laborious but far from being, as commonly thought, impossible).

9 Ochse (1977:588) writes: 'A plantation of pepper is very exacting as regards quality of the soil and care to be bestowed'.

sekali'; asa belega' dibantun-ditupi', mati' (Pepper gardens cannot be ignored at all; if [you] momentarily stop weeding and caring for, [it] dies).

Indeed, the Kantu' say, though perhaps not without hyperbole, 'If you ignore pepper for even ten days, it will die' (Padoch 1982:113). This poses a problem for the Kantu' during the seasons of peak labour demand in their swiddens (planting, weeding, and harvesting). The Kantu' say only a household that has adult children in it, which is capable of splitting up its labour force, can cultivate pepper. The timing constraints of pepper cultivation have another consequence as well: with respect to the market. It takes a minimum of three years for a pepper garden to start producing, so there is a three-year time-lag in the cultivator's response to market conditions (De Waard 1964:24).[10] And once a crop of pepper has ripened, it must be harvested: pepper cannot be stored on the vine (Padoch 1982:113).[11] The time-lag in initial production, and the inability to delay harvesting, place the pepper cultivator at the mercy of market prices (which is not the case with rubber). Cramb (1993:222) says that reliance on pepper, compared with other cash crops, leads to greater peaks when prices are high and deeper lows when prices are low. More generally, this time-lag introduces elements of volatility and uncertainty into the marketing of this crop, which have characterized it throughout history.[12]

History of the pepper trade

Black pepper is today the most important spice in the world in terms of usage and value (Duke and Du Cellier 1993:394), and it has been globally

10 In contrast, the negligible cost of starting or stopping rubber tapping (in an existing rubber garden) permits the Kantu' to respond adroitly to fluctuating market prices (Dove 1993b:139-40). This capability, coupled with the minimal costs of establishing rubber and the ability to self-exploit household labour, even permits the Kantu' to respond inversely to price fluctuations (that is, to increase production as prices fall or decrease as prices rise) (Dove 1993b:143-4, 1996:43).

11 Although pepper cannot be stored on the vine, Duke and Du Cellier (1993:398) say that it can be stored off the vine – and thus withheld from the market – for several years without loss of quality. Instead of empowering small producers, it appears, as Duke and Du Cellier go on to suggest, this characteristic is largely exploited by international speculators (De Waard 1964:24), which merely adds to the problems faced by the smallholder. Andaya (1993b:79) gives an historic example, 'Pepper can be stored for several years without deteriorating, and by the middle of the seventeenth century, warehouses in Europe were filling up as sellers tried to keep the price high and as consumers turned to ginger as a substitute. By 1652 Europe was said to be glutted with pepper, the surplus sufficient for at least three years.'

12 This is reflected in Chau Ju-kua's comment on the fluctuation in the cost of the 'great abundance of pepper' from thirteenth-century Central Java (Su-ki-tan): 'At the right season and in good years, twenty-five taels of "trade money" will buy from ten to twenty packages of pepper, each package holding fifty pecks. In years of dearth or times of disturbance, the same sum will buy only half of that amount' (Hirth and Rockhill 1911:83). Schrieke (1966:56) writes of these disturbances: 'It happened repeatedly that because of troubles between *negeri* [kingdoms] in the interior there would be no pepper shipped to the market in Jambi'. Burkill (1966, II:1779) attributes fluctuation in (nineteenth-century) pepper production to variation in political conditions, variation in market prices, and the inherent ecological dynamic that necessitated periodic relocation of the pepper gardens.

important since classical times: a number of writers have noted the acquaintance with pepper of classical scholars, including Pliny, Dioscorides, and Theophrastus, dating back to the fourth century BC (Burkill 1966, II:1776; Flückiger and Hanbury 1879:576; Watt 1972:264). Pepper is native to the Western Ghats of India (Burkill 1966, II:1776). It was introduced from India to Indonesia near the beginning of the first millennium AD (and Indonesia is still a major producer, along with India, Malaysia (Sarawak), and Brazil); Burkill (1966, II:1779) suggests that Hindu colonists migrating from India brought pepper to Indonesia between 100 BC and AD 600. Wolters says that there is some evidence that western Indonesia may have been exporting pepper to China before AD 400: in a Chinese translation undertaken in 392 of the *Sutra of the Twelve Stages of the Buddha*, five maritime kingdoms are enumerated and the account of one, called She-yeh – which has been plausibly identified as Java[13] explicitly mentions that 'This land produces long pepper and black pepper' (Wolters 1967:66-7, 183).[14] Hirth and Rockhill (1911:223, note 2) suggest that the first Chinese author to mention pepper as a product of the East Indies was Chóu K'ü-féi, an assistant sub-prefect in Kui-lin, in his Sung topography, the *Ling-wai Tai-ta*, published in 1178. Pepper from a number of places in Java also is mentioned in the *Chu-fan-chï* (literally, 'A description of barbarous peoples'), a trade handbook compiled in 1225 by the Superintendent of Maritime Trade in Fukien, Chau Ju-kua. Indeed, it may be largely due to the trade in pepper that Chau Ju-kua ranks Shö-p'o (Java) second after Ta-shï (the realm of the Arabs), of all 'the wealthy foreign lands which have great store of precious and varied goods' for trade with China (Hirth and Rockhill 1911:23).[15] Reid (1993:12) suggests that Ming trading missions and territorial expansion early in the fifteenth century stimulated pepper production in Indonesia and its trade to China.[16] Marco Polo also mentions

13 Wolters (1967:184) believes that Lévi's (1918:83) identification of She-yeh with Java is 'reasonable'.

14 In the Chinese text used by Wolters, the common referent for 'long pepper' is *Piper longum*, which is native to India but not Java. If Java is indeed the location of the kingdom of She-yeh, then the plant to which the text refers is probably *Piper retrofactum*, which in appearance and use resembles *Piper longum* and has been traded from Java to China since ancient times (Burkill 1966, II:1775, 1782).

15 The decisive role of pepper in this ranking is suggested, in part, by the fact that China officially eventually had to prohibit trade with Java because purchases of pepper were draining China's currency reserves (Hirth and Rockhill 1911:78, 82).

16 Reid (1993:12) actually writes that the Ming trade missions 'were probably responsible for the introduction of Indian pepper plants to northern Sumatra'. This would give a much later date for this introduction than that given by other writers cited in the text. Reid (1993:32-3) subsequently gives the following timetable for the dissemination of pepper, but in a context in which he implies it is the rise of production centres and not actual introduction that he is talking about: 'Round pepper (*Piper nigrum*) had [...] spectacular travels – from southern India to the tip of northern Sumatra around 1400, the Malayan peninsula by 1500, the western coast of Sumatra by about 1550, the inland Minangkabau districts of Sumatra, western Java, and southern Sumatra by 1600, and southern Borneo by 1630'.

black pepper as part of the 'surpassing wealth' of Java, during his visit there at the end of the thirteenth century (Polo 1969:241).

A critical shift in the historic development of the pepper trade involved the displacement of Chinese buyers by European buyers. Up until 1500, almost all Indonesian pepper was being exported to China (Reid 1993:144). But by 1600 more than half of the much-increased Southeast Asia pepper crop was being taken to the Middle East and Europe (Reid 1993:144; Andaya 1993b:43-5). The goal of dominating the pepper trade was one of the principal motives behind the expansion of first the Portuguese and later the Dutch and English into Asia (Andaya 1993b:45).[17] By the early modern period, pepper was the most important Southeast Asian export by a factor of ten (Reid 1995:100). Schrieke (1966:50) cites the *Daghregister* (1625:202) as saying '[The pepper trade was] now as earlier the most important thing in the commerce [of the Dutch Company]'. During the sixteenth and seventeenth centuries, when the *Hikayat* was written, pepper went through a major boom and bust: prices reached peak levels during the period 1616-1650 (and production continued to rise through 1670), but then prices declined by one-half between 1650 and 1653, and by the 1670s they had dropped to just one-quarter of what they had been in the 1640s (Reid 1993:299; Andaya 1993b:79-80).[18]

Much of this history is reflected in pepper's terminology. The English term 'pepper' derives from the Sanskrit name for one of India's pepper plants, *pippali* (*Piper longum*) or 'long pepper' (Monier-Williams 1899: 628).[19] Over time, as this term came into English and the other European languages, its referent shifted from long pepper to round or black pepper.[20] The Sanskrit name for black or round pepper (*Piper nigrum*) is *maricha* (Monier-Williams 1899:790). Black pepper is called *miricha* and *mrica* in old and modern Javanese respectively (Zoetmulder 1982, I:1143; Horne 1974:384); but it is called *lada* in Indonesian and Malay (Echols and Shadily 1992:321; Wilkinson 1959, II:636-7), Sundanese, and many of the languages of Sumatra and Borneo (as in the case of the Kantu', Richards 1981:174). Burkill (1966, II:1776-7) suggests that *lada* was originally a term for peppers indigenous to the archipelago (*Piper cubeba* L. and *Piper*

17 Watt (1972, VI:264) writes that 'The trade in pepper [...] during the Middle Ages was certainly the most important branch of commerce between Europe and the East'.

18 This was not the end of pepper cultivation, which experienced a number of boom-bust cycles during the subsequent three centuries. By the time of Marsden's study of West Sumatra in the 1770s, for example, pepper was again of sufficient importance for him to write: 'Of those productions of Sumatra, which are regarded as articles of commerce, the most important and most abundant is pepper. This is the object of the East India Company's trade thither, and this alone it keeps in its own hands.' (Marsden 1975:129.) In addition, the trade to China was not necessarily affected by the boom-bust cycles of the European trade (Andaya 1993b:122).

19 The Sanskrit *pippali* is derived from the term for the sacred fig tree (*Ficus religiosa* L.), *pippala*, and its fruit (Monier-Williams 1899:627-8).

20 This suggests that India's westward pepper trade was first based on *Piper longum* and then only later on *Piper nigrum* (see Burkill 1966, II:1774, 1777).

retrofactum Vahl, both of which were ancient trade commodities and are still called by this term today), and that it was adopted, in the aforementioned languages, for the incoming black pepper. The term used for black pepper in the *Hikayat*, *sahang*, which is still used in contemporary Banjarese (Hapip 1977:155), is also an old Javanese (Kawi) word (Ras 1968:593; Wilkinson 1959, II:999; Zoetmulder 1982, II:1596).

The Banjar Kingdom

According to the *Hikayat*, the Banjar kingdom was founded by a trader, who on his deathbed said: 'In my heart I still take pride in considering myself as nothing but a prominent merchant' (Ras 1968:229, 231, 267-8).[21] The importance of trade to the kingdom is reflected in the fact that the presence of foreign traders is cited throughout the *Hikayat* as a sign of a healthy kingdom: the stereotypical line is 'The country was bustling and prosperous, and foreign traders also came in great numbers' (Ras 1968:335, 231, 373).[22] The participation of the large trading community in periodic dynastic conflicts is recounted in the *Hikayat* (Ras 1968:413); and when the Banjar capital is periodically relocated – whether because of dynastic conflict or land accretion between it and the sea (Ras 1968:192-6) – the relocation of the traders is explicitly mentioned as well (Ras 1968:371, 405). The *Hikayat* variously describes the 'foreign traders' as Acehnese, Balinese, Buginese, Biaju, Chinese, Johorese, Hollanders, Javanese, Madurese, Makassarese, Malaccans, Malay, Minangkabau, Patanis, Sumbawanese, and those from Banten, Jambi, Gujerat, Keling, Macao, Palembang, and Tuban (Ras 1968:263, 371, 431). This diversity was in the best interests of Banjar; but it was contested by the Dutch and English, who thought that their own best interest lay in monopolizing the trade of Banjar (and other major trade entrepôts). Dutch and English machinations notwithstanding, the Banjar rulers were quite successful well into the colonial period in maintaining a free trade policy (even when they signed treaties to the contrary, as they did with the Dutch in 1635). Thus, Suntharalingam writes of Banjar's response at the beginning of the eighteenth century to actions by the chief of the English trading mission that had led to open hostilities with Banjar:

21 Banjarese have a reputation to this day as one of the archipelago's trading groups. Peluso (1983:99) writes, 'Like those of Chinese and Bugis heritage [...] the Banjarese had a knack for trade'. The dispersion of Banjarese traders around the archipelago was amplified by the capitulation of the Banjar sultanate to the Dutch at the end of the Banjarmasin War of 1859-1863 (Peluso 1983:98). See Hudson (1967, Chapter 3 – Historical Sketch of the Southeast Barito Basin) for a brief introduction to Banjar history. See Dunn (1975), Wheatley (1959), and Wolters (1967) on early trade in the Indo-Malay region.

22 The importance of Banjar to the foreign (European) traders is reflected in the fact that it is one of the first five place-names in Borneo to appear on the earliest European maps (Cleary and Eaton 1992:42).

'Despite Landen's [the English chief's] behaviour, the Sultan showed little animosity towards the British traders. Indeed, he appeared anxious that the English should continue trading in his dominions, as he realised that foreign trade was vital for the prosperity of the state. The Banjarmasin rulers adopted the policy of trying to encourage as many foreign nations to trade at Banjarmasin as possible, provided they did so peacefully.' (Suntharalingam 1963:43.)

The memory of this period of international fame and exchange is remarkably fresh among contemporary Banjarese: villagers in the today remote upland Banjarese village of Rantau Balai told the author that Martapura (one of the historic capitals of Banjar) was once known as 'little Singapore'.[23]

The Banjar kingdom's trade initially involved forest products, the oldest trade goods of the archipelago.[24] Ras (1968:189, 198) suggests that natural abundance in such products, coupled with navigable waterways for bringing them out to the coast,[25] were critical determinants of the devel-

23 Since Singapore was not founded until early in the nineteenth century (1819), Martapura could not have been known as 'little Singapore' until much later than the period under discussion here, the seventeenth century. Nevertheless, the contemporary claim that Martapura was once known by this term still represents a contemporary memory of the region's historic place in international trade. (I am grateful to Barbara Andaya for this insight.)
The memory of a time when the region figured prominently in international affairs is retained in other ways as well. For example, the villagers of Rantau Balai retain a legend of a 'Chinese Rajah of Brunei' who attacked and defeated an early, pre-Islamic Banjar ruler, then was attacked and defeated in turn by 'Majapahit', whereupon he 'fled' (*lari*) to, 'fasted' (*betapah*) atop, and 'disappeared' (*hilang*) on nearby Mount Payangan. A similar story is recounted in the *Silsilah Kutai*, a Malay chronicle from East Kalimantan, which tells of a Chinese prince who comes to Kutai by ship, loses a cockfight, and flees and settles in the interior (Ras 1968:82); Ras (1968:120-1) notes that a visiting Chinese prince is a recurring figure in many Malay and Javanese stories. The same villagers also retain a memory of the ruler of South Kalimantan formerly paying tribute (*bayar upeti*) to the ruler ('Rajah Betawi') of Java (just as they say Indonesia as a whole once paid tribute to the Dutch). These comments have something of the flavour of the unique historical consciousness of Dayak inhabitants of the mountain hinterlands of Southeast Kalimantan studied by Tsing (1993:74): 'The official model portrays Meratus as caught in an archaic prehistory, before national historical time. Mountain-Meratus leaders turn this around to show the state as timeless. And within state timelessness, these leaders imply, Meratus subjects can sustain an improvised time of social networking and provisional community building.'

24 Recent work on non-timber forest products has suggested that their potential economic value is much greater than had been thought (Peters, Gentry and Mendelsohn 1989; Dixon, Roditi and Silverman 1991), and that the importance and historical depth of their trade are correspondingly great. Trade in forest products has an especially long history in Southeast Asia, with early records of it (beween western Indonesia and China) dating from the fifth century (Wolters 1967), if not considerably earlier (Von Heine-Geldern 1945; Dunn 1975). Cleary and Eaton (1992:60) say that 'to conceptualize the important jungle trade as an unsophisticated and anachronistic part of the "traditional" economy, is both misleading and inaccurate [...]. The trade was economically and socially sophisticated, as well as ecologically balanced.' They suggest that the Dutch and English interest in Banjar's pepper trade also would have stimulated its trade in forest products.

25 The historic concern for navigable waterways and transportation in the region is reflected in the story told by the villagers of Rantau Balai, about how a former Banjarese Rajah, from the pre-Dutch era, tried to cut channels between two loops of the Riam Kanan River and also between two loops of the Pa'au River (a tributary of the Riam Kanan). The two attempts are said to have ultimately failed due to encountering bedrock formations. The villagers of Rantau Balai believe that two long gulleys stretching inland from the two aforementioned rivers – both

opment of the Banjar kingdom in southeastern Borneo.[26] The importance and diversity of this trade is reflected in the rich, evocative lists that appear throughout the *Hikayat* of goods sent to other kingdoms as gifts or tribute.[27] Several examples follow (Ras 1968:255, 427):[28]

'So Wiramartas was sent to the king of China [to ask for men able to make statues of bell-metal], taking with him ten diamonds, forty pearls, forty emeralds, forty red

---

of which were shown to the author – mark the sites of these attempts. They refer to them, both located upstream from their village, as either *lobang rajah* (the crater of the rajah) or *bekas rajah* (the remnants of the rajah). The villagers explicitly say that the purpose of these diggings had nothing to do with irrigation, for example, but was intended to *mempercepat jalan* (quicken the way). See Hudson (1967:76) on inter-river linkages built in the Banjar region by the Dutch at the end of the nineteenth century. The historical focus – not on agriculture but on transport and trade – is reflected in such things as another site upriver from the village of Rantau Balai, called Gelang Rajah (the bracelet of the rajah), which is said to have been a royal diamond works dating from the same *jaman* (era) as the two 'craters of the rajah'.

26 Forest products continue to be important to the Banjarese even today. In the contemporary highland Banjarese village of Rantau Balai, during the slack agricultural period between weeding and the harvest, most people leave the village to seek tradable products in the forest, including diamonds, gold, dammar, rattan, fish, and *kementing/kebenting* (*kemiri* or *Aleurites moluccana*) – which is mostly the same list of goods that the Banjar kingdom was sending abroad three and four centuries ago. See Tsing (1993:55-6) on the forest-product trade of the nearby Meratus Dayak: 'Meratus hike or raft down to these [Banjar] markets, where they sell rattan, rubber, peanuts, mung beans, ironwood, incense woods, wax, and numerous other minor crops and forest products'.

27 Groeneveldt's (1960) list (cited in Hudson 1967:64) of the trade products sought by China in Southeast Borneo at this time is similar: 'rhinoceros horns, peacocks, parrots, gold dust, crane crests, wax, rattan-mats, [chillies], dragon's blood, nutmegs, deer hides, and so on'.

28 Dragon's blood, dammar, mats, pepper, coral, pearls, civet, and bees-wax – all mentioned in tribute lists in the *Hikayat Banjar* – each merit separate sections in the second part of Chau Ju-kua's *Chu-fan-chï* (Hirth and Rockhill 1911:197, 199, 220, 226, 229, 234, 238). One other commodity mentioned in the *Hikayat*'s tribute lists, gold, is discussed in Chau Ju-kua's section on P'o-ni (Borneo), where he says that gold vessels are both used and traded, and trade-gold is used in barter (Hirth and Rockhill 1911:155-6, 158). Chau Ju-kua identifies Indonesia as the source, or one of the sources, of most of these goods: thus, he identifies kingdoms in Java, Sumatra, or Borneo as the source of the mats, pepper, coral, pearls, civet, and bees-wax (Hirth and Rockhill 1911:77-8, 81, 83-5, 220, 222, 230 note, 234, 239 note). Hirth and Rockhill (1911:198 note) themselves identify Borneo and Sumatra as the source of the 'ordinary' dragon's blood (*hüékié*) traded to China. This Indonesian product appears to have partially replaced in the China trade a similar product from Ta-shï (Arabia) (Hirth and Rockhill 1911:197-8), a process that Wolters (1967) claims underlay the development of Indonesia's China trade. Fifteenth-century Chinese sources refer to 'blood dessicate', an alternative term for dragon's blood, produced in Banjarmasin (Wolters 1967:124). In some of these cases, Javanese and Sumatran ports likely served only as places for trans-shipment of Bornean products, for as Wheatley (1959:33) writes, 'Authors all too often described commodities as natural products of regions from which they were only re-exports'. See Wolters' (1967, Chapter 9) argument that the term for China's so-called Persian (Po-ssu) trade actually referred more to the value of the products involved than to their geographic origin in Persia. An obvious example of this was the inclusion of trade cloth from India in tribute set from Shê-p'o (Java) to China in AD 430 (Wolters 1967:151). Chau Ju-kua himself notes that this is what happened with some of the woven mats that he describes in his trade manual: 'The mats called *yé-sin-tién* come from Tan-jung-wu-lo [Banjar]. The foreign traders carry them to San-fo-tsi [Palembang], Ling-ya-mön [Lingga?] and Shö-p'o [Java] for trade' (Hirth and Rockhill 1911:220).

corals, forty rubies, forty opals, forty loads of beeswax, forty bags of dammar,[29] a thousand coils of rattan, a hundred gallons of honey, and ten orang-outangs.'

'The mission [to request military assistance from Demak for a dynastic struggle] was entrusted to Patih Balit who took with him a gift of homage consisting of one thousand coils of rattan, one thousand dish-covers, ten loads of beeswax, one thousand bags of hard dammar, ten diamonds and ten loads of dragon's blood.'[30]

The way that these lists are used to signify and summarize relations with other kingdoms testifies to the importance of forest products in the self-identity of Banjar.[31]

Schrieke suggests that this forest product trade, important though it was, left Banjar a relative backwater, and that this changed only with the development of pepper cultivation and trade.[32] He writes:

'Southern Borneo was not by any means so important for Java in those days, because the only goods to be obtained there were forest products – such as dammar, dragon's blood, wax, *myrabolans* [sic][33] for the batik industry, rattan and wicker work – which could also be imported from elsewhere, Palembang and Timor, for example. Banjarmasin in actuality became a center of importance only in the first part of the seventeenth century, when the amount of pepper grown was increasing every year thanks to the Chinese for whom the Bantam market was insufficient or inaccessible, overwhelmed by the European demand as it was.' (Schrieke 1966:29-30.)

Pepper was a relatively late addition to Banjar's products. Lindblad (1988: 31) suggests that Hindu immigrants first brought pepper to Martapura (later one of the Banjar capitals) only in the fifteenth century. As a result of a combination of factors at the beginning of the seventeenth century – 1. the crippling of the Central and eastern Javanese ports (Jepara, Ceribon, Tuban, Madura, Surabaya) by Mataram and the displacement of not just their trade but their traders to Banjar, 2. the monopolist policies of Aceh, 3. the international competition for Sumatran pepper, and 4. the Dutch

29 'Dammar' or damar is a generic term for resin from a variety of dipterocarps, especially of the genera *Shorea*. It was used traditionally for torches, caulking for boat seams, and as a glue, incense, fumigant, and medicine (De Beer and McDermott 1989:38).

30 'Dragon's blood' is a resin obtained from several species of rattan of the genus *Daemonorhops*. It has been sought in trade for use as both a medicine and a dye (Burkill 1966, I:758-60).

31 The importance of these same forest products also is seen in the identity that the region took on in the eyes of others: Cleary and Eaton (1992:40) note that southern Borneo was known to the fourteenth century Javanese as Puradvipa or diamond land.

32 Hudson (1967:65) writes: 'It was the introduction of commercial pepper cultivation into the Banjar sultanate in the seventeenth and eighteenth centuries, rather than the conversion to Islam in the sixteenth, that led to the growth of Banjarmasin's importance in the Indonesian trade sphere and the expansion of Banjar-Malay peoples throughout the whole Banjar region of the Southeast Barito basin at the expense of the indigenous Dajak population'.

33 Myrobalans is the trade name for fruits of a variety of tannin-rich species of the genus *Terminalia*, which are used in the manufacture of tannin for leather, dye for cloth, and ink (Burkill 1966, II:2173-80).

closing of many other ports to traders of other nations (Schrieke 1966:60-1, 67; Hudson 1967:65-6) – there was an efflorescence in the Banjar trade in pepper.[34] This led the Dutch to attempt to set up a factory in Banjarmasin in 1606 and the English to do the same in 1615 (Suntharalingam 1963:37-8).[35]

Interpreting key elements in the Hikayat Banjar

With this background on historic and contemporary pepper cultivation in mind, King Ampu Jatmaka's speech in the *Hikayat Banjar* now can be examined in detail.

Jambi and Palembang

The first sentence of this passage runs as follows: 'And let not our country plant pepper as an export-crop, for the sake of making money, like Palembang and Jambi.'

Andaya (1993b:43, 1995:169) suggests that pepper was first cultivated in Sumatra in the fifteenth century for export to China.[36] By 1545 Jambi already was known to the Portuguese as a pepper producer (Andaya 1993a:97, 1993b:45). In 1615 English and Dutch ships first reached Jambi and established trading posts; and despite intermittent battles with one another, they were able by the second quarter of the century to displace both the Chinese and Portuguese from their dominant positions in Jambi's pepper trade (Andaya 1993a:103, 1993b:43-56). The pepper that was the subject of this trade was cultivated, for ecological reasons, in the upstream regions (Andaya 1993b:17-8), from whence it was gathered and then traded

34 Hudson (1967:66) writes: 'Banjarmasin had begun planting and exporting pepper on a small scale at the beginning of the seventeenth century in response to the needs of Chinese traders who were finding it increasingly difficult to secure pepper at Patani and Bantem [Banten], the ports previously utilized'.

35 The initial ventures by both the Dutch and the English were short-lived, establishing a pattern of in-and-out moves that prevailed for the next two centuries (Lindblad 1988:8; Suntharalingam 1963:37-8). Cleary and Eaton (1992:44) note that during the two centuries between 1615 and 1814, the Dutch and the English each concluded four separate formal agreements with Banjar regarding trade in general and pepper in particular.

36 Schrieke (1966, I:53) says that pepper production in Pidie on Sumatra's east coast was 'mentioned as early as 1416 in the Chinese records'. An earlier date for pepper cultivation in Sumatra does not seem to be out of the question, however. Chinese trade manuals from the twelfth and thirteenth centuries clearly show that pepper was abundant all over Java by then. We also know from these same chronicles that Jambi sent missions to China in the last quarter of the eleventh century (and Palembang's relations with China date from the beginning of the tenth century) (Hirth and Rockhill 1911:62, 66 note 18). Given the importance of pepper to China, and its abundance in Java, it seems unlikely that it took – in effect – five centuries for the Sumatran kingdoms to take advantage of this opportunity to grow pepper for the same market. However, it must be noted that when Chau Ju-kua compiled his *Chu-fan-chï* in 1225, he does not list pepper among the numerous 'native products' of San-fo-ts'i (Palembang) (which sufficed to rank it third, after Arabia and Java, among the richest sources of trade goods for China) (Hirth and Rockhill 1911:23, 61).

Figure 4. Kantu' pepper garden and adjoining secondary forest

to the foreigners under the auspices of the downstream state. This linkage of upriver and downriver in the pepper trade was initially very successful and, coupled with the decline of other pepper-producing regions as a result of competing efforts to monopolize the pepper trade both by indigenous kingdoms (Aceh and Banten) and by colonial powers (Dutch and English) (Schrieke 1966:55), it made Jambi into the second city of Sumatra (Andaya 1993a:99). In part because of the market glut and precipitate price declines cited earlier from Reid (1993), this linkage broke down (Andaya 1993a:104). The relation of the European traders with the downriver half of the state became more problematic, and this made relations between upriver and downriver more problematic as well (Andaya 1993a:104). It ended, by the end of the seventeenth century, in rejection of pepper by those upriver people and the decline in political-economic importance of those downriver (Andaya 1993a:109-12).

The dramatic rise and fall of Jambi, and the internal and external factors that drove it, obviously had a powerful psychological impact on Banjar – and doubtless on other kingdoms in the region involved in pepper production, or perhaps in any commodity production at all – who saw it as an object lesson for themselves. The validity of this lesson is borne out by the fact that other kingdoms (including Banjar) experienced the same sort of cyclic commitment(s) to pepper as Jambi did. Thus, Raffles (1978:131) blames the 'oppressive, unprincipled, and impolitic' policies of the Dutch for the disappearance of pepper cultivation from Java and Sumatra in the early nineteenth century,[37] and Crawfurd (1856:65) uses the same argument for the near-extinction of pepper cultivation during the same period in Southeast Borneo.

Expensive food, poor crops, hot vapours

The next line in the passage from the *Hikayat Banjar* reads as follows: 'Whenever a country cultivates pepper all food-stuffs will become expensive and anything planted will not grow well, because the vapours of pepper are hot'.

Note, first, that this passage is specifically about pepper (as opposed to all crops, or all cash crops). The case of the contemporary Kantu' demonstrates that the constraints that apply to pepper need not apply to other cash crops, such as rubber. This point is driven home by other injunctions in the *Hikayat*, which actually contain preferred lists of crops to be planted instead of pepper: for example, 'What people should definitely cultivate with energy is rice, maize, yams, taro and bananas' (Ras 1968:331). The

[37] Raffles's commentary must be weighed in the light of his own agenda, which was to depreciate the prior Dutch rule in the East Indies and, thereby, rationalize the need for English rule. As Bastin writes in his Introduction to the Oxford University Press reprint of Raffles's work, 'Undoubtedly one of the main reasons for writing the book was Raffles's desire to publicize his activities in Java and to contrast the benevolence of his own measures with the "tyrannical and rapacious" policies of the Dutch colonial regime' (Raffles 1978:xvii).

specific focus on pepper also is made evident by the absence from the *Hikayat* of proscriptions of any other crops or commodities besides pepper; there is no proscription on gathering the highly commercialized forest products. In the absence of any admonitions in the *Hikayat* regarding trade in these commodities, the critique of pepper is all the more marked.

As regards the reference to expensive foodstuffs: in the case of Jambi, intensive involvement in pepper cultivation did lead to a greater reliance on imported rice (Andaya 1993b:66). The reference to 'expensive' here probably does not just allude to reliance on foodstuff markets, however, but to the conditions of this reliance, based on the flexibility versus inflexibility of one's agricultural strategy. As the case of the contemporary Kantu' suggests, the relatively long maturation period of pepper makes it harder for farmers to respond to short-term market fluctuations, and this inevitably means that they will often have to trade a crop that the market is not favouring for one that the market is favouring – and paying higher prices (that is, receiving less in the trade) in consequence (Andaya 1993a:104, 1993b:79). As a result, the farmer in Jambi developed a preference for shorter-maturation crops like rice, cotton, tobacco, and gambier over pepper (Andaya 1993a:114). By planting such crops, the farmers would stand a better chance of getting foodstuffs – and also cloth, which was the major good traded for pepper[38] – on good terms.

The remaining portion of this line from the *Hikayat* states that anything planted will not grow well, 'because the vapours of pepper are hot'.[39] One reason that nothing else will grow well with pepper is its environmental impact.[40] Pepper is one of Southeast Asia's most environmentally demanding export crops. As the case of the contemporary Kantu' demonstrates, pepper cultivation is associated with soil erosion and impoverishment and grassland succession. On the one hand, pepper cultivation compares favourably with the normal swidden cultivation of food crops, since pepper gardens may be cultivated eight to ten years in succession, whereas swiddens are normally cultivated for just one or at most two years. On the other hand, the longer cropping period in the pepper garden is only attained, in part, by the importation of nutrients from surrounding lands (which thereby 'subsidize' the garden). In addition, whereas a multiple-year forest fallow will restore the former swidden and permit its recultivation, this does not suffice in the case of the pepper garden. For all of these latter reasons, Brookfield et al. (1990:497) blame

38 The *Hikayat* elsewhere describes the 'good state' of society as one in which 'food *and* clothing were very cheap' (Ras 1968:335, emphasis mine).

39 The concept of pepper's 'vapours' has some historical depth. Thus, Chau Ju-kua's trade manual contains the following comment on pepper cultivation in Central Java: 'The pepper gatherers suffer greatly from the acrid fumes they have to inhale, and are commonly afflicted with headache' (Hirth and Rockhill 1911:83).

40 In many parts of Asia, nutrient-demanding and environmentally stressful crops are termed 'hot', whereas those less demanding and more benign are termed 'cold' (Kurin 1983).

pepper for the first sustained 'attack' on the region's upland forests. One casualty of this attack is diminished land and resources to grow other crops well, just as the *Hikayat* suggests.

Malice and government disorder

The next sentence in the passage under study reads as follows: 'That [pepper cultivation] will cause malice all over the country and even the government will fall into disorder'.

The cultivation of pepper in Jambi clearly led to ill-will when involvement in the colonial trade obliged downriver to intensify its exactions (and on worse terms) from upriver, and the latter's awareness of their own best interests led them to resist. The malice was spawned, therefore, by deviation from what Scott (1976) calls the 'moral economy', in which basic subsistence is held sacrosanct.[41] This deviation was perhaps an inevitable development from collaboration of the native state with the colonial powers, given the latter's interest in a set of values antithetical to the moral economy.

> 'The Jambi experience encapsulates a basic problem of many coastal rulers who found themselves caught up with a European power. Any alliance between them was bound to increase tensions as the Europeans attempted to manipulate or recast traditional upstream-downstream relations to attain specific commercial goals.' (Andaya 1993a:119-20.)

The Europeans were interested in monopolizing the trade with the coastal powers, to maximize their profits.[42] Coastal states like Jambi responded by trying to resist them, while at the same time seeking in turn to monopolize trade with their own upriver hinterland. Similarly, Saleh (1976:212-3) says that the Banjar aristocracy resisted attempts by Westerners to enforce a monopoly with them but enforced their own monopoly with the pepper peasantry.[43]

Government disorder enters with this new system of values. Abandonment of the moral economy gives every faction in the coastal courts a self-interest in striking their own deals with both the European traders and the upriver producers; and the Europeans had even greater incentives to do the same.[44] Hudson writes:

41 Thus Andaya (1995:183) writes: 'Underlying all these developments was a growing sense of social unease because of what modern researchers term "food insecurity" '.

42 The Dutch established monopolies on the pepper trade in Palembang in 1662 and in Jambi in 1679 (Andaya 1993b:244).

43 Reid (1993:250) writes: 'Banjarmasin's ruler began in the 1660s to force growers to sell to his agents at low prices in order to fulfil his contract with the VOC and still make a profit' (*Generale Missiven* 1960-1988, III:422, 455).

44 The spirit that governed the actions of some of the Europeans involved is summed up in this quotation by Schrieke (1966, I:65) from Jan Pieterszoon Coen, Governor-General of the Dutch Company in 1619-1623 and 1627-1629: 'There is nothing in the world [...] that gives man better right than might and power added to right.'

> 'It should not be imagined [...] that there was any sort of unity behind Banjar policy. The court was ridden by factionalism and regionalism, with individuals and policies contending to guide the kingdom. There were shifting "internationalist" factions that sought to gain power with the support of one foreign power.' (Hudson 1967:70.)

Indeed, Saleh (1976:215) interprets the *Hikayat*'s mention of 'hot vapours' as a metaphor for inter-family power rivalries, political intrigues, group conflicts and usurpations. European traders both contributed to and took advantage of these turbulent political waters, to advance their own agendas, typically by tipping the balance of power in favour of a given faction in exchange for a monopoly on the pepper trade.[45]

Rural pretensions

The next line in the *Hikayat* reads as follows: 'The rural people will become pretentious towards the townsfolk if pepper is grown for commercial purposes, for the sake of money.'

The interpretation of this important line poses a number of pithy questions, beginning with the question: Who are the so-called 'rural people'?

Definition of 'Sakai'

There are four references to 'rural people' in three (of the four) pepper passages in the *Hikayat*: in two of these references, 'rural people' is acceptably translated from *orang desa* in the original (Ras 1968:330, 374):

> 'Orang kota tiada diupamai oleh orang desa'
> (The rural population will not think highly of the townsfolk) (pp. 330-1)[46]
> 'Maka orang kota tiada ditakuti oleh orang desa'
> (For the rural people will not fear the townsfolk any more) (pp. 374-5)

But in the other two references, 'rural people' is less clearly translated from *Sakai* or *orang sakai* (Ras 1968:264, 330):

> 'Orang sakai pun banyak barani pada orang kota'
> (The rural people [Sakai] will become pretentious towards the townsfolk) (pp. 264-5)
> 'Orang yang kaparak pada raja itu tiada ditakuti oleh sakai yang bersahang itu'[47]
> (The functionaries from the capital will not be respected by the people in the rural areas [Sakai] who grow pepper) (pp. 330-1)[48]

The term *Sakai* has the following denotations and connotations in Malay:

45 Pepper contracts were integral bargaining chips in all conflicts (Andaya 1993b; Schrieke 1966, I:62-3).

46 *Diupamai* literally translates as 'taken as example' (Ras 1968:612).

47 Another version of the *Hikayat* gives *Saki* and *Sakie* as alternate wordings for *sakai*, capitalized as is customary for ethnic labels in the *Hikayat* (Ras 1968:240 note 6, 330 note 37).

48 *Kaparak*, from *parak*, translates literally as 'persons who during the audience-meetings have their seats nearest to the king' (Ras 1968:579).

'subject, dependent, of subject peoples in contrast to the ruling race, aborigines who do not speak Malay' (Wilkinson 1959, II:1002). In his own glossary, Ras (1968:593) translates *sakai* as 'the tribal or rural people subjected to the rule of a political centre (*nagara*), and so in a certain way the opposite of *rakjat*'. Ras (1968:589) translates *rakyat* as 'the (Malay) subordinates or subjects, i.e. the (male) members of the ruling race who were in principle entitled to attend the royal audience meetings on Saturdays, in contradistinction to the *sakai*, i.e. the (Dayak) tribal people subjected to Malay rule'.[49] *Sakai* does not, thus, simply denote 'rural people'; and it specifically does not denote Banjar.[50]

The referent for *sakai* is the Dayak, therefore, who were distinguished from the Banjarese – and thus meriting this term – by religion, culture, and political marginality (as they still are today, Tsing 1993).[51] The initial role of the Dayak in the Banjar kingdom was to supply the coast with the interior forest products that were its chief trade goods up to the end of the sixteenth century (Hudson 1967:67). It appears that the Dayak expanded on this role, when pepper was introduced, to be Banjar's first pepper cultivators. Hudson (1967:56, 67) writes: 'The historical evidence seems to indicate that pepper cultivation in the Hulu Sungai was in the hands of Dajak swidden agriculturalists until the closing decades of the seventeenth century'. Pepper would have fitted fairly well into the extensive, mobile,[52] low-density system of swidden cultivation then prevalent; and the trade-based, non-agricultural Banjar kingdom would have had few alternatives in any case. As Reid (1993:35) writes, 'Pepper cultivation was [...] developed in a particularly sparsely populated region of seventeenth-century Banjarmasin (southern Borneo) that had no previous tradition of pepper or even of intensive rice cultivation'. The lack of a large, regional agricultural population available for pepper cultivation is reflected in the fact that slaves also played a significant role in this system:[53] Reid (1993:35), citing Speelman (1670:112), writes that 'One of Makassar's

49 Most of the references to *rakyat* in the *Hikayat* involve audiences before the king or the king's professions of responsibility for his 'subjects'; and most of the references to *sakai* involve waging war on, exacting tribute from, or sending instructions to the upland tribal peoples.

50 The distinction between *sakai* and *rakyat* reveals additional nuances in the *Hikayat*, as in one of the pepper passages (Ras 1968:331), in which concern is expressed that 'The functionaries from the capital will not be respected by the [Sakai] who grow pepper', and that 'The [*rakyat*] have no difficulty in providing for themselves'. The costs and benefits of royal policy are being contemplated here for not one but two distinct populations.

51 Andaya (1993a:120) notes that throughout the Malay-Indonesian archipelago, forest products were gathered by non-Moslem peoples upstream who traded them to Moslem peoples downstream.

52 The contemporary swidden system of the Meratus Dayak of this region is notable for its mobility (Tsing 1993).

53 On the other hand, slaves played a prominent role in the system of pepper cultivation in Jambi (Andaya 1993b:80, 96), despite the presence of a larger, interior agrarian population, just because the labour demands of pepper – especially as state demands for productions accelerated – are onerous, and also therefore unpleasant.

Figure 5. Preparing 'burnt earth' fertilizer

major exports to the region was "male and female slaves fitted for labour in the pepper-gardens".'[54] (The use in the *Hikayat* of the term *sakai* as opposed to *rakyat* for the pepper growers also makes sense if one of its referents was non-Banjarese slaves, in addition to non-Banjarese tribesmen.)

Dayak 'pretensions' and the Banjar response

The passage under discussion continues to say that 'The rural people will become pretentious'. The traditional Dayak role in the Banjar Kingdom, as collectors of forest products, was critically important; and this importance would have been further enhanced by the Dayak's initial involvement in pepper cultivation. The actual versus realized extent of this importance was the site of an ongoing contest between Dayak and Banjar. In a manner that typified the policy of coastal states towards interior, resource-producing populations all over the archipelago, including Jambi,[55] Banjar sought to strengthen its negotiating position and weaken that of the Dayak by insulating them from contacts with other traders or outsiders (Hudson 1967:55).[56] As Vlekke (1961:202) writes, 'The Sultans of Banjarmasin feared nothing so much as direct contact between the Europeans and the Dayaks, whom they cruelly exploited and oppressed'. Hudson (1967:65) even suggests that some of the periodic relocations of the Banjar capital (for example, from Negara to Banjarmasin) may have been motivated at least partly by an attempt to minimize contacts between foreign merchants and the Dayak people of the interior who produced the export commodities. The contest between Banjar and Dayak would have intensified with the introduction of pepper and the European traders, as the signing of trade agreements with the European powers obliged Banjar to structure the terms of exchange between the interior resource-producing populations and the coastal trading population more and more in favour of the latter (and the Europeans).[57] This contest between interior Dayak and

54 Hudson (1967:56) reports that Banjar even raided the Dayak populations themselves for slaves, although he does not say whether Dayak slaves were put to work in pepper gardens.

55 Andaya (1993b:46-7) writes that the Portuguese tried to prevent the late-coming English and Dutch from competing for Jambi's trade by keeping secret the route from the coast to Jambi's capital 120 kilometres upriver adding that one means employed to this end (ultimately without success) was to deliberately place Palembang, Jambi's neighbour, in Java on Portuguese maps.

56 This practice has continued down to contemporary times. Brosius (n.d.:17), referring to the relations between Penan hunter-gatherers and the agricultural Kayan-Kenyah in Sarawak, writes: 'The presence of a Penan band in an area meant access to forest products and to the income generated by trade in those products. Longhouse [Kayan-Kenyah] aristocrats were proprietary about "their" Penan, and jealously guarded their prerogatives to trade with certain groups.'

57 This also held true in Jambi, of which Andaya (1993b:244) writes as follows:
'Following the English and Dutch arrival [...] the bonding [between upstream and downstream] created by the exchange of gifts, the acceptance of mutual obligations, and putative kinship ties was frequently undermined by the commercial tensions injected into the marketplace. These tensions were in turn exacerbated by competition among Europeans and their

coastal Banjar (as between upriver and downriver in Jambi) belongs to a wider set of developments associated with the historic expansion of trade relations in the archipelago, which transformed (and then often re-transformed) power relations at a number of different levels, including those of intra-island (as in Borneo and Sumatra), inter-island, and between indigenous states and the European powers.[58]

The principal expression of 'pretension' by the Dayak was resistance of one sort or another.[59] In East Kalimantan (Peluso 1983),[60] as in Jambi (Andaya 1993a:112, 1993b:137), this resistance sometimes took the form of open aggression towards the coastal kingdom. However, in Southeast Kalimantan, although some Dayak groups like the Ngaju were both feared and used as mercenaries by the Banjar (Ras 1968:22, 50, 52),[61] open aggression appears to have been less common. Rather, the major response of these Dayak seems to have been passive resistance to pepper cultivation, perhaps coupled with flight from Banjar control (both of which responses also were documented in Jambi, Andaya 1993a:102, 107-8, 114, 1993b:136). This response has been documented for other Dayak, at other times and

---

resistance to absorption into the familial network that lay at the heart of indigenous economic activity.'

58 Regarding the transformation of relations between upriver and downriver, or interior and coast, Schrieke (1966:27-8) writes as follows, for Java: 'The great flowering [in the fifteenth century] of the eastern Javanese coastal towns was destined eventually to have a disintegrative effect [...]. The coastal regents, who after all already enjoyed a very large amount of freedom under Javanese royal sway, became more and more assured of their own power under the influence of the growing economic importance of their provinces; they became autonomous coastal potentates more and more independent of the weak bearers of the royal dignity they were outgrowing. The struggle of the commercial towns against the agrarian interior for hegemony [...] was then able to make its beginning.'
Regarding the effect of the expansion of trade relations on the power balance not just within, but between, the archipelago's major islands, Schrieke (1966:29) writes about how successive Javanese kingdoms used their rice surplus to dominate the trade with – and also political economy of – the rest of the archipelago: 'The position of economic preponderance which Java thus occupied in the archipelago found its expression in the recognition of political supremacy on the part of its dependencies scattered everywhere'.
Another level of interaction involved the actions of the European trading powers, and the way their interventions suppressed the prominence of some places in the archipelago and raised (often indirectly) the prominence of others. Of course, the forces operating at all of these different levels often interacted, as they did in the seventeenth-century rise to prominence of Banjarmasin (as argued already).

59 To return to the discussion of moral economy from the previous section, there is a history of peasant resistance, even rebellion, throughout Southeast Asia whenever this morality is violated by the state (Scott 1976).

60 Peluso (1983:28, 30, 32, 61, 70) describes in eastern Borneo a loose and shifting political-economic relationship between upriver Dayak gatherers of forest products, the downriver Malay Kutai Sultanate, and Bugis traders, with the sultanate at first mediating between the Dayak and Bugis, and then the Bugis mediating between the sultanate and the Dayak – a situation further complicated by the arrival of the Dutch who, as Peluso notes, 'conveniently chose to recognize the sovereignty of those who recognized them'.

61 Interestingly, the *Hikayat* in one place lists the Biaju (or Ngaju) as one of the 'foreign traders' resident in Banjar (Ras 1968:371).

places, with other cash crops;[62] and it appears to be based on a strong cultural commitment to subsistence rice culture.[63] Thus, Hudson (1967:66) writes that: 'Although they [the Dayak in Southeast Borneo] were willing to collect forest products and grow a little pepper to trade for Chinese goods, they were not willing to allow these secondary pursuits to interfere with the primary task of rice cultivation.'[64]

The Banjar response to the Dayak resistance to pepper cultivation was not simply heightened exactions and oppression, as happened in Jambi and in many other cash-crop producing areas, but displacement of Dayak by Banjar. Hudson (1967:67) writes:

> 'As mere part-time cultivators, the Dayak were not able to meet the increasing export needs nor were they willing to give up their traditional way of life to become full-time commercial agriculturalists. The result was that elements of the Banjar-Malay population began to move inland from the coast and to displace the Dayak in the best interior pepper growing region which was in the upper Negara drainage above Amuntai.'

The result was a replacement, during the seventeenth and eighteenth centuries, of part-time Dayak pepper producers, operating within the overall framework of a system of swidden agriculture, by full-time Banjar pepper producers, operating within a system of sedentary agriculture. Hudson (1967:67-8) writes:

> 'By the end of the eighteenth century, pepper was being cultivated in more or less permanent plots in the Hulu Sungai, and there were inland regional pockets with relatively high population densities, from which we may infer that Dayak, with their shifting cultivation, had given way to Banjar-Malay sedentary agriculturalists.'

This displacement of Dayak by Banjar was accompanied by Banjar slave-raids, land acquisition, and coercive proselytizing in the Dayak regions (Hudson 1967:56).

62 See Dove (1996) regarding Dayak resistance to complete commitment to rubber cultivation in West Kalimantan.

63 Ironically, the new value system that the interior Dayak were resisting also imperilled their coastal foes. Schrieke (1966, I:228) writes (with reference to Banda in particular), that when 'ever-expanding commerce gave birth in the course of time to a trading class, the power of money destroyed the autocratic authority of the coastal potentates and the traditional aristocratic class prejudices'.

64 Hudson (1967:66, note 25) adds: 'This Dajak reluctance to becoming overly dependent on a cash or trading economy at the expense of subsistence agriculture continues to the present day. It is one of the economic and psychological characteristics that sets the majority of the Southeast Barito Dayak off from the contiguous Banjars.'

Figure 6. Pepper drying on the verandah of a Kantu' longhouse

Private consumption

The next two lines in the *Hikayat* passage read:

> 'If people grow pepper it should be about four or five clumps per head, just enough for private consumption. Even four or five clumps per head will cause much vapour, owing to the great number of people involved, let alone if it is grown extensively as a crop; then the country inevitably would be destroyed.'

The phrase 'private consumption' should probably be interpreted as referring not to pepper consumption but to the wider issue of household subsistence. This interpretation is supported by the fact that this phrase is structurally opposed, within the overall passage on pepper, to the earlier phrase regarding planting pepper as an export crop. Thus, we have first an injunction against planting as an export crop, followed by approval (albeit reluctant) for planting pepper to guarantee subsistence. Further light is shed on this opposition by looking at the original Banjarese text of this earlier phrase: the reference to 'export crop' turns out to be the translation of *dagangan negri* (Ras 1968:264). This term has a much narrower referent than just export or trade (which is in fact a slightly misleading translation of it in this context): it properly translates as 'state trade' (referring to the sort of state involvement in export that Jambi had in its collaboration with the European traders). This translation is supported by Saleh's (1976:208) suggestion that 'By the end of the seventeenth century pepper was being cultivated in all the regions of Banjar, mostly on the big appanage lands of the king, the royal family and the ruling class'. Thus, two different types of pepper production and trade are distinguished in this passage: production and trade carried out by and on behalf of the state is discouraged, while production and trade oriented towards household subsistence is approved. This interpretation is consistent with the interpretation made in the previous section of the injunction against cultivating pepper for commercial purposes, 'for the sake of money': not all pepper cultivation is proscribed, and not all cultivation for market, just certain types.[65]

Andaya (1995:185) suggests that the traditional system of pepper cultivation in Jambi (before the sixteenth- to seventeenth-century boom) involved cultivation of a few stakes[66] of *pohon wang*, 'money trees' (citing Forbes 1885:135), to meet not all of the household's needs but just its

[65] Thus, the *Hikayat* is urging (with pepper) a return to a different type of market-orientated economy, not to a mythical pre-market 'natural economy' (Roseberry 1991:223), which would have been a strange outcome from a trading state like Banjar.

[66] Andaya (1995:175) cites the *Hikayat*'s recommendation of five to ten clumps as an indication of how many the 'few stakes' might actually be. In fact, the *Hikayat* recommendations range from 'a few pepper trees per head' to 'about four or five clumps per head', and to 'about ten or twenty plants per head', or 'some ten or twenty plants only', (Ras 1968:265, 331, 375, 443). Burkill (1966, II:1778), writing of pepper growing in the sixteenth century and earlier, says: 'Pepper was probably grown all through Malaysia at that time in a very haphazard way. Later a more intensive cultivation came in, with special plantations, in the place of a vine or two near the homestead.'

periodic, non-subsistence needs for cash. This means that pepper would have been just one part of a wider composite household economy (or 'portfolio'), likely comprising a subsistence sector of swidden (and perhaps also irrigated or swamp-based) cultivation of rice, and a market-oriented sector of forest-product gathering and cultivation of export crops like pepper. Precisely this sort of economic setting – which still characterizes much of the agricultural economy in outer Indonesia today, as indicated by the earlier discussion of Kantu' pepper and rubber cultivation – is indicated by the use of the term 'money trees' for pepper, since it would seem to make sense only against a non-monetary subsistence background. Hudson (1967:67) suggests that this also characterized the initial system of pepper cultivation in Banjar: 'The earliest commercial pepper was grown by Dayak as a part of the swidden complex, a situation that continued to the end of the seventeenth century in some interior regions'. This Dayak system of pepper cultivation represents the one that is permitted in the *Hikayat*; it is replaced, when sedentary Banjar cultivators displaced the Dayak (as described in the preceding section), by the state-trade type of system that is enjoined in the *Hikayat*. The contest between these two systems, whether waged within or between groups, is a *Leitmotiv* of the history of commodity production in Indonesia (see Dove 1996).

Discussion and conclusions

A notable omission from the *Hikayat* is the failure to mention the Dutch in the pepper passages, despite the very salient role that they played in the affairs of the Banjarese sultanate during the century in which the *Hikayat* was written, and played specifically out of their interest in pepper. Dutch interest in controlling the pepper trade in southeastern Borneo led them to destroy the capital of the sultanate in 1612 and finally win concession of a monopoly on the pepper trade in 1635 (Suntharalingam 1963:37).[67] A second notable omission is the failure to include pepper in any of the detailed lists of gifts, tribute, and trade that are reproduced throughout the *Hikayat* (Ras 1968:255, 305, 325, 363, 427, 441), until the very end of the *Hikayat*, after the Dutch bombardment of Banjarmasin, when a mission to Mataram is described (Ras 1968:483): 'They went to offer the diamond Misim together with a quantity of pepper, rattan, dish-covers and beeswax.'

Regarding, first, the omission of pepper from the commodity lists: it seems highly unlikely that pepper was not actually present in gift and tribute (and simply trade) shipments before this date, especially in the

67 The follow-up to this treaty was, again, unsatisfactory for the Dutch. The sultan did not honour the Dutch monopoly, and he had the entire staff of the Dutch trading post killed in 1638 (Lindblad 1988:9).

case of China, which we know to have had a strong appetite for pepper (and a shipment of tribute to which is described in the *Hikayat*, Ras 1968:255). The fact that pepper only appears in these shipments after the coming of the Dutch symbolically associates pepper with the Dutch era. It is perhaps a symbolic statement to the effect that pepper took on a significance during the Dutch era which neither it nor any other commodity had ever had before. The Dutch made pepper into something qualitatively different from all other trade commodities; and relations with the Dutch over pepper were qualitatively different from all other trading relations. Pepper, in this sense, was not part of the trade of Banjar before the coming of the Dutch. This politically charged interpretation of pepper is supported by the fact that pepper first appears in a shipment of goods to Mataram intended (as the preceding comment from the Sultan of Mataram implies) to plumb its political intentions vis-à-vis Banjar. The outcome of the three-sided relationship between Banjar, Mataram, and the Dutch eventually drove Banjar into the arms of the Dutch in 1635, for protection from Mataram, but at the price of granting the Dutch a monopoly on its pepper trade (Suntharalingam 1963:35).[68]

The second notable omission of any mention of the Dutch in the *Hikayat*'s pepper passages is related to this explanation of the omission of pepper from the tribute lists. The coming of the Dutch is described in the *Hikayat* as follows (Ras 1968:465):

> 'Then, about two years later [after Marhum Panambahan had recommended, but in vain, that they move the Banjar capital to a safer location], the Hollanders came. Four ships anchored south of Pulau Kambang and bombarded the town of Bandjar. There was great consternation among the Bandjarese.'

The most significant aspect of this passage, from the standpoint of this analysis, is its location within the *Hikayat*: it follows all four of the major passages warning against the planting of pepper.[69] The implication is that the coming of the Dutch follows the problematic circumstances surrounding pepper cultivation and is, therefore, to some extent explained by these same circumstances. That is to say, just as the previous paragraph shows how the Dutch brought pepper – in its politically charged sense – to Banjar, so now we see that pepper – again in its politically charged sense – brought the Dutch to Banjar (just as pepper once brought the Dutch to Jambi). This interpretation is supported by a metaphor that is used by

68 Banjar was able to conclude peace with Mataram just two years later, at the price of sending tribute to Mataram, which allowed it to then renege on its agreement with the Dutch (Suntharalingam 1963:37-8).

69 There are two prior mentions of 'Hollanders' in the *Hikayat*: one in the listing of foreign styles of dress that they should avoid (Ras 1968:265) and a second in the listing of foreign traders who visited the kingdom (Ras 1968:263) – but the latter apparently referred to a purely trade-orientated visit by one or more Dutch ships, and not to a visit by warships with a political agenda such as is described later on in the *Hikayat* (Ras 1968:465).

Marhum Panambahan in his speech to the court two years before the arrival of (and bombardment by) the Dutch, in which he recommends the relocation of the capital:

> ' "I propose that we move the capital to somewhere on the Mangapan river. [...] for it is like a banana-tree in front of one's gate, too many people take an interest in it. Since this place lies near the sea it is an easy prey for an enemy. We had better move elsewhere." At that time none of the *dipatis* [governors] was willing to move because it would give too much trouble.' (Ras 1968:463.)

As it turns out (predictably, of course, since the purpose of the *Hikayat* is in part to glorify Banjar's rulers), the capital does indeed prove to be like a banana tree in front of the gate, and it is the Dutch who take an interest in it. Since it is Banjar's pepper production and trade that is of most interest to the Dutch, we can see how pepper is to the kingdom as the banana tree is to the household. The *Hikayat*'s injunctions against pepper cultivation thus represent a (failed) attempt to remove this 'banana-tree in front of the gate'.

This interpretation of the banana tree image is reinforced by a contemporary shamanic story from the Meratus Dayak of southeastern Kalimantan (Tsing 1993:77):

> 'The Banjar king at Kayu Tangi built his palace in a grove of thorned bamboo, where it could not be seen by invaders. But the Dutch shelled the grove with pieces of silver. When people saw the silver, they cut down the bamboo to gather the money. Then the Dutch captured the palace, and the king was forced to disappear at the headwaters of the Barito river.'

Kayu Tangi, near Martapura (Figures 2, 3), was the site of the fourth, and last, Banjar *kraton* (palace) and capital (although in fact it was Banjarmasin, the third capital, that the Dutch attacked in 1612, and it was Kayu Tangi to which they eventually fled, Ras 1968:55, 77).[70] Tsing (1993:77) writes, 'I don't know whether they [such stories] are "true"; but some are telling parables of political relations in the region'. And indeed, many of the elements that I have been discussing in the *Hikayat* are found in Tsing's story as well. The emphasis in Tsing's story on the desirability of not being seen, of invisibility, is obviously analogous to the *Hikayat*'s passage on the banana tree in front of the gate: insofar as relations with the colonial powers were concerned (at that time and place), it was best *not* to attract attention. In both stories, economic activity draws what proves to be the fatal attention: in the *Hikayat*, it is cultivation of a banana tree, which I have been reading as pepper cultivation; and in Tsing's story, it is gathering of money from the Dutch, which we can read as responding to

70 I say 'eventually', because after fleeing from Banjarmasin, the Banjarese settled at Pamakuan, Tambangan, and then Batang Banju, for ten years apiece according to the *Hikayat* (Ras 1968:465-9), before finally settling at Kayu Tangi.

the economic opportunities – like expanded pepper cultivation – brought by the Dutch. In both cases, it is local misreading of the relationship with outsiders (and its dangers), and local misdeeds, that lead to their downfall. The two stories share these similarities because they both represent the local, Kalimantan perspectives on contact with outsiders, foreigners, Europeans; the two stories also differ in some respects, however, because one represents the perspective of the coastal state whereas the other represents the perspective of the interior tribesman. For example, whereas the *Hikayat* focuses on the attraction of local resources (the banana tree), Tsing's story focuses on the resources of the outsider (the Dutch money); and whereas the *Hikayat* focuses on the greed of outsiders for resources (the banana tree) not their own, Tsing's story focuses on greed of the local people for the outsiders' money. Tsing's story thus attributes the fall of the Banjar capital to local greed, fed by the cunning of outsiders; whereas the *Hikayat* attributes it to local failure to take sufficient precautions against the greed of outsiders.

If the *Hikayat* offered such good advice to the Banjarese, why was it ignored? Why did the state trade in pepper continue (as it obviously did)?[71] There are several possible answers to this question. One is simply that good advice is often ignored. The *Hikayat* presents an example of this in the passage on the banana tree, where the court rejects Marhum Panambahan's advice to move the capital because 'it would give too much trouble'. The evidence that there is a difference of opinion within the court in matters such as this would also have been a factor. Moreover, given such differences of opinion the pressures from the Dutch would have been difficult to resist. (Whereas it is true that the world system does not completely dictate the terms of incorporation of local systems, it also is true that it usually plays from a position of great strength vis-à-vis local systems.) Perhaps most important, however, the *Hikayat* may not have articulated this warning until it was too late. The last rewriting of the *Hikayat* was carried out around 1663, half a century after the Dutch established their first trading post in the Banjar capital, and nearly three decades after the treaty that granted the Dutch a monopoly on the pepper trade. Accordingly, it may be more useful to see the pepper passages in the *Hikayat* not as a warning before-the-fact but as an after-the-fact commentary.[72] In that case, the more appropriate question is not: Did the

71 Suntharalingam (1963:47) estimates that the annual exports of pepper from Banjarmasin averaged just over one thousand tonnes at the beginning of the eighteenth century; Raffles (1978, I:215) puts this figure at the beginning of the nineteenth century at 'about twelve to fifteen hundred tonnes'; Lindblad (1988:31) puts the figure at one thousand tonnes early in the twentieth century; and contemporary government statistics put the production of South Kalimantan at 948 tonnes (in 1992) (Kantor Statistik Propinsi Kalimantan Selatan 1992:24).

72 Thus, Hudson's (1967:70) suggestion that 'the compiler' of the *Hikayat* was a member of the Banjar court's 'isolationist faction' is, although on the right track, wrong. The *Hikayat* statements on pepper were not so much an articulation of an isolationist position, as an articulation of what the implications were for Banjar society of isolation versus non-isolation.

Banjarese heed the warning? But rather: Was the warning prescient? And the answer is: Yes.

Acknowledgements

I first carried out research in the interior of Borneo from 1974 to 1976 with support from the National Science Foundation (Grant GS-42605) and with sponsorship from the Indonesian Academy of Science (LIPI). I gathered additional data during six years of subsequent work in Java between 1979 and 1985, making periodic field trips to Kalimantan, with support from the Rockefeller and Ford Foundations and the East-West Center, and with sponsorship from Gadjah Mada University. A recent series of field trips to Kalimantan, beginning in 1992, have been supported by the Ford Foundation, the United Nations Development Programme, and the John D. and Catherine T. MacArthur Foundation, and with sponsorship from BAPPENAS and Padjadjaran University. Initial analysis of these data was supported by a fellowship in Yale University's Program in Agrarian Studies. The author is grateful to Barbara Andaya, Peter Boomgaard, Carol Carpenter, and the participants in the 'Environmental History of Indonesia' workshop for constructive comments on an earlier draft of this paper, to John Cusick for assistance with library research, and to Helen Takeuchi for editorial assistance. Earlier findings from this study were presented at the 91st Annual Meeting of the American Anthropological Association in San Francisco, 2-6 December 1992, and at the Annual Meeting of the American Institute of Biological Sciences (Society for Economic Botany) in Honolulu, 9-13 August 1992. I am especially grateful to Peter Boomgaard, Freek Colombijn, David Henley, and the Koninklijk Institut voor Taal-, Land- en Volkenkunde for the invitation to the June 1996 workshop on the 'Environmental History of Indonesia,' for which this contribution was prepared. The author alone is responsible for the analysis presented here.

References

Andaya, B.W.

1993a 'Cash cropping and upstream-downstream tensions; The case of Jambi in the seventeenth and eighteenth centuries', in: A. Reid (ed.), *Southeast Asia in the early modern era; Trade, power, and belief,* pp. 91-122. Ithaca: Cornell University Press.

1993b *To live as brothers; Southeast Sumatra in the seventeenth and eighteenth centuries.* Honolulu: University of Hawaii Press.

1995 'Women and economic change; The pepper trade in premodern Southeast Asia', *Journal of the Economic and Social History of the Orient* 38-2:165-90.

Bartlett, H.H.

1957 *Fire in relation to primitive agriculture and grazing in the tropics; Annotated bibliography.* Vol. 2. Ann Arbor: Department of Botany, University of Michigan.

Beer, J.H. de, and M.J. McDermott
1989 *The economic value of non-timber forest products in Southeast Asia; With emphasis on Indonesia, Malaysia and Thailand*. Amsterdam: Netherlands Committee for IUCN/WWF.

Blacklock, J.S.
1954 'A short study of pepper cultivation with special reference to Sarawak', *Tropical Agriculture* [Trinidad] 31:40-56.

Brookfield, H. et al.
1990 'Borneo and the Malay Peninsula', in: B.L. Turner et al. (eds), *The earth as transformed by human action; Global and regional changes in the biosphere over the past 300 years*, pp. 495-512. Cambridge: Cambridge University Press.

Brosius, J.P.
n.d. 'Prior transcripts, divergent paths; Resistance and acquiescence to logging in Sarawak, East Malaysia'. [Manuscript.]

Burkill, I.H.
1966 *A dictionary of the economic products of the Malay Peninsula*. Kuala Lumpur: Ministry of Agriculture and Cooperatives. 2 vols. [First edition 1935.]

Cleary, M. and P. Eaton
1992 *Borneo; Change and development*. Singapore: Oxford University Press.

Cramb, R.A.
1993 'Shifting cultivation and sustainable agriculture in East Malaysia; A longitudinal case-study', *Agricultural Systems* 42:209-26.

Crawfurd, J.
1856 *A descriptive dictionary of the Indian islands and adjacent countries*. London: Bradbury and Evans.

Dixon, A., H. Roditi and L. Silverman
1991 *From forest to market; A feasibility study of the development of selected non-timber forest products from Borneo for the U.S. market*. Cambridge: Project Borneo. 2 vols.

Dove, M.R.
1986a 'Peasant versus government perception and use of the environment; A case-study of Banjarese ecology and river-basin development in South Kalimantan', *Journal of Southeast Asian Studies* 17-1:113-36.

1986b 'The practical reason of weeds in Indonesia; Peasant vs. state views of *Imperata* and *Chromolaena*', *Human Ecology* 14:163-90.

1993a 'A revisionist view of tropical deforestation and development', *Environmental Conservation* 20:17-24, 56.

1993b 'Smallholder rubber and swidden agriculture in Borneo; A sustainable adaptation to the ecology and economy of the tropical forest', *Economic Botany* 47-2:136-47.

1996 'Rice-eating rubber and people-eating governments; Peasant versus state critiques of rubber development in colonial Indonesia', *Ethnohistory* 43-1:33-63.

Drewes, G.W.J. (ed. and translator)
1979 *Hikajat Potjut Muhamat; An Achehnese epic*. The Hague: Nijhoff. [KITLV, Bibliotheca Indonesica 19.]

Duke, J.A. and J.L. Du Cellier
1993 *CRC handbook of alternative cash crops*. Boca Raton, Fl.: CRC Press.

Dunn, F.L.
1975 *Rain-forest collectors and traders; A study of resource utilization in modern and ancient Malaya*. Kuala Lumpur: Malaysian Branch of the Royal Asiatic Society. [Monograph 5.]

Echols, J.M. and H. Shadily
1992 *Kamus Indonesia-Inggris; An Indonesian-English dictionary*. Third edition. Jakarta: Gramedia.

Flückiger, F.A. and D. Hanbury
1879 *Pharmacographia; A history of the principal drugs of vegetable origin, met with in Great Britain and British India*. Second edition. London: Macmillan.

Forbes, H.O.
1885 *A naturalist's wanderings in the eastern archipelago; A narrative of travel and exploration from 1878 to 1883*. London: Sampson Low.

Generale Missiven
1960-88 *Generale Missiven van Gouverneurs-Generaal en Raden aan Heren XVII der Verenigde Oostindische Compagnie*. Edited by W.P.Coolhaas [vol. 1-8] and J. van Goor [vol. 9] 's-Gravenhage: Nijhoff. [RGP, Grote Serie, 9 vols.]

Groeneveldt, W.P.
1960 *Historical notes on Indonesia and Malaysia; Compiled from Chinese sources*. Jakarta: Bhratara.

Hapip, A.D.
1977 *Kamus Banjar-Indonesia*. Jakarta: Pusat Pembinaan dan Pengembangan Bahasa, Departement Pendidikan dan Kebudayaan.

Heidhues, M.F. Somers
1992 *Bangka tin and Mentok pepper; Chinese settlement on an Indonesian island*. Singapore: Institute of Southeast Asian Studies.

Heine-Geldern, R. von
1945 *Prehistoric research in the Netherlands Indies*. New York: Southeast Asia Institute.

Hirth, F. and W.W. Rockhill (eds and translators)
1911 *Chau Ju-kua; His work on the Chinese and Arab trade in the twelfth and thirteenth centuries, entitled Chu-fan-chï*. St. Petersburg: Imperial Academy of Sciences.

Horne, E.C.
1974 *Javanese-English dictionary*. New Haven: Yale University Press.

Hudson, A.B.
1967 *Padju Epat; The ethnography and social structure of a Ma'anyan Dajak group in Southeastern Borneo*. [PhD thesis, Cornell University, Ithaca.]

Kantor Statistik Propinsi Kalimantan Selatan
1992 *Kalimantan Selatan dalam angka 1992*. Banjarmasin: Kantor Statistik Propinsi Kalimantan Selatan.

Kurin, R.
1983 'Indigenous agronomics and agricultural development in the Indus basin', *Human Organization* 42-4:283-94.

Lévi, S.
1918 'Pour l'histoire du Ramayana', *Journal Asiatique* (Jan.-Fèv.):1-160.

Lindblad, J.T.
1988 *Between Dayak and Dutch; The economic history of Southeast Kalimantan, 1880-1942*. Dordrecht: Foris. [KITLV, Verhandelingen 134.]

Marsden, W.
1975 *The history of Sumatra*. Kuala Lumpur: Oxford University Press. [Reprint of the third edition (1811), introduced by John Bastin.]

Monier-Williams, M.
1899 *A Sanskrit-English dictionary [...]*. Oxford: Clarendon Press.

Noorlander, J.C.
1935 *Bandjarmasin en de Compagnie in tweede helft der 18de eeuw*. Leiden: Dubbeldeman. [PhD thesis, Rijksuniversiteit Leiden.]

Ochse, J.J.
1977 (with R.C. Bakhuizen van den Brink) *Vegetables of the Dutch East Indies; Survey of the indigenous and foreign plants serving as pot-herbs and side-dishes*. Amsterdam: Asher. [First edition, *Indische groenten*, Buitenzorg: Archipel, 1931.]

Padoch, C.
1982 *Migration and its alternatives among the Iban of Sarawak*. The Hague: Nijhoff. [KITLV, Verhandelingen 98.]

Peluso, N.
1983 Markets and merchants; The forest product trade of East Kalimantan in historical perspective. [M.Sc. thesis, Cornell University, Ithaca.]

Peters, C.M., A.H. Gentry, and R.O. Mendelsohn
1989 'Valuation of an Amazonian rain forest', *Nature* 339:655-6.

Polo, Marco
1969 *The travels of Marco Polo*. New York: Airmont.

Potter, L.M.
1988 'Indigenes and colonisers; Dutch forest policy in South and East Borneo (Kalimantan) 1900 to 1950', in: J. Dargavel, K. Dixon, and N. Semple (eds), *Changing tropical forests; Historical perspectives on today's challenges in Asia, Australasia and Oceania*, pp. 127-53. Canberra: Centre for Resource and Environmental Studies, Australia National University.

Raffles, T.S.
1978 *The history of Java*. Kuala Lumpur: Oxford University Press. 2 vols. [First edition London: Black, Parbury and Allen, and Murray, 1817.]

Ras, J.J.
1968 *Hikajat Bandjar; A study in Malay historiography*. The Hague: Nijhoff. [KITLV, Bibliotheca Indonesica 1.]

Reid, A.
1993 *Southeast Asia in the age of commerce 1450-1680. Vol. 2: Expansion and crisis*. New Haven/London: Yale University Press.
1995 'Humans and forests in pre-colonial Southeast Asia', *Environment and History* 1:93-110.

Richards, A.
1981 *An Iban-English dictionary*. Oxford: Clarendon Press.

Roseberry, W.
1991 *Anthropologies and histories; Essays in culture, history, and political economy*. New Brunswick: Rutgers University Press.

Saleh, M.I.
1976 'Pepper trade and the ruling class of Banjarmasin in the seventeenth century', in: *Papers of the Dutch-Indonesian historical conference, held at Noordwijkerhout, The Netherlands, May 19-22*, pp. 203-21. Leiden/Jakarta: Bureau of Indonesian Studies.

Schrieke, B.
1966 *Indonesian sociological studies; Selected writings of B. Schrieke*. Part one. Second edition. The Hague: Van Hoeve. [First edition 1955.]

Schwaner, C.A.L.M.
1853-54 *Borneo; Beschrijving van het stroomgebied van den Barito en reizen langs eenige voorname rivieren van het zuid-oostelijk gedeelte van dat eiland in de jaren 1843-1847*. Amsterdam: Van Kampen. 2 vols.

Scott, J.C.
1976 *The moral economy of the peasant; Rebellion and subsistence in Southeast Asia*. New Haven: Yale University Press.

Snouck Hurgronje, C.
1888 'Nog iets over de Salasila van Koetai', *Bijdragen tot de Taal-, Land- en Volkenkunde* 37:109-20.

Speelman, C.
1670 'De handelsrelaties van het Makassaarse rijk volgens de notitie van Cornelis Speelman uit 1670', in: J. Noorduyn (ed.), *Nederlands historische bronnen III*, 1983, pp. 96-121. Amsterdam: Verloren.

Suntharalingam, R.
1963 'The British in Banjarmasin; An abortive attempt at settlement 1700-1707', *Journal of Southeast Asian History* 4-2:33-50.

Tsing, A.L.
1993 *In the realm of the diamond queen; Marginality in an out-of-the-way place*. Princeton: Princeton University Press.

Vlekke, B.H.M.
1961 *Nusantara; A history of Indonesia*. Revised edition. Bruxelles: Manteau, Jakarta: Soeroengan.

Waard, P.W.F. de
1964 'Pepper cultivation in Sarawak', *World Crops* 16-3:24-30.

1989 'Piper nigrum L.', in: E. Westphal and P.C.M. Jansen (eds), *Plant resources of South-East Asia; A selection*, pp. 225-30. Wageningen: PUDOC.

Watt, G.
1972 *A dictionary of the economic products of India*. Delhi: Gordhan. [7 vols. in 10 parts. First edition Calcutta: Superintendent of Government Printing, 1889-96.]

Wheatley, P.
1959 'Geographical notes on some commodities involved in Sung maritime trade', *Journal of the Royal Asiatic Society* 32-2:1-140.

Wilkinson, R.J.
1959 *A Malay-English dictionary*. London: Macmillan. 2 vols.

Wolters, O.W.
1967 *Early Indonesian commerce; A study of the origins of Srivijaya*. Ithaca: Cornell University Press.

Zoetmulder, P.J.
1982 *Old Javanese-English dictionary*. The Hague: Nijhoff. 2 vols. [KITLV.]

ROBERT CRIBB

Birds of paradise and environmental politics in colonial Indonesia, 1890-1931

The history of environmental politics in Indonesia falls into two distinct episodes. The more recent dates from the mid-1970s when, for the first time since independence in 1945, the Indonesian government began to show concern for environmental protection and, more important, began to undertake measures to stem the rate of environmental destruction in the archipelago. Within a decade, the environment had become a significant issue on the Indonesian political agenda, a topic of dispute between government ministers, a field of rapid legal development, a significant element in the changing business-government relationship, and the inspiration for hundreds of non-governmental organizations working within the limited democratic space permitted by the New Order government.[1] The earlier episode was much more modest in scope and until recently has gone virtually unnoticed by scholars. During the last half century of colonial rule, the Dutch authorities in Indonesia also saw a kind of environmentalist light and undertook a series of measures to preserve from destruction and permanent loss what they saw as important elements in the ecology of the Indies.[2]

Contests over the environment are nothing new in Indonesian history. Forests and forest products have long been a subject of conflict between various social groups, especially rulers and ruled, as a matter of simple competition over power and wealth.[3] Locals and outsiders have competed over access to fish and other marine resources. And both indigenous and outside powers have seized parts of the archipelago for the sake of their natural resources. We can distinguish the environmental politics of twentieth-century Indonesia, however, from the more general category of resource politics in two important respects. First, the arguments for

1 On environmental politics in Indonesia since the 1970s, see Cribb 1988, 1990; Smith 1992; MacAndrews 1994.

2 To my knowledge, the only systematic attempt to describe and explain these policies is Boomgaard 1993. Further notes to this most valuable article refer to the published version; my reading of the article, however, is based on an English-language version kindly provided to me by the author.

3 On the history of competition over forests in the Indies, see Peluso 1992 and Boomgaard 1994.

Aru hunters shooting the great bird of paradise (Wallace 1890:opposite page 337).

environmental protection have arisen primarily out of a modern, scientific understanding of the world. Thus the argument for conservation rests on an understanding that the extinction of a species is possible and that this can happen both by direct extermination and by destruction of its habitat. The argument for controlling pollution rests on an awareness of the long-term and short-term consequences of different kinds of industrial and biological emission. The argument for forest preservation rests on knowledge of hydrography and plant physiology and of the consequences of forest removal for climate and rainfall. This is not to say that the information used to back environmentalist arguments has been or is now necessarily accurate; in any era, scientific propositions are subject both to scientific error for various reasons and to the limitations of whatever paradigms may be current at the time. Rather, the debate was conducted in an idiom in which scientific truth appeared to be the decisive factor in winning or losing the argument. Although sections of the contemporary environmental movement are sympathetic to nonscientific Western romanticism and arguments from indigenous peoples about the need to preserve humankind's spiritual harmony with nature, such arguments are peripheral to the main environmentalist case.[4]

The second distinctive feature of the environmentalist argument is its appeal to posterity. A significant part of the moral appeal of the environmentalist argument comes from its promise to ensure that future generations will have much the same access to the benefits of natural environment as we do. In this respect, environmentalism lies firmly in the tradition – once Western, now global – of confidence in human progress. Although the broad environmental movement contains a doomsday element that believes that only radical change can deliver humankind from environmental disaster, the core of the movement sees environmental degradation as another challenge in human history that the correct application of foresight and science can solve, and that this can be done at a cost acceptable to those living in the present.

Especially because of its association with expanding scientific knowledge, the history of environmental protection has often had a triumphalist tone. In many societies of the world, environmentalist arguments are exercising steadily more power. Projects such as the Snowy River irrigation scheme in Australia or the Delta Works in the Netherlands, which both public and policy-makers widely regarded as unambiguous social goods at the time they were carried out, would face determined resistance on environmental grounds if they were proposed today. By contrast, it would be hard to find a significant constituency to argue that the creation of, say, Yellowstone or Serengeti National Parks was a mistake. The history of

[4] The issue of non-scientific philosophies of the human relationship with and within nature is a complex one. For a recent study focusing on Asia, see Bruun and Kalland (1995); a valuable older study examining divergent philosophical trends in the West is Passmore (1974).

environmental protection, therefore, tends to be told in heroic terms of far-sighted individuals who overcame official and public inertia and obstruction to stem the rate of environmental destruction for the sake of humankind. This picture contrasts with the more sombre range of dilemmas facing environmentalists today: Should the pollution hazards of coal-fired power stations be preferred to the radioactive hazards of nuclear power? Should the well-managed national parks of southern Africa be permitted to help fund themselves by allowing the export of ivory? Should humans deliberately cause the extinction of the smallpox virus?

In this chapter, I wish to draw attention not only to the existence of environmental politics in Indonesia in the early part of the twentieth century, but also to suggest that dilemmas of the kind outlined above were also characteristic of earlier stages of environmental movements. The episode I discuss is the issue of protecting birds of paradise in the Netherlands Indies, an issue which arose towards the end of the nineteenth century and which received close government attention especially during the first three decades of the twentieth century. We shall see that the debate raised a wide range of issues, especially in the areas of scientific knowledge and long-term and short-term economic and social costs, and that the nature of the debate was not dramatically different from many contemporary environmental discussions.

The term 'bird of paradise' normally refers to any of about forty-five species of bird of the family *Paradisaeidae* indigenous to New Guinea and adjacent islands (including Halmahera and the Aru Islands). Although their closest relatives are believed to be crows and ravens, birds of paradise are known for their bright and ornate plumage which, as in many bird species, is borne exclusively by the males. The sole function of the plumes appears to be for display to females during elaborate courtship rituals, and their sheer exuberance has made them a useful case-study in the theory of evolution by sexual selection (Diamond 1981). Westerners making first contact with the indigenous peoples of New Guinea during recent centuries found bird-of-paradise pelts and feathers widely used for adornment, and this practice is presumably an ancient one. The trade of feathers and pelts to Asia may also be ancient; kingfisher and peacock feathers were exported from Java to China possibly as early as the sixth century.[5] By the sixteenth century, however, the plumes were certainly being traded to the Middle East and when the Spanish expedition of (the then deceased) Ferdinand Magellan reached the Moluccas (Maluku) in

5 Hall 1992:196. Wolters (1967), however, does not mention a trade in feathers or pelts, though he does briefly discuss the use of bird plumage for adornment by indigenous peoples in Borneo (p. 177). The fact that cloves were exported from Ternate and Tidore, a few dozen kilometres from the habitat of some bird-of-paradise species, as early as 1700 AD, means that there is no practical obstacle to pelts having been traded at this early stage. See Andaya 1993:1-2; Swadling 1996:53-9.

1521 the King of Bacan entrusted them with two preserved bird-of-paradise pelts, along with some slaves and a consignment of cloves, as a gift for Emperor Charles V (Westermann 1947:15; Swadling 1996:62).

Antonio Pigafetta, chronicler of the Magellan expedition, reported that local people called the birds 'bolondinata', which is generally interpreted as a rendering of '*burung dewata*' or bird of the gods.[6] Both this Malay name and its English equivalent arose from the fact that the pelts were only ever seen without feet, wings or viscera. The standard practice of the hunters was to remove these parts, to seal the carcass with gum or resin and then to smoke it. In the finest specimens, these operations were carried out so skilfully that the cuts were generally invisible and the story grew that birds of paradise never alighted. 'They flie [wrote the Dutch navigator J.H. van Linschoten in 1598] alwaies into the Sunne, and keepe themselues continually in the ayre [...] for they haue neither feet nor wings, but onely head and bodie, and the most part tayle.'[7] The males, according to a seventeenth-century report, had a hollow in their backs in which the females laid their eggs and raised their young.[8] The legend still had enough power in 1760 to lead the Swedish naturalist Carolus Linnæus to give the species that he examined the scientific name *Paradisaea apoda* (literally 'paradise creature without legs') (Westermann 1947:16-7).

For three centuries after the initial European contact with birds of paradise, the trade appears to have remained at a fairly low level and the annual harvest of pelts probably did little damage to bird populations.[9] The Raja Ampat Islands in particular traded pelts to the west, as well as delivering them as tribute to their overlords, the rulers of Tidore (Andaya 1993:99, 103, 192, 205). Dutch records from the second quarter of the nineteenth century show a sudden expansion of the trade in the mid-1830s followed by a decline which may or may not have been due to diminishing stocks.[10] British naturalist A.R. Wallace, who worked in the

6 Crawfurd 1856:54. Slightly later visitors also recorded the name '*manuk dewata*', with the same meaning. This term has entered English in the somewhat bastardized form of 'manucode' as the common name for Australian species of the family *Paradisaeidae*. See Yule and Burnell 1903:94-5.

7 Quoted in Yule and Burnell 1903:95.

8 *Evelyn's Diary*, 1645, quoted in Yule and Burnell 1903:95. Stories of this kind may also have had their origin in the desire of bird-hunters to discourage and distract those who might otherwise have been tempted to seek the birds in the wild for themselves. Wallace records the suspicion and obstruction which one of his assistants faced when he attempted to collect birds in the hinterland of Sorong. See Wallace 1890:438.

9 In the absence of detailed studies of the birds of paradise before the nineteenth century, we exclude the possibility that scarce species may have become extinct as a result of human activity. Hunting by Austronesian communities who had settled in vulnerable island ecosystems on Madagascar and Hawaii and in New Zealand certainly led to the extinction of some bird species.

10 The figures below give the recorded value of pelts exported from Ternate from 1832 to 1850. The price of a single pelt in Ternate in 1839 was *f* 2.50, but since the price of pelts

archipelago from 1854 to 1862, reported his impression that most species of bird of paradise were much more difficult to obtain than they had been twenty years earlier, but he attributed this phenomenon to the fact that Dutch officials had been using the authority of the Sultan of Tidore to instruct villagers to collect rare species for scientific collections and as souvenirs. The villagers, he suggested, resented this activity, for which they were not paid, and reacted by pretending that the birds had died out (Wallace 1890:439-40).

During the nineteenth century, however, the pace of exploitation seems to have increased gradually. The Raja Ampat Islands declined as a source of pelts, their place being taken initially by the Aru Islands and later by the New Guinea mainland, though this change may have been due to the fact that the two species which were most in demand, *Paradisaea apoda* and *P. minor* (the Greater and Lesser Birds of paradise respectively, but sometimes known collectively as yellow birds of paradise) were most common in Aru and on the mainland. Batavia and Singapore began to develop as significant centres for the trading and transit shipment of pelts. Prices began to rise: a string of pelts which fetched *f* 50 in Ternate in 1839 had risen in value to *f* 100-120 by 1875.[11] The character of the hunt also began to change. Until the middle of the nineteenth century, birds of paradise were hunted mainly by local people. Their favoured technique was to build hides in trees which were known to be used by courting birds of paradise and to shoot them with blunt-headed arrows. The stunned birds would then fall to the ground where they could be collected by other villagers without damaging the pelt in any way (Wallace 1890:422). After the pelts had been preserved, they were traded or delivered downstream or along the coast to medium-scale political and commercial centres. In the second half of the nineteenth century, however, the hunt seems to have been dominated increasingly by semi-professional hunters from the Moluccas, especially Ternate and Geser, who moved from region to region and delivered the pelts to market through independent Chinese and Bugis traders. These hunters for the most part used firearms, because there was now a market for feathers as well as entire birds and hunters could afford some spoilage of carcasses for the sake of a greater kill rate. From 1891, the regular steamship service of the KPM provided an additional means of

---

probably went up in times of high demand, the fluctuation in the number of birds killed may not have been as great as these figures suggest.

1832	*f* 1218	1836	*f* 10223	1845	*f* 311	1848	*f* 154
1833	*f* 2829	1837	*f* 5235	1846	*f* 90	1849	-
1834	*f* 5171	1844	*f* 500	1847	*f* 24	1850	*f* 37
1835	*f* 9065						

Source: Westermann 1947:17.

For figures on the trade in birds, bird skins and feathers (which thus include crowned pigeons, parrots and other species), see Boomgaard 1996.

11 Westermann 1947:18. Part of this price increase was a consequence of inflation.

exporting the pelts, and the hinterland of the ports of Hollandia, Ansus, Kokas, Mimika, and Merauke all emerged as prime hunting areas.[12]

The actual hunting appears to have been done mainly by Papuans. Sometimes they hunted independently for pelts to trade for tools, food, and other goods; more commonly, they hired firearms and ammunition from hunters from Ternate, Ambon, Seram, Geser, and South Sulawesi, paying off the fee in pelts. These hunters, however, were generally dependent on Chinese or European middlemen, who provided them with hunting equipment, weapons (often out-of-date muzzle-loaders) and ammunition, as well as an advance to cover incidental costs. Colonial officials noted an appreciable turnover rate of hunters who feared to return to the Moluccas because they could not pay their debts or had been killed by Papuans (Westermann 1947:20-4).

The demand behind this expansion of trade came from the European and American fashion industries, in which bird-of-paradise feathers were widely used, mainly for the decoration of ladies' hats. From the early nineteenth century, milliners had begun to use an enormous range of materials, from owls' heads, swallows' wings, and the preserved bodies of entire robins and humming-birds, to egret and ostrich plumes (Doughty 1975:16-22; Swadling 1996:83-8). In the second half of the nineteenth century, movements against the feather trade began to develop in the United States and Britain. These movements drew on several elements in the Western tradition. On the one hand, they reflected a willingness to project human characteristics onto animals and to give animals the consideration which should be due to human beings. Passionate accounts were published depicting the alleged brutality of the slaughter of birds and drawing attention to the fact that many species were hunted during the nesting season, so that the young often starved to death after the killing of their parents. Other authors presented more utilitarian arguments, pointing to the value of birds to agriculture as eaters of insects which might otherwise prey on crops. And increasingly in the course of the nineteenth century attention was drawn to the risk that some or even most birds might become extinct (Doughty 1975:31-50; Swadling 1996:92).

The fact of extinction had intrigued scientists ever since geological research had uncovered the fossils of animals no longer known to exist (Axelrod 1967:1-4; Raup 1984:1-4). Darwin's ideas of evolution, which took the scientific world by storm in the middle of the nineteenth century, explained that extinction was a natural part of the evolutionary process and suggested that no particular value should be attached to the preservation of individual species. The social variant of Darwinism was similarly complacent about the disappearance of 'less advanced' ethnic

[12] See Crawfurd 1856:55; *Paradijsvogels* 1919:338-9; Westermann 1947:18, 22; De Wit 1914: 441-5.

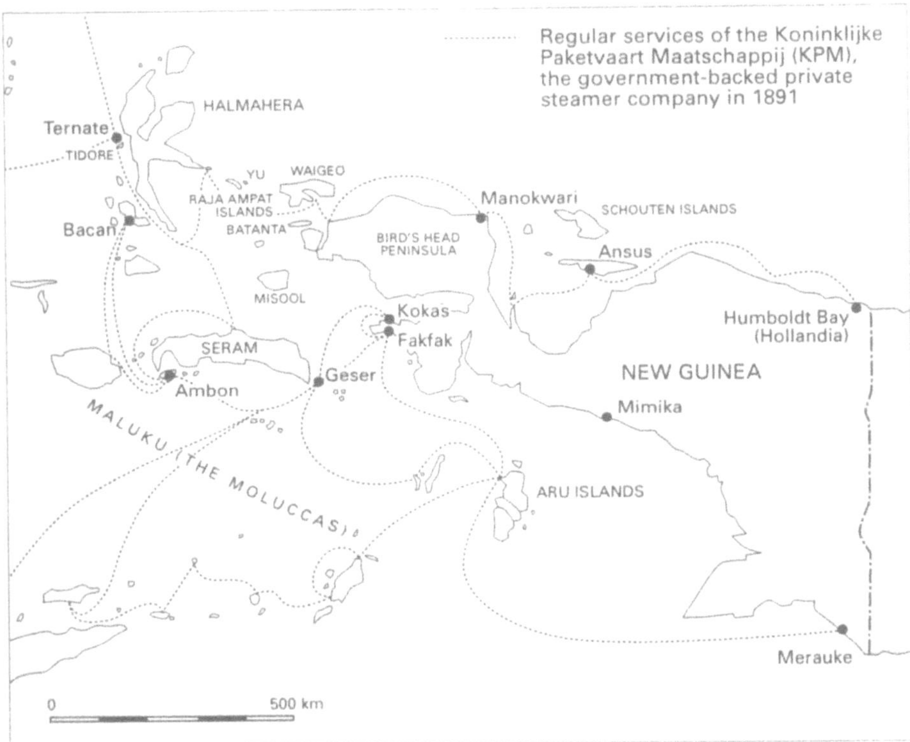

The Moluccas and New Guinea, 1891

groups. As the evolutionary paradigm worked its way through scientific practice, however, scientists increasingly recognized a number of 'problem cases', physical features of birds and animals which could not easily be explained in terms of natural selection. Complex organs such the ear and the eye fell into this category, as did ornamentation which appeared to have no practical function. A scientific consensus gradually developed, therefore, that for future research purposes it was important to preserve all species and particularly important to preserve species which demonstrated the creative power of the evolutionary process. This scientific abhorrence of extinction coincided with a more general growth in ideas of natural beauty and a desire to ensure that coming generations would also be able to marvel at creatures such as the rhinoceros and the elephant.

In the early centuries of the European expansion overseas from the fifteenth century onwards, a number of extinctions were known to have occurred, including the extermination of the dodo on Mauritius by 1690 (Fuller 1987:122) and of Steller's sea cow in the Bering Sea in about 1768. In the second half of the nineteenth century, however, a much wider range of species appeared to be suddenly at risk as a result of human rapacity. Attention focused initially on the spectacular large animals of southern Africa, where spreading European settlement accompanied by large-scale hunting had greatly diminished the herds of grazing animals such as elephant and springbok which were once a striking feature of the landscape. One species, the blue antelope, was apparently extinct by 1800 (Mackenzie 1988:89). As early as 1858, the colonial government in Cape Town issued a government notice for the 'Preservation of elephants and buffaloes' (Grove 1987:27; Mackenzie 1988:203), but the limited measures of the nineteenth century were insufficient to prevent the extinction of the quagga (a relative of the zebra) in about 1872. Many other species appeared to be close to extinction. Similar alarm began to develop in the United States, where the American bison was brought close to the point of extinction and the passenger pigeon was pushed over the brink.

In each of these cases, (near-)extinction appeared to be a consequence of unrestrained hunting. The settlers of South Africa and of the American West hunted for food, for meat and skins to sell, and to remove grazing animals which competed with their own cattle and sheep. The pace of settlement in both regions would probably have been much slower if settlers had not had access to this source of income and capital as they developed more sedentary farming activities. In both cases, the destruction of habitat certainly played a greater role than opinion of the day conceded in reducing animal populations, but hunting certainly accelerated the process. In the last decade of the nineteenth century, the large number of bird-of-paradise pelts reaching dealers in London and Paris suggested that those birds too might soon disappear from the face of the earth.

The first recorded warning that continued hunting of birds of paradise might lead to the extinction of one or more species came in an 1890 article by the former Resident of Ternate, F.S.A. de Clercq,[13] who predicted bleakly: 'Now that the birds are almost never found along the coast and the killing has moved into the interior, it will not be long before nothing remains of these most glorious products of Creation, which are a delight to ornithologists and a wonder to the whole world'. In 1894, probably prompted both by this warning and by Dutch press reports reaching the Indies in the overseas mail editions of newspapers such as the *Nieuwe Rotterdamsche Courant*, the Governor-General, Jhr C.H.A. van der Wijck asked the Residents of Ternate and Amboina (Ambon), whose territories covered the habitat of the birds of paradise in the Indies, to report on the nature of the trade and to advise whether protective or regulatory measures were needed. The Resident of Ternate concluded that New Guinea was so vast and so little explored that no risk of extinction was likely to emerge for several decades, but the Resident of Ambon reported that bird-of-paradise species limited to smaller islands such as Waigeo and the Aru archipelago might indeed be in danger of extermination.[14] Both the question and the replies were significant. Only a few years before, the colonial government had dismissed out of hand a request by the Landbouw-vereeniging (Agricultural Association) in Sukabumi in West Java for measures to protect insectivorous birds from hunting, commenting that it saw 'no value' in such a measure.[15] Now, however, senior figures in the colonial government had absorbed enough of the spirit of the age to take for granted that extinction was an undesirable possibility and that government action might be necessary and able to avert it.

Conservationist arguments in the late-twentieth century are marked by a strong sense of urgency, but the same cannot be said of the deliberations of the Indies government a century ago. More than ten years were to elapse before the birds of paradise received even perfunctory protection and not until 1931 was all hunting of birds of paradise prohibited. The main reason for this slow progress appears to have been that the colonial government felt itself on unknown territory both scientifically and administratively. As the contradictory advice received from the two Residents in 1894 indicated, there was a great deal of scientific uncertainty about the degree to which birds of paradise were threatened with extinction. Moreover, although pioneering conservation measures had been introduced in other parts of the world in the course of the nineteenth century, neither the technical detail of those measures nor the extent to which they had succeeded in their aims was widely known. Time and time again,

13 Cited in Van Houten 1896:19-20.

14 ARA MvK, verbaal 11-10-1897, no. 30, J. van Oldenborgh (Resident Ternate en Onderhoorigheden) to Governor-General, 13-4-1894; G.W.W.C. Baron van Hoëvell (Resident Ambon) to Governor-General, 27-6-1894.

15 ARA MvK, Oost-Indische Besluiten 29-4-1887 no. 21.

therefore, the intention to do something to protect birds of paradise ran aground on these two rocks of uncertainty, only to be pushed back into the policy mainstream by pressure from outside the Indies.

Thus, no sooner had the colonial government decided in 1894 that there was no need for the time being to take any action[16] than new cries of environmentalist alarm over the hunt for birds of paradise were brought to its attention. In November 1985, the Minister of Colonies in The Hague received a letter, signed by the executive of the Bond ter Bestrijding eener Gruwelmode (Association to Combat a Revolting Fashion) and others, deploring the 'plunder hunt' (*roofjacht*) of what they called 'the most beautiful birds in the world' and urging the minister to prevent it.[17] This letter was followed within weeks by the appearance of an article by M.C. Piepers, an amateur entomologist and a former official in the Indies Department of Justice, arguing that birds of paradise and a number of other animals and plants faced extinction if protective measures were not taken. Piepers urged the colonial government to give legal protection to endangered species and to establish nature reserves on the model of Yellowstone National Park in the United States. P.J. van Houten (1896) warned publicly that the accelerating rate of hunting might lead to the extinction of one or more species of bird of paradise; other shrill voices of alarm addressed the Dutch press and Parliament[18] and in 1896 the Minister of Colonies put a series of mild requests for information to the colonial government. The question of whether the birds were at risk of extinction now receded in importance; there seems to have been a consensus for the time being that even if extinction was not imminent, the current level of hunting was unsustainable and ought to be reduced. The problem was: how could this be done?

During 1896 and 1897 the colonial government and officials in the ministry in The Hague canvassed and discarded a wide range of possibilities. They considered a complete ban on the export of bird-of-paradise feathers and pelts from the Indies, but rejected this idea because they felt that hunting needed only to be limited, not ended, and because they feared that a full-scale ban would simply encourage smuggling. The Resident of Ternate suggested an export tax on the pelts, but others pointed out that this would do nothing to limit the hunt; there were also technical problems linked to the fact that New Guinea was not yet formally part of the Netherlands Indies customs zone. Other suggested imposing a ban or at least heavy licence fees on the import of firearms into New Guinea, but

16 ARA MvK, verbaal 11-10-1897, no. 30, H.J.W. van Lawick van Pabst (secretaris, Dept van Binnenlandsch Bestuur) to Governor-General, 4-9-1894

17 ARA MvK, verbaal 7-1-1896, no. 37, Bestuur Bond ter Bestrijding eener Gruwelmode, redactie *Androcles*, and bestuur Vereeniging 'Ornis' to Minister of Colonies, 4-11-1895.

18 ARA MvK, verbaal 18-2-1896, no. 14, Minister of Colonies to Governor-General, 18-2-1896; ARA MvK, verbaal 11-10-1897, no. 30, Philornithes (1897?), *Het vogelvraagstuk, opgedragen aan de leden der Eerste en Tweede Kamer van de Staten-Generaal en de Nederlandsche Vrouwen* (Amsterdam: Bond ter Bestrijding van den Vogelmoord).

this idea too raised the spectre of smuggling. It was also pointed out that local farmers needed firearms to protect their fields from wild animals. Some suggested introducing a closed season for hunting, or banning it in specified regions, but these ideas faced formidable practical obstacles. In 1896, the colonial government still had no permanent administrative presence on the New Guinea mainland and it would have been impossible to enforce any kind of ban. Even a ban on hunting in the Aru Islands, the Resident of Ambon argued, would require a warship to be permanently stationed in the waters of the Arafura Sea.[19]

Amid the gloom that these objections produced, however, the Governor-General saw one ray of light. Ever since they established a territorial empire in the Indies, the Dutch had been accustomed to profiting from the collection of wild products such as edible birds' nests by means of a *pacht* (lease), or revenue farm. Under this system, the right to collect a product in a particular region was put out to tender and was normally awarded to the highest bidder who could convince the colonial authorities of his ability to manage the *pacht*. The system was also used for the distribution of opium and salt and the management of gambling and prostitution. By the late nineteenth century many authorities believed that the *pacht* system was inefficient and exploitative, and the main *pachten* were in the process of being abolished. For managing the bird-of-paradise trade, however, the *pacht* seemed to have practical advantages. With hunting permitted to continue, there would be less risk of smuggling. By dividing the New Guinea coast and adjacent islands into five or six *pacht*-districts, the government believed it could give a continuing place in the trade to virtually all the larger European concerns which had already built up interests in the trade. Because it would be dealing with no more than a few *pachters* (game tenants), the government could more easily apply limits to the number of birds exports, thus providing greater protection to the birds. The proceeds from the *pacht* would help to pay for the extension of colonial administration in New Guinea. And, finally, the Resident of Ternate reported that the Sultan of Tidore had successfully introduced a *pacht* on the hunting of Golden Doves (probably *Ptilinopus luteovirens*) on the island of Yu, near Gebe in the northern Moluccas.[20]

The idea of applying a *pacht* system as a kind of territorial advance guard for the colonial administration, however, was a break with tradition, and the government in Batavia decided to wait for the more detailed, expert advice on the matter which they hoped would come from the two *controleurs* (district officers) who took up posts in Manokwari and Fakfak in 1898. When they reported in 1899, both *controleurs* wrote forcefully in

19 For these discussions, see ARA MvK, verbaal 30-8-1898, especially Governor-General to Minister of Colonies, 6-9-1897 and G.F. de Bruijn Kops (Resident Ambon) to Governor-General 22-5-1897.

20 ARA MvK, verbaal 30-8-1898 no. 8, Governor-General to Minister of Colonies, 6-9-1897.

favour of controlling the hunt, not just because of the risk of exterminating the birds but because of the consequences of the wholesale introduction of firearms into New Guinean society. On the other hand, both outlined serious practical problems with the implementation of a *pacht* in areas beyond the administrative control of the Netherlands Indies: it would be very difficult, both of them pointed out, to provide the environment of law and order needed for an effective *pacht* and difficult, too, to keep the *pachter* (franchise-holder) honest. The Resident of Ternate added a further warning of the risk that the *pacht* might come into the hands of Chinese entrepreneurs from Singapore, who would be likely to engage in smuggling on the side and who would not act at all as an advance guard for Dutch authority.[21]

While hesitating over the idea of a *pacht*, the colonial government also turned to Dr J.C. Koningsberger, appointed in January 1898 as agricultural zoologist attached to the Buitenzorg (now Bogor) Botanic Gardens, for scientific advice on the extent of the danger to birds of paradise. Koningsberger's advice, however, added to the uncertainty. He shared the general conviction that birds of paradise should not be allowed to become extinct, but he repeated the argument that there was not the slightest risk of extinction on the New Guinea mainland because of the vast extent of territory and the difficult terrain. In the Aru Islands, however, and on the northern islands of Waigeo, Batanta, and Misool, he argued that local species were indeed in danger of extinction and that a local ban on hunting should be enforced.[22]

Koningsberger's main interest, moreover, was not in birds of paradise but in the ecological relationship between environment and agriculture. He was concerned especially with the insect pests of agriculture on Java and with the role that birds could play as agents of biological control.[23] He wanted regulations which would give protection to useful and interesting birds while permitting people to hunt species harmful to agriculture. His own research stressed the complexity of the relationship between plants, birds, insects, and other animals and he emphasized that it was impossible to draw up a list of birds which merited protection and those which should be exterminated. Instead, he preferred that the regional administrators[24] should make their own local decisions, bearing in mind

21 ARA MvK, verbaal 28-7-1905, no. 16, Van Oosterzee (wd controleur Manokwari) to Resident Ternate, 4-7-1899; Kroesen (controleur West- en Zuid-Nieuw-Guinea) to Resident Ternate 6-9-1899; Horst (Resident Ternate) to Directeur Binnenlandsch Bestuur, 9-101900.

22 ARA MvK, verbaal 28-7-1905, no. 16, J.C. Koningsberger to Directeur, 's Lands Plantentuin, no date [June 1898].

23 See Koningsberger 1898, 1901-09, 1903.

24 The Netherlands Indies was divided administratively into about thirty-six regions. Most were headed by Residents, but a few were headed by Governors or by Assistant-Residents. The collective term for these officials, *gewestelijke hoofden*, I have translated in this chapter as 'regional administrators'.

that a bird which was useful to agriculture in one region might be a pest to farmers in another, and that a species in danger of extinction on one island might be abundant elsewhere. The Governor-General did in fact write to the regional administrators in early 1900 asking them to take whatever steps they could to prevent the killing of birds with blowpipes, but the main effect of Koningsberger's intervention was to introduce the idea that the protection of birds of paradise should take place within the framework of a more general ordinance to protect interesting and useful birds.[25]

After further contributions to the debate from the regional administrators and the Raad van Indië (Council of the Indies), the Director of Justice, S.J. Lagerweij, finally produced a draft ordinance for the protection of birds throughout the colony in November 1901. The ordinance effectively proposed to create three categories of birds. One group was to be given full protection from killing or possession, living or dead, while a second group was not to be hunted with firearms, blowpipes or catapults (these weapons were thought to be the preferred instruments of sport hunters, as opposed to the bows and arrows, snares, birdlime, and so on of traditional and subsistence hunters). Finally, there was a third category of birds whose hunting could be permitted by the regional administrators either without restriction or for specified times and places. Birds of paradise were to be placed into this category, so that regional administrators could decide the extent to which hunting could be permitted. In the spirit of the age, those found infringing the regulations were subject to a fine or to detention, up to eight days in jail in the case of Europeans, and up to three months in a labour camp in the case of natives.[26]

As drafts of the new ordinance circulated within the colonial administration, however, officials quickly became aware of the anomaly that birds of paradise, supposedly the most endangered birds in the Indies, were placed in the category given least protection. Back to the drawing board, therefore, went the Director of Justice to produce a draft ordinance specifically regulating the bird-of-paradise hunt. Hunting on Waigeo, Misool, and Batanta was now to be banned altogether, while hunters elsewhere were required to purchase an annual licence for *f* 25 and were permitted to hunt only during an open season of at least six months which was to be declared by the regional administrator.[27]

Originally prepared in September 1903, the draft ordinance, amended in various small ways, finally made its way to the Ministry of Colonies in June 1905. The Minister, A.W.F. Idenburg, however, worrying about the

25 ARA MvK, verbaal 28-7-1905, no. 16, P.C. Arends (Directeur Binnenlandsch Bestuur) to Governor-General, 25-11-1898; Governor-General to Minister of Colonies, 29-3-1904.

26 The draft ordinance is to be found in ARA MvK, verbaal 28-7-1905, no. 9. See also the 'Toelichting op de concept-ordonnantie houdende verbodsbepalingen op het dooden van vogels in Nederlandsch-Indië', 28-11-1901.

27 See the draft ordinance in ARA MvK, verbaal 28-7-1905, no. 16.

costs to the colonial treasury if the ordinance were enforced and about the effect on attitudes to authority if it were not, referred the draft back to Batavia for comments on cost and feasibility.[28] In 1901, the Dutch administrative presence in New Guinea had been further extended with the establishment of a post at Merauke in the south, but in 1906 the Resident of Ternate pronounced the draft to be 'practisch onuitvoerbaar' (unworkable) and the colonial government did not hasten to reply to the Minister.[29] Instead, the administration urged the Resident to use the authority which he already possessed to limit the hunt by requiring licences for the carrying of firearms within his jurisdiction.[30] The bird-of-paradise ordinance finally was ambushed in 1906 by a suggestion from the Director of Agriculture, Melchior Treub, to include the birds in a planned general law on hunting and wildlife protection which had been in gestation since about 1900.[31] The idea of special measures to protect birds of paradise was driven further into the background by new suggestions in 1907 for a *pacht* system in New Guinea, coming this time from H. Colijn, a prominent contributor to colonial debates and a later Netherlands Prime Minister. Unlike the earlier *pacht* proposals, which had focused exclusively on birds of paradise, Colijn's suggestions involved introducing six general trading monopolies over defined regions of New Guinea. Colijn's aim appears to have been to bring a greater element of order into what had become something of a turbulent frontier province by delegating to private companies some of the authority which the colonial state itself was unable or unwilling to exercise (Colijn 1907:46-50). The proposal, however, aroused vehement opposition, especially from independent traders who saw monopolies as antithetical to their interests. As the practical implications of Colijn's proposals sank in, too, colonial officials began to doubt that monopoly would achieve economic development where open competition had not and they feared that the monopolies would be profitable mainly at the expense of the indigenous peoples of the region. Colijn's ideas were discussed in the Dutch Parliament but were rejected.[32]

So, more than ten years of deliberation on the life and death of birds of

28 ARA MvK, verbaal 21-11-1905, no. 7, Minister of Colonies to Governor-General, 21-11-1905; ARA MvK, verbaal 1-9-1908, no. 62, Koningsberger to Governor-General, 13-8-1906.

29 ARA MvK, verbaal 2-2-1909, no. 15, Directeur Landbouw to Governor-General, 23-8-1907.

30 ARA MvK, verbaal 10-6-1909 no. 5, Resident Ternate to Governor-General, 21-1-1909. According to Westermann (1947:23, 25), the licence requirement was introduced in 1905; this was a local measure and it receives only oblique mention in the Ministry of Colonies files which I consulted. Westermann also states that the colonial government applied export tariffs to bird-of-paradise pelts, but it is likely that this refers to a local tax, such as the so-called 'vogelbelasting' ('bird tax') collected in South New Guinea (Gonggrijp 1934:1139) rather than to a true tariff which would have required the approval of the metropolitan government.

31 ARA MvK, verbaal 2-2-1909, no. 15, A3 (Eerste bureau), 'Maatregelen tot tegengaan van de uitroeiing van paradijsvogels', 15-12-1908.

32 Westermann 1947:26; for objections to Colijn's proposals, see ARA MvK, verbaal 14-1-1908, no. 18 and ARA MvK, verbaal 18-2-1908, no. 31.

paradise had come to nothing. This lack of result seems puzzling at first. Colonial officials for the most part appreciated the risk of extinction and, although they discounted the more frantic claims that the birds were on the point of disappearance, they were willing to take measures to avert loss of species before the crisis was upon them. On the other hand, they saw the bird-of-paradise trade as a useful part of the economy of eastern Indonesia and did not want to restrict it more than was necessary to protect the birds and to avert undesirable social consequences of the hunt. They had in mind a model of sustainable harvesting, but were unable to deliver it because they could not devise a satisfactory balance between the need to protect the birds, the economic benefits of the trade and the administrative capacity of the Netherlands Indies government. Unwilling to put a flawed ordinance on the books, they preferred to make do with no ordinance at all.

During the decade of fruitless reflection from 1897 to 1907, the colonial administration had enjoyed the luxury of policy making in a relative political vacuum. After the initial flurry of attention to the possible extinction of birds of paradise in 1894-1897, there was no significant external pressure on the Indies to produce quick results. By 1907, however, external pressure began again to inject a greater sense of urgency into the issue. At the very end of the 1906, the Nederlandsche Vereeniging tot Bescherming van Dieren (Netherlands Society for the Protection of Animals) wrote to the Minister of Colonies pointing out that nothing had been achieved during the previous decade and urged swift action. In May 1908, the Society wrote again, pointing out that the Indies was now out of step with international opinion and referred extensively to measures taken to protect endangered species in other parts of the world.[33] The Dutch faced an embarrassing contrast with the colonial administrations in the eastern half of New Guinea. As early as 1891, the German Neu Guinea Compagnie in northeastern New Guinea had introduced a licence system for the hunting of birds of paradise 'in the interests of the preservation of these noble birds' (Sack and Clark 1979:70), while in 1894 the British in Papua (southeastern New Guinea) introduced an ordinance which protected rare birds of paradise in the d'Entrecasteaux Islands.[34] In 1908, the British were to introduce a Wild Birds Ordinance which extended the ban over the whole colony (Westermann 1947:34; Swadling 1996:267). Much had also been done in other parts of the world: in particular, Britain, France, Germany, Italy, Spain, Portugal, and the Congo had signed a treaty in 1900 pledging to cooperate in protecting the wildlife of their Central African colonies (*Correspondence* 1906:86-91; Mackenzie 1988:207-

[33] ARA MvK, verbaal 1-9-1908, no. 62, Nederlandsche Vereeniging tot Bescherming van Dieren to Minister of Colonies 31-12-1906 and 30-5-1908.
[34] ARA MvK, verbaal 24-3-1896, no. 14, Minister of Colonies to Governor-General, 24-3-1894; Swadling 1996:264-6.

9), and a wide range of countries and colonies had introduced nature protection measures of one kind or another (Swadling 1996:93-6). The Netherlands Society for the Protection of Animals also suggested that failure to control the trade in New Guinea was endangering good relations with the Germans, because unlicenced Malay hunters were crossing the border to hunt illegally in German territory, knowing that they could sell their catch legally in Dutch New Guinea.

The most significant pressure on the Netherlands, however, came from Britain, where protection of birds had become a matter of widespread national interest. In 1909, Britain began to take steps to legislate against the import of the plumage or skins of wild birds and approached the Dutch to enquire whether they would be willing to take part in an international convention resulting in a treaty to ban the trade.[35] The proposal filled Dutch officials with dismay. A complete ban on the hunt had been one of the options originally considered when protection of birds of paradise was first raised, but it had been rejected because no officials saw a complete ban as necessary and because they saw the trade as a useful element in the commercial development of western New Guinea. A decade later, the commercial arguments for allowing the trade to continue were even stronger: in 1907 the recorded value of pelts exported from the Indies was *f* 694,829 and the trade was clearly of major importance in the east of the colony. Simply opposing the treaty, however, would be unwise, the colonial ministry felt, because it would expose the Netherlands to accusations of being concerned only with profit and not with higher moral awareness and civilization. Acceding to the treaty, on the other hand, was likely to require considerable expenditure to police the ban. Instead, the Netherlands decided to go along with the idea of a treaty but to press for the toleration of trade in pelts and feathers where the hunt was properly regulated.[36]

In the event, neither the treaty nor the British ban went ahead at this stage, but the Dutch authorities had been warned. Their strategy required first of all that the Indies put its house in order by regulating the hunt. The Ministry in The Hague initially hoped to argue that the licencing of firearms was a sufficient measure, but a gloomy report from the Resident of Ternate in 1909 soon disabused them of this idea. The Resident reported that he had no legal grounds to refuse anyone a licence and that virtually all birds were under threat.[37]

These considerations probably hastened the promulgation of the Indies'

[35] ARA MvK, verbaal 22-5-1909, no. 51, British ambassador (Acton) to Jhr R. de Marees van Swinderen, 13-3-1909.

[36] ARA MvK, verbaal 22-5-1909, no. 51, A3 (Eerste bureau), 'Internationale regeling tot voorkoming van de uitroeiing van siervogels, met name van paradijsvogels', no date.

[37] ARA MvK, verbaal 22-5-1909, no. 51, A3 (Eerste bureau), 'Internationale regeling tot voorkoming van de uitroeiing van siervogels, met name van paradijsvogels', no date; ARA MvK, verbaal 10-6-1909, no. 5, Resident Ternate to Governor-General, 20-1- 1909.

first formal measures to protect birds of paradise, when the Ordonnantie tot Bescherming van sommige in het wild levende Zoogdieren en Vogels (Ordinance for the Protection of Certain Wild Mammals and Birds) came into force in 1910.[38] The ancestral draft of this ordinance had been written in 1898 (*Koloniaal Verslag* 1898:83) but it had undergone a bureaucratic odyssey even longer than that of the bird-of-paradise ordinance. In its final version, the 1910 ordinance gave blanket protection to all wild mammals and birds in the colony, and then made a number of exceptions to that protection. What were considered to be noxious species – tigers, which endangered human life, monkeys (including orang-utan) which were seen as robbers of gardens, along with kingfishers, ricebirds and barbets seen as predatory on fishponds, ricefields or orchards – could be killed without restriction. So could traditional game such as wild pig, pigeons, and water-fowl. The ordinance also partly excepted deer, elephants, birds of paradise, parrots, and a few other species from protection by giving regional administrators the right to permit the hunting of any or all of them (*Indisch Staatsblad* 1909, no 497).

As a measure to protect the birds of paradise, the 1910 ordinance had little practical significance. There was legal uncertainty over whether the ordinance applied to the north and west of western New Guinea,[39] which was considered to be tributary to the Sultan of Tidore, and therefore not automatically subject to regulations which applied in the so-called direct-rule territories. The ordinance was also weak on controlling trade in bird-of-paradise feathers and pelts. And it left protection to the discretion of the regional administrator. The then Resident of Ternate appears to have been determined to do what he could to protect the birds, but the measure was not a convincing one for international audiences.

Ineffective though the 1910 ordinance was in practical terms, it was instrumental in establishing the framework within which further conservationist measures were considered. The essence of that framework was that hunting was entirely acceptable as long as it did not put any valued

38 *Indisch Staatsblad* 1909, nos 497 and 594 and 1910, no. 337. The regulations were issued on 14-11-1909, but came into effect only on 1-7-1910. Reptiles, including snakes and crocodiles were not covered by the ordinance; nor were butterflies.
The first conservationist measure in the Indies was probably the colonial government creation of a nature reserve at Tjibodas (now Cibodas) on the slopes of the volcanic Mount Gede in West Java in 1889. The reserve was created as an extension of the Botanical Gardens at Buitenzorg (now Bogor) and there was no suggestion at the time that it might be the first in a colony-wide system of reserves (Boomgaard 1993:316; Went et al. 1945)

39 For much of the last half century of colonial rule, Dutch New Guinea was divided into three administrative regions, called North, West and South New Guinea. I have used the capitalized terms in this paper to refer to these divisions. The whole of Dutch New Guinea was also sometimes called West New Guinea, but I have here used the terms western New Guinea in that context. The region's internal administrative structure underwent several changes in the twentieth century. At times parts or all of the region fell under the jurisdiction of colonial authorities in Ambon or Ternate, at other times parts or all were separate jurisdictions in their own right.

species of bird or mammal at risk of extinction. The economic importance of hunting – either for trade or to protect crops and life – might influence the level of protection given to a species but in the last instance priority would be given to protecting prized species from extinction. During the next twenty years, until the hunting of birds of paradise and the trade in pelts was banned completely, economic interests jostled with scientific arguments for influence on policy. As we shall see, the changing economic importance of the bird-of-paradise trade steadily diminished the arguments against protection, but at the same time scientific arguments in favour of protection became steadily more powerful, especially because they tended to encourage a step-by-step process of gradually increasing protection.

The first step in this incremental process was the issue of a supplementary ordinance in 1912 which brought birds of paradise under the 1910 ordinance by extending its coverage to the parts of New Guinea considered as tributary to Tidore.[40] This new ordinance also gave the birds of paradise their first formal protection, by declaring a closed season for hunting (which also applied to parrots and crowned pigeons of the genus *Goura*) for the five months from November to March each year.[41] In addition, hunters were required to pay *f* 25 for a licence, which was valid for only one year at a time and which gave them the right to use one gun. The firearms had to be turned in at a government administrative post at the end of the hunting season and would only be released on payment of a further licence fee at the start of the next season. A further provision restricted the licences to traders who had been settled in New Guinea for at least a year; the intention here was to force those who made a profit from the hunt to contribute to the island's commercial development, as well as to exclude itinerant hunters who might be less concerned about sustainability.

The closed season was unambiguously aimed at conservation. No-one was entirely sure when the birds of paradise bred, but the idea of giving them a respite from slaughter for several months of the year seemed a sensible way to give them some opportunity to reproduce. The increased licence fee on the other hand reflected the government's desire to share in what it saw as the substantial profits being made from the hunt. In fact, the fee probably increased hunting pressure rather than diminishing it, because hunters needed to catch more birds to achieve the same profit. The residence requirement was introduced for two reasons. First, for much the

40 *Indisch Staatsblad* 1911, no. 473; the regulation came into effect on 1-1-1912. Gerdes Oosterbeek 1918:176; Westermann 1947:27. The 1911 ordinance (*Indisch Staatsblad* 1911, no. 473) is of slight further historical interest as it is an early, minor example of a Dutch inclination to treat western New Guinea as having a status different from other parts of the archipelago, on the grounds of its biogeographical distinctiveness. This inclination, of course, culminated in the abortive attempt to develop the region as a separate colony from 1949 to 1962.

41 Westermann 1947:24. Parrots were in demand both for their feathers and for the aviary trade; crowned pigeons were hunted locally for food and both feathers and entire heads were exported.

same reasons as Colijn had proposed a monopoly system, the government wanted to concentrate the trade in the hands of people who would take a long view, both of the sustainability of the hunt and of their relations with the Papuans. The authorities wanted to exclude what they saw as rough adventurers who would sweep into a region, slaughter its birds, defraud its people and move on. Instead they hoped that traders, forced to stay on the New Guinea coast for five months when the birds of paradise were out of season, would search out other forest products and so stimulate a more diverse economy.

In fiscal terms, the new rules were a clear success. The Ternate regional government issued 1880 licences in 1912, a useful source of income for one of the Indies' poorer regional governments (Westermann 1947:29). Whether the regulations attracted a better class of hunter and trader is unproven; complaints about the behaviour of the hunters certainly continued. The regulations, however, had no appreciable impact on the level of hunting. The value of export tariffs levied on all bird pelts in 1911 had been *f* 101,344, rising to *f* 216,828 in 1912, and falling only slightly to *f* 215,041 in 1913.[42] Economics, it appeared, still had the upper hand. Yet two years later, in 1914, the Resident of Ternate announced that he would do all in his power to prevent the extinction of birds of paradise in his residency and he limited to six the number of bird-of-paradise species which could be hunted, as well as ending the hunt altogether on the Raja Ampat and Schouten Islands. In the remainder of the residency the hunting season was reduced from seven to six months (Westermann 1947:35).

These practical measures of protection were the work of a conservationist official who was in a position to act, but in the four years since 1910 environmental issues had also received far more public and international attention than previously. The Nederlandsch-Indische Vereeniging tot Natuurbescherming (Netherlands Indies Society for Nature Protection) had been founded in 1912, and it began almost immediately to lobby the colonial government for the creation of nature reserves and the protection of birds of paradise. The society was a predominantly European association with a small leavening of Javanese aristocrats and was dominated by scientists. It provided a forum which amplified and directed the general concern for environmental protection which the 1910 ordinance had demonstrated. It also quickly developed international links, not just with the metropolitan Dutch Vereeniging tot Beheer van Natuurmonumenten (Association for the Maintenance of Nature Reserves) and its redoubtable treasurer, P.G. van Tienhoven, but also with the Swiss naturalist and anthropologist Paul Sarasin, who had undertaken important research in Sulawesi (Celebes) around the turn of the century (see Sarasin

42 ARA MvK, verbaal 22-3-1922, no. 59, Overzicht van de gedurende de jaren 1911 t/m 1919 op vogelhuiden geheven uitvoerrechten.

and Sarasin 1905, 1905-06). Sarasin was responsible for organizing the first international campaign against whaling, but he was also in close contact with the American Audubon societies which had begun an effective campaign for the protection of endangered American birds and which in 1913 and 1914 succeeded in obtaining a ban on the import of all bird pelts and feathers into the United States and Canada (Van Heijnsbergen 1988:79). If any decline in the rate of hunting took place between 1912 and 1913, the American ban probably deserves more credit than the Netherlands Indies regulations of 1912.

This international network was significant because it contributed to the respectability of the colonial conservationists and because it added to the political pressures favouring conservation. In the early part of the twentieth century, the Netherlands was a slightly uneasy colonial power in Southeast Asia. Suggestions were in the air that the Dutch might give up parts of the Indies (and western New Guinea was one of the parts most often mentioned) to one of the new great powers, Germany or Japan, in the interests of international equity. Although these suggestions never became serious proposals, they added to the desire of the Netherlands Indies government to be seen to fulfil its international obligations. When S.H. Koorders, chairperson of the Indies Nature Protection Society, therefore, wrote to the Governor-General in 1913 that German intentions to ban the bird-of-paradise trade in their colony in northeastern New Guinea were being held up by Dutch tolerance of the trade, the rebuke carried a slight hint of international threat (Westermann 1947:29).

The colonial establishment, moreover, although wishing to keep the hunt going, still accepted that extinction was undesirable, thus leaving itself open to being convinced at any time that extinction had become a real risk. The international connections were important because of the practical examples they provided of animals which had once seemed so numerous that extinction was out of the question but which had nonetheless become extinct. Perhaps the most powerful example of the age was the passenger pigeon (*Ectopistes migrarius*) of North America, once found in vast flocks in the Midwest. Hunted for food and sport, its numbers dwindled and the last individual died in a Cincinnati zoo in 1914. This case, like that of the American bison, the great auk, and the dodo showed the dangers of complacency and demonstrated that species had to be protected in significant populations to be assured of survival. The debate favoured the incremental protection of birds of paradise because it was always plausible to believe that only a little extra protection at only a little extra cost might make all the difference to the survival of the species.

World War I led to something of a respite for the birds of paradise: the value of exported pelts fell from *f* 701,960 to *f* 464,634 from 1913 to 1914. After the hostilities ended, however, the trade emerged with renewed vigour. For the years 1918-1922, the export figures for bird pelts and feathers (predominantly but not exclusively bird of paradise) were as follows:

	Pelts	Value (*f*)	Feathers (kg)	Value (*f*)
1918	60,760	795,958	352	78,510
1919	121,284	2,073,957	1100	98,872
1920	61,892	1,758,969	257	80,621
1921	97,766	1,643,893	550	75,972
1922	68,434	1,013,779	1096	290,757

Source: Westermann 1947:20.

This unprecedented volume of trade led to a renewed international campaign to end the bird-of-paradise trade. Within the colony the campaign was spearheaded by Dr K.W. Dammerman, head of the zoological museum at the Buitenzorg gardens and one of the leading figures in the Netherlands Indies Society for Nature Protection. During 1920 he conducted a rather polemical debate in print and in government memoranda with the Resident of New Guinea, C. Lulofs, who scathingly attacked the 'sentimental objections' of British and American ornithological societies to the hunt.[43] In the Netherlands itself, the zoological society Natura Artis Magistra set up a committee to promote a ban on the export of birds and bird parts from the colonies (Westermann 1947:37) and shortly afterwards the British government announced that the import of bird feathers and pelts (except from ostriches and eider ducks) would be prohibited throughout the British Empire from 1922. This announcement brought the bird-of-paradise debate to a head within official circles, for the colonial government now had to decide whether or not to appeal to the British government to exempt the birds of paradise and, if so, how to argue the case.

The argument for continuing the hunt developed along the following lines. The main birds hunted for their pelts, the two yellow species, were also the most common and widespread. Proponents of hunting also noted that less than half the presumed habitat of the birds was in fact being worked by hunters and they observed that many areas of the island were still so difficult to reach that the cost of hunting birds there was likely to exceed the price they would fetch in local markets.[44] To this they added technical arguments based on the population biology of the birds themselves. The plumes which were prized for trade were used in courtship rituals. Even if all the males in their prime were removed from a local

43 ARA MvK, verbaal 21-10-1921, no. 119, C. Lulofs (Resident, New Guinea) to Governor-General, 10-8-1920; K.W. Dammerman (chef van het Zoölogisch Museum) to Directeur, 's Lands Plantentuin, 29-10-1920; Lulofs 1917; Westermann 1947:36.

44 Westermann 1947:30; ARA MvK, verbaal 22-3-1922, no. 59, Minister van Koloniën to P.G. van Tienhoven, 3-3-1922.

population by hunting, the proponents maintained, the species could still reproduce itself. The evidence of the hunters, which cannot be regarded as totally disinterested, was that males began to breed at one year of age, but did not attain their full splendour until at least four.[45] If this were so, then, the future of the species was not at risk. To this argument was added the claim that the hunt had become necessary for the basic subsistence of many Papuan communities, who had few other items to trade for basic tools. If the bird trade were stopped, it was argued, many Papuans would suffer a decline in living standards.[46] This argument was also developed in an ecological direction with the assertion that if the hunt were stopped, then the development of Dutch New Guinea would move inexorably in the direction of a plantation economy, resulting in the wholesale clearance of jungle and the possible extinction of a great many more species.

The opponents of hunting responded that not enough was known of the distribution or population biology of the birds to assert that hunting did not damage the species. They pointed out that although breeding by males which had not reached their full adult plumage was not unknown amongst birds, it was uncommon and was not the case in the closest relatives of the birds of paradise. The testimony of the hunters to the contrary was seen as untrustworthy. To the more economic arguments, the opponents of hunting responded that opportunistic Ternate hunters were not on the whole a civilizing force for Papuan societies. They introduced alcohol, guns, and disease, as well as influencing the economic balance of power in previously isolated communities. 'The fact that there are any people left in New Guinea is certainly no thanks to the bird trade', commented an Assistant-Resident as early as 1906 (Westermann 1947:24).

The instinct of many historians would be to argue for the primacy of economic arguments over considerations of ethical duty or scientific significance. The stress which both sides laid, however, on the question of extinction meant that once evidence began to accumulate that at least some
species of bird of paradise might be seriously affected by hunting, the
idiom of protection could be relatively easily accepted. The technica
arguments over the mating habits of birds of paradise also worked to th
advantage of the totally unrelated crowned pigeons. Having asserted th
the hunt for male birds of paradise alone could not endanger the pop

45 ARA MvK, verbaal 21-10-1921, no. 119, C. Lulofs (Resident, New Guinea) to Gove
General, 10-8-1920; ARA MvK, verbaal 26-4-1922, no. 31, Board of Trade, Plumag
Advisory Committee, Report of evidence given by Dr De Beaufort and Mr F. 's Jaco
regard to birds of paradise and crown pigeons, 8 March 1922; Westermann 1947:31; Lulo
Current scientific knowledge supports this argument in a slightly different way: it appe
the presence of a dominant male may suppress the development of display plu
otherwise mature males in the vicinity. When a dominant male is removed, therefore,
is likely to be taken almost immediately by another fertile male. See Diamond 1981:258.

46 ARA MvK, verbaal 21-10-1921, no. 119, Resident, New Guinea to Governor-Gen
1920; Westermann 1947:30-1.

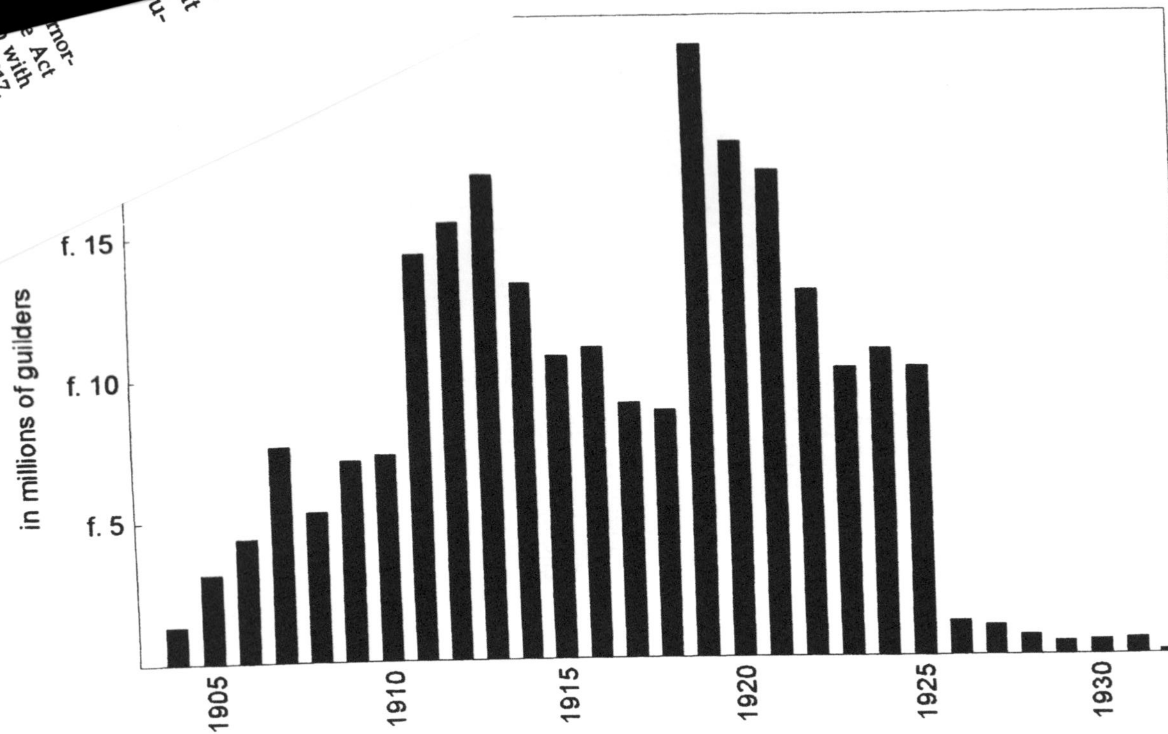

Table 1. Value of birds, bird skins and feathers exported from the Netherlands Indies, 1904-1932 (Boomgaard 1996). Figure for 1912 is corrected from original sources.

lation, the authorities were receptive to the argument that crowned pigeons, of which both males and females were hunted, did indeed need protection.[47]

Reviewing these arguments, the colonial government decide to press a case with the British. In order to show, however, that its arguments took good account of the conservationist case, it began to strengthen the protection given to species with the strongest claim to being endangered. From 1921, only muzzle-loading firearms could be used for hunting in North and West New Guinea,[48] and in 1922 the hunting of all but the two yellow species of bird of paradise was banned (*Staatsblad* 1922, no 748). The willingness of the colonial government to accept risk of extinction as sufficient grounds for further protection had worked once again to negate the influence of economic considerations. The show of concern, however, was in vain: the British authorities refused to make an exception even of the yellow birds of paradise.

The process of incremental protection continued from 1922. In 1923, a local tax of 10% (called a *telrecht*) was put on the value of all pelts exported from North and West New Guinea, and a similar tax was introduced in the south in 1927. The hunting ban on most species was reinforced with an export ban in 1924 and, in the same year, at the request of the Australian government, the hunt was banned in the vicinity of the border with Australian Papua and New Guinea, in order to discourage illegal hunting in Australian territory. In 1925, the Bird's Head Peninsula was closed to all hunting. There were further limits in hunting territory in 1926 and 1927, and the hunt was ended for good, in legal terms at least, in 1928. Finally, in 1931, a new Dierenbeschermings-verordening (Animal Protection Ordinance) came into effect, giving full protection to all birds of paradise, along with many other species (Westermann 1947:37-40). By this time, however, the risk to the birds of paradise had greatly diminished as a result of declining interest in pelts from the hat trade. The wearing of dead birds of paradise on one's head was no longer accepted in polite circles in Europe and America.[49]

The forty-year conflict over bird-of-paradise hunting marks the start of environmental politics in Indonesia in two ways. First, in contrast with earlier resource politics where the main issue was the allocation of scarce natural resources, the debate over birds of paradise had at its heart a commitment by all parties to preserve the birds for future generations. In this case, the benefits which future generations were expected to get was

47 ARA MvK, verbaal 21-10-1921, no. 119, Resident of New Guinea to Governor-General, 10-8-1920.

48 ARA MvK, verbaal 22-3-1922, no. 59, 'Uittreksel uit het Register der Besluiten van den Resident van Nieuw Guinea', 14-3-1921.

49 Illicit trade in birds of paradise as a curiosity has continued. See Special Correspondent 1990 and Muller 1991:67.

aesthetic and scientific, rather than economic, but the idea of the present generation's responsibility to ensure the quality of life of tomorrow's generation was central to the argument. Second, the debate was technocratic and technical: judgements of the risk of bird-of-paradise extinction were based on an understanding of the population biology of the birds. Although today's scientific judgement would probably be that habitat clearing represented a far greater danger to the birds than did even massive hunting – thanks mainly to the polygyny of the plume-bearing males and the presence of a reserve population of unplumed fertile males – the weight of scientific evidence in the first decades of the century suggested that hunting had to be stopped. The importance of science in the argument did not prevent sentiment from playing a major role as well, and the fact that almost none of the participants in the debates had seen or were likely to see a bird of paradise in the wild seems to have added to the air of altruism with which the debate was conducted. This combination of science and sentiment, as in contemporary environmental campaigns, put the economic arguments in favour of bird-of-paradise hunting at a great disadvantage.

The course of the debate resembled modern environmental politics in two further respects. First, public opinion in developed countries played a powerful role in pressing local authorities into action. There was a good deal of interest in protecting birds of paradise among colonial officials, but pressure from the Netherlands and Britain was instrumental in galvanizing that interest into action. And, secondly, the idea of maintaining a sustainable harvest from the wild, which underpinned the first attempts to devise a protection ordinance, proved to be practically and politically unattainable. As in the more recent cases of whales, elephants and seals, scientific uncertainty about population biology, the desire to prevent smuggling and illicit hunting, and the elevation of birds of paradise to become conservationist symbols produced an irresistible pressure towards total protection. Ironically, the idea that nature could survive only in isolation from humankind has probably turned out to be an obstacle in the long run to ideas for the sustainable management of the environment.

Abbreviations

ARA	Algemeen Rijksarchief, Den Haag
MvK	Archive of the Ministerie van Koloniën (ARA)

References

Andaya, L.Y.
1993 *The world of Maluku; Eastern Indonesia in the early modern period.* Honolulu: University of Hawaii Press.

Axelrod, D.I.
1967 *Quaternary extinctions of large mammals*. Berkeley/Los Angeles: University of California Press.
Boomgaard, P.
1993 'Protection de la nature en Indonésie pendant la fin de la période coloniale (1889-1949)', in: J. Pouchepadass (ed.), *Colonisations et environnement*, pp. 307-44. Paris: Société Française d'Histoire d'Outre-Mer.
1994 'Colonial forest policy in transition, 1865-1916', in: R. Cribb (ed.), *The late colonial state in Indonesia; Political and economic foundations of the Netherlands Indies, 1880-1942*, pp. 117-37. Leiden: KITLV Press. [Verhandelingen 163.]
1996 (with the assistance of R. de Bakker), *Forests and forestry 1823-1941*. Amsterdam: Royal Tropical Institute. [Changing Economy in Indonesia 16.]
Bruun, O. and A. Kalland (eds)
1995 *Asian perceptions of nature; A critical approach*. Richmond: Curzon Press.
Colijn, H.
1907 *Nota betreffende de ten aanzien van Nieuw-Guinea te volgen gedragslijn*. Batavia: Landsdrukkerij.
Correspondence
1906 *Correspondence relating to the preservation of wild animals in Africa*. Vol. I. London: His Majesty's Stationery Office.
Crawfurd, J.
1856 *A descriptive dictionary of the Indian islands and adjacent countries*. London: Bradbury and Evans.
Cribb, R.
1988 *The politics of environmental protection in Indonesia*. Clayton: Centre of Southeast Asian Studies, Monash University. [Working Paper 48].
1990 'The politics of pollution control in Indonesia', *Asian Survey* 30:1123-35.
Dammerman, K.W.
1929 *Preservation of wild life and nature reserves in the Netherlands Indies*. [Weltevreden?]: Fourth Pacific Science Congress, Java.
Diamond, J.M.
1981 'Birds of paradise and the theory of sexual selection', *Nature* 293:257-8.
Doughty, R.W.
1975 *Feather fashions and bird preservation; A study in nature protection*. Berkeley: University of California Press.
Fuller, E.
1987 *Extinct birds*. London: Viking.

Gerdes Oosterbeek, W.F.
1918 'Jacht', in: *Encyclopaedie van Nederlandsch-Indië*, vol. 2, pp. 173-7. 's-Gravenhage: Nijhoff, Leiden: Brill.

Gonggrijp, G.F.E.
1934 *Geïllustreerde encyclopaedie van Nederlandsch-Indië*. Leiden: Leidsche Uitgeversmaatschappij.

Grove, R.
1987 'Early themes in African conservation; The Cape in the nineteenth century', in: D. Anderson and R. Grove (eds), *Conservation in Africa; People, policies and practice*, pp. 21-39. Cambridge: Cambridge University Press

Hall, K.R.
1992 'Economic history of early Southeast Asia', in: N. Tarling (ed.), *The Cambridge history of Southeast Asia. Volume 1: From early times to c. 1800*, pp. 183-275. Cambridge: Cambridge University Press.

Heijnsbergen, P. van
1988 'Het ontstaan van internationale geïnstitutionaliseerde samenwerking ten aanzien van het natuurbehoud', *Vakblad voor Biologen* 68:79-81, 85.

Houten, P.J. van
1896 *Nota aangeboden aan het bestuur der Maatschappij ter Bevordering van het Natuurkundig Onderzoek der Nederlandsche Koloniën*. Amsterdam: De Bussy.

Koloniaal Verslag
1898 *Koloniaal Verslag 1898*. 's-Gravenhage: Algemeene Landsdrukkerij.

Koningsberger, J.C.
1898 *Eerste overzicht der schadelijke en nuttige insecten van Java*. Batavia: Kolff. [Mededeelingen uit 's Lands Plantentuin 22.]
1901-09 *De vogels van Java en hunne oeconomische beteekenis*. Batavia: Kolff. 2 vols. [Mededeelingen uit 's Lands Plantentuin 50; Mededeelingen uitgaande van het Departement van Landbouw 7.]
1903 *Ziekten van rijst, tabak, thee en andere cultuurgewassen die door insecten worden veroorzaakt*. Batavia: Kolff. [Mededeelingen uit 's Lands Plantentuin 64.]

Lulofs, C.
1917 'De paradijsvogeljacht', *Tijdschrift voor het Binnenlandsch Bestuur* 53:654-61.

MacAndrews, C.
1994 'Politics of the environment in Indonesia', *Asian Survey* 34:369-80.

Mackenzie, J.M.
1988 *The empire of nature; Hunting, conservation and British imperialism*. Manchester: Manchester University Press.

Muller, K.
1991 *Indonesian New Guinea; Irian Jaya*. Berkeley/Singapore: Periplus.

Paradijsvogels
1919 'Paradijsvogels' in: *Encyclopaedie van Nederlandsch-Indië*, vol. 3, pp. 338-9. 's-Gravenhage: Nijhoff, Leiden: Brill.

Passmore, J.
1974 *Man's responsibility for nature; Ecological problems and Western traditions.* London: Duckworth.

Peluso, N.L.
1992 *Rich forests, poor people; Resource control and resistance in Java.* Berkeley: University of California Press.

Piepers, M.C.
1896 'Door welke maatregelen kan tot eene rationeele bescherming der inheemsche dieren- en plantenwereld in Nederlandsch-Indië worden gekomen?', *Tijdschrift voor Nederlandsch-Indië* 25(nieuwe serie), I:38-72.

Raup, D.M.
1984 'Death of species', in: M.H. Nitecki (ed.), *Extinctions*, pp. 1-19. Chicago/London: University of Chicago Press.

Sack, P. and D. Clark (eds and translators)
1979 *German New Guinea: The annual reports.* Canberra: Australian National University Press.

Sarasin, P. and F. Sarasin
1905 *Reisen in Celebes ausgeführt in den Jahren 1893-1896 und 1902-1903.* Wiesbaden: Kreidel.
1905-06 *Versuch einer Anthropologie der Insel Celebes.* Wiesbaden: Kreidel.

Smith, S.L.
1992 *The politics of Indonesian rain forests.* Clayton: Centre of Southeast Asian Studies, Monash University. [Working Paper 76.]

Special Correspondent, A
1990 'Plucking profits from Irian Jaya', *Inside Indonesia* 24:24-5.

Swadling, P.
1996 *Plumes from paradise; Trade cycles in outer Southeast Asia and their impact on New Guinea and nearby islands until 1920.* Boroko: Papua New Guinea National Museum.

Wallace, A.R.
1890 *The Malay archipelago; The land of the orang-utan and the bird of paradise.* New edition. London: Macmillan.

Went, F.W., K.W. Dammerman, W.M. Docters van Leeuwen and P. Dakkus
1945 'The Tjibodas biological station and forest reserve', in: P. Honig and F. Verdoorn (eds), *Science and scientists in the Netherlands Indies*, pp. 403-16. New York: Board for the Netherlands Indies, Surinam and Curaçao.

Westermann, J.H.
1947 *Fauna en natuurbescherming in Nederlandsch Nieuw-Guinea.* Amsterdam: Nederlandsche Commissie voor Internationale Natuurbescherming. [Mededeelingen 13.]

Wit, A. de
1914 *Natuur en -menschen in Indië.* Amsterdam: Maatschappij voor Goede en Goedkope Lectuur.

Wolters, O.W.
1967 *Early Indonesian commerce; A study of the origins of Srivijaya.* Ithaca: Cornell University Press.

Yule, H. and A.C. Burnell
1903 *Hobson-Jobson; A glossary of Anglo-Indian words and phrases, and of kindred terms, etymological, historical, geographical and discursive.* Edited by W. Crooke. London: Murray.

Notes on contributors

Peter Boomgaard is Director of the Royal Institute of Linguistics and Anthropology (KITLV) in Leiden. He also holds a chair in the economic and ecological history of Southeast Asia at the University of Amsterdam. He is chief editor of the *Changing economy in Indonesia* series and author of *Children of the colonial state; Population growth and economic development in Java, 1795-1880* (1989).

Harold Brookfield is Emeritus Professor of Human Geography at the Research School of Pacific and Asian Studies of the Australian National University. His recent publications include (together with Lesley Potter and Yvonne Byron) *In place of the forest; Environmental and socio-economic transformation in Borneo and the eastern Malay Peninsula* (1995). He is co-editor of *Southeast Asia's environmental future; The search for sustainability* (1993).

Freek Colombijn is a post-doctoral researcher at the Royal Institute of Linguistics and Anthropology (KITLV) in Leiden. He obtained his PhD from Leiden University and published his thesis, *Patches of Padang; The history of an Indonesian town in the twentieth century and the use of urban space*, in 1994.

Robert Cribb is Research Professor at the Nordic Institute of Asian Studies in Copenhagen. He writes on various aspects of Indonesian history in the twentieth century, including the history of environmental politics. His recent publications include the *Historical dictionary of Indonesia* (1992) and (together with Colin Brown) *Modern Indonesia; A history since 1945* (1995).

Michael R. Dove is Director of the Program on Environment at the East-West Center in Honolulu. He is author of *Swidden agriculture in Indonesia; The subsistence strategies of the Kalimantan Kantu'* (1985) and editor of *The real and imagined role of culture in development; Case studies from Indonesia* (1988).

David Henley currently works as a researcher at the Royal Institute of Linguistics and Anthropology (KITLV) in Leiden. He studied geography at

Cambridge University and obtained his PhD at the Australian National University. His thesis was published in 1996 as *Nationalism and regionalism in a colonial context; Minahasa in the Dutch East Indies.*

Bernice de Jong Boers studied cultural anthropology at the University of Utrecht and is now a PhD student at the Royal Institute of Linguistics and Anthropology (KITLV) in Leiden. Her research focuses on the environmental history of Nusa Tenggara, particularly Sumbawa.

Jeya Kathirithamby-Wells, formerly Professor of Asian History at the University of Malaya, now lectures and researches at Cambridge University. Author of numerous articles, she contributed to Volume I of the *Cambridge history of Southeast Asia* (1992) and co-edited *The Southeast Asian port and polity; Rise and demise* (1990) with John Villiers.

Han Knapen studied cultural anthropology at the University of Nijmegen and is now a PhD student at the Royal Institute of Linguistics and Anthropology (KITLV) in Leiden. He is engaged in the study of ecological change in Southeast Kalimantan.

Masyhuri works as a researcher at the Centre for Economic and Development Studies of the Indonesian Institute of Sciences (LIPI) in Jakarta. He studied at Gadjah Mada University and the University of Indonesia before obtaining his PhD at the Free University of Amsterdam in 1995 with a thesis entitled *Pasang surut usaha perikanan laut; Tinjauan sosial-ekonomi kenelayanan di Jawa dan Madura, 1850-1940.*

Jan Willem Nibbering is a rural development expert. He obtained his PhD at the Australian National University with a thesis entitled *Hoeing in the hills; Stress and resilience in an upland farming system in Java* (1992).

Lesley Potter is Associate Professor and Head of the Geography Department at the University of Adelaide. She is author (together with Harold Brookfield and Yvonne Byron) of *In place of the forest; Environmental and socio-economic transformation in Borneo and the eastern Malay Peninsula* (1995). She is currently working on a history of people and forests in Southeast Asia.

Anthony Reid is Professor of Southeast Asian History at the Research School of Pacific and Asian Studies of the Australian National University. His recent publications include *Southeast Asia in the age of commerce, 1450-1680* (2 volumes, 1988 and 1993) and, as editor, *Indonesian heritage; Early modern history* (1996).

Index

www.ingramcontent.com/pod-product-compliance
Ingram Content Group UK Ltd.
Pitfield, Milton Keynes, MK11 3LW, UK
UKHW041143300926
15471UKWH00011B/111

9 789067 181242